Guide to the
HVAC/R CERTIFICATION AND COMPETENCY TESTS

Robert Featherstone
Oakland Community College

Jesse Riojas
Oakland Technical Center

Prentice Hall

Upper Saddle River, New Jersey
Columbus, Ohio

Library of Congress Cataloging-in-Publication Data

Featherstone, Robert.
   Guide to the HVAC/R certification and competency tests / Robert Featherstone, Jesse Riojas.—1st ed.
   p. cm.
   ISBN 0-13-010694-1
   1. Heating—Problems, exercises, etc. 2. Air conditioning—Problems, exercises, etc. 3. Ventilation—Problems, exercises, etc. 4. Mechanics (Persons)—Certification—United States. I. Riojas, Jesse R. II. Title.

TH7015.F43 2001
697'.0076—dc21
                                                                                          00-023084

Vice President and Publisher: Dave Garza
Editor in Chief: Stephen Helba
Executive Editor: Ed Francis
Production Editor: Christine M. Buckendahl
Production Coordination: Megan Hill, Lithokraft II
Design Coordinator: Robin G. Chukes
Cover Designer: Tom Mack
Production Manager: Matthew Ottenweller
Marketing Manager: Jamie Van Voorhis

This book was set in Times by Lithokraft II and was printed and bound by Victor Graphics, Inc. The cover was printed by Victor Graphics, Inc.

Copyright © 2001 by Prentice-Hall, Inc., Upper Saddle River, New Jersey 07458. All rights reserved. Printed in the United States of America. This publication is protected by Copyright and permission should be obtained from the publisher prior to any prohibited reproduction, storage in a retrieval system, or transmission in any form or by any means, electronic, mechanical, photocopying, recording, or likewise. For information regarding permission(s), write to: Rights and Permissions Department.

10 9 8 7 6 5 4 3 2
ISBN: 0-13-010694-1

PREFACE

The HVAC/R industry has progressively become more technical. With new technology, industries are requiring their employees' skills be updated to match the current changes. HVAC/R competency/certification tests are directed toward students and technicians at all skill levels. As a result, national certification competency exams are being offered by many organizations, including the Air Conditioning and Refrigeration Institute (ARI), Air Conditioning Contractors of America (ACCA), Refrigeration Service Engineer Society (RSES), and the ESCO Institute.

The *Guide to the HVAC/R Certification and Competency Tests* is designed to help students and technicians pass competency and certification exams. This book is divided into two areas: a Study Guide and a Question Section. The Study Guide and the Question Section cover System Components and Tools; Electrical Theory and Application; Air Flow Components and Duct Fabrication; and Indoor Air Quality and Safety. The Question Section covers one additional topic: Service and Troubleshooting.

The student/technician will benefit most by reading the Study Guide first. Then, the student should turn to the Question Section for the related statement. For example, the first statement in question 1 of System Components and Tools (Hand Tools) reads "A torx driver has a star shaped head that is used on torx type screws." The student would then turn to the same section in the Questions and try to answer question 1.

In Section Four of the Study Guide, an EPA Certification Study Guide has been provided to assist the student in successfully passing Type I, II, III, and Universal Recover/Recycle Certification Examinations.

It is the hope of the authors that all information provided in this book will assist the student in successfully passing certification exams offered by any HVAC/R national organization.

Acknowledgments

Thanks to:

- Dave Arscheene, Owner of Dave's Remodeling Co., Mt. Clemens, Michigan, for assistance with Basic Construction in Section 1.
- Cathy Artley, School to Work Facilitator, Oakland Technical Center, Clarkston, Michigan, for assistance with organization and all the miscellaneous work done on the manuscript.
- Richard D. Featherstone, Jr., Fire Marshall for Trenton, Michigan, for assistance with Safety in Section 4.

We also thank the reviewers of this text for their helpful comments and suggestions: Ricky A. Black, Northeast State Technical Community College; David Eishen, Cedar Valley College; and Herb Haushan, College of DuPage.

CONTENTS

Part 1: Study Guide

Section 1 — System Components and Tools Study Section

Hand Tools	3
Basic Construction	4
Refrigeration Theory	5
Air Conditioning Components	7
Refrigerants	9
Lubricants	10
Compressors	11
Air Cooled Condensers	12
Water Cooled Condensers	13
Indoor Evaporator Coils	13
Metering Devices	14
Service Valves	15
Leak Detection	15
Recovery Equipment	16
Soldering and Brazing	16
Manifold Gauges	18
System Evacuation	18
Refrigerant Charging	19
Defrost Systems	19
Fundamentals of Gas Combustion	21
Gas Furnace Controls and Components	21
Combustion Air	23
Vents and Chimneys	24
Gas Piping	31
Oil Furnaces	33
Heat Pumps	35

Section 2 — Electrical Theory and Application Study Section

Electrical Theory	41
Magnetism	42
Direct Current	42
Series Circuits	43
Parallel Circuits	43
Alternating Current	43
Power Distribution	44
AC Motors	45
Shaded-Pole Motors	47
Split Phase Motors	47
Capacitor Start Induction Run Motors (CSIR)	47
Permanent Split Capacitor Motor (PSC)	48
Capacitor Start Capacitor Run Motor (CSR)	48
Three-Phase Motors	48
Electrical Controls	48
Solid-State Electronics	55
Wiring Layouts and Electrical Diagrams	56
Electrical Meters	62
Electrical Troubleshooting	63

Section 3 — Air Flow Components and Duct Fabrication Study Section

Air Flow Tools and Measurements	65
Blowers and Fans	67
Air Duct Systems and Fabrication	70
Flexible (Flex) Duct	72
Fiberglass Ductboard	72
Air Side Components	73
Air Filters	74
Humidifiers	75
Temperature and Humidity	75

Section 4 — Indoor Air Quality (IAQ) and Safety Study Section

Indoor Air Quality (IAQ)	79
Personal Safety	81
Refrigerant Safety	81
Soldering and Brazing Safety	83
Electrical Safety	83
EPA Certification Study Guide	84
Core Section	84
Type I Certification	86
Type II Certification	86
Type III Certification	88

Part 2: Questions

Section 1 — System Components and Tools Questions

Hand Tools 91
Basic Construction 93
Refrigeration Theory 94
Air Conditioning Components 97
Refrigerants 101
Lubricants 103
Compressors 105
Air Cooled Condensers 105
Water Cooled Condensers 106
Indoor Evaporator Coils 108
Metering Devices 109
Service Valves 110
Leak Detection 111
Recovery Equipment 112
Soldering and Brazing 113
Manifold Gauges 115
System Evacuation 116
Refrigerant Charging 117
Defrost Systems 118
Fundamentals of Gas Combustion 120
Gas Furnace Controls and Components . . . 121
Combustion Air 124
Vents and Chimneys 126
Gas Piping 131
Oil Furnaces 135
Heat Pumps 138

Section 2 — Electrical Theory and Application Questions

Electrical Theory 143
Magnetism 144
Direct Current 145
Alternating Current 146
Power Distribution 148
AC Motors 149
Electrical Controls 154
Solid-State Electronics 161
Wiring Layouts and Electrical Diagrams 161
Electrical Meters 169
Electrical Troubleshooting 170

Section 3 — Air Flow Components and Duct Fabrication Questions

Air Flow Tools and Measurements 173
Blowers and Fans 177
Air Duct Systems and Fabrication 178
Flexible (Flex) Duct 182
Fiberglass Ductboard 182
Air Side Components 185
Air Filters 186
Humidifiers 187
Temperature and Humidity 188

Section 4 — Indoor Air Quality (IAQ) and Safety Questions

Indoor Air Quality (IAQ) 191
Personal Safety 194
Refrigerant Safety 195
Soldering and Brazing Safety 198
Electrical Safety 199
EPA Certification Core Section 201
Type I Certification 204
Type II Certification 205
Type III Certification 207

Section 5 — Service and Troubleshooting Questions

Air Conditioning Troubleshooting 209
Gas Furnace Troubleshooting 220
Oil Furnace Troubleshooting 231
Heat Pump Troubleshooting 236
Air Distribution Troubleshooting 242

Part 3: Answers

Section 1 — System Components and Tools 249
Section 2 — Electrical Theory and Application 255
Section 3 — Air Flow Components and Duct Fabrication 258
Section 4 — Indoor Air Quality (IAQ) and Safety 260
Section 5 — Service and Troubleshooting . . . 263

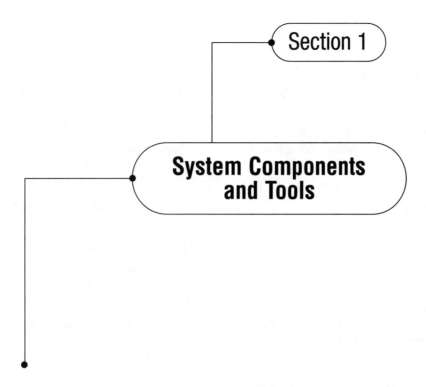

Section 1

System Components and Tools

This section contains information that will help technicians pass certification exams. It provides emphasis on refrigeration and air conditioning theory as well as a description of the systems' mechanical components. It also covers the mechanical components of a gas/oil furnace and heat pump. With a thorough understanding of this section, the technician will be better prepared and more skilled in mechanical troubleshooting.

HAND TOOLS

Hand tools that technicians use on a daily basis can range from the most basic to precision made tools. The technician must be sure that the tools used are correct for the job. He or she must also keep the tools in good working condition. The following are types of tools that may be used in the HVAC/R field.

A **torx driver** has a star shaped head and is used on torx type screws.

A **nut driver** has a hexagon shaped head and is used to remove a screw with a six point hex head shape.

An **adjustable wrench** has a variable open end that is used to remove a nut or bolt.

An **allen wrench** is used to loosen or tighten a set screw.

A **box wrench** is a closed end wrench that is used to tighten or loosen nuts and bolts.

The **flare nut wrench** is used to tighten or loosen a flare type mechanical fitting on a refrigeration or air conditioning system.

An **open end wrench** is used to tighten or loosen a nut or bolt from the side.

The **pipe wrench** is used to assemble or disassemble threaded pipe.

A **service valve wrench** is used to manipulate service valves on a refrigeration or air conditioning system. The most common sizes are $\frac{3}{16}"$, $\frac{1}{4}"$, $\frac{5}{16}"$, and $\frac{3}{8}"$.

A **socket wrench** is used to fit on top of a nut or bolt. Common socket drives are $\frac{1}{4}"$, $\frac{3}{8}"$, and $\frac{3}{4}"$.

A **Torque wrench** is used to apply a specific amount of torque (force × length arm) to a nut or a bolt. Using a torque wrench will prevent over- or under-tightening.

Tongue and groove pliers are used to hold objects of many shapes and sizes. They are also used to loosen and tighten threaded pipe.

A **level** is used to test for uniformity of work. When held in a horizontal position, it will show if the work is level. If it is held in a vertical position it will show if the work is plumb.

A **carpenters square** is an "L" shaped tool that is used to measure materials and check for uniformity.

A **crosscut saw** is used to cut wood against the grain. It has 6 to 12 teeth per inch that bend away from the saw blade in an alternating pattern.

A **ripsaw** is used to cut wood with the grain. It has 4 to $5\frac{1}{2}$ teeth per inch that bend away from the saw blade in an alternating pattern.

A **hacksaw** is primarily used to cut metal. It has 18 to 32 teeth per inch. Please note that cutting copper tends accumulate copper filings in the saw teeth.

A **keyhole saw** is used for cutting drywall or wood. It has 8 to 10 teeth per inch.

A **single cut file** has one row of teeth.

A **double cut file** has two rows of teeth that crisscross each other.

There are four types of **files**: the rasp, the rough cut file, the smooth file, and the finish file. The **rasp** is used to start a smoothing process on extremely rough cut wood. The **rough cut file** is used on rough cuts of wood, plastic and metal. The **smooth cut file** is used to continue the smoothing of a material after the use of a rough cut file. The **finish file** will complete the final smoothing process.

A **strait cut snip** is designed for straight and wide radius cuts in sheet metal. The maximum recommended thickness for the strait cut snip is 18 gauge metal.

A **right cut snip** is designed for straight and tight right radius cuts in sheet metal. The maximum recommended thickness for the right cut snip is 18 gauge metal.

A **left cut snip** is designed for straight and tight left radius cuts in sheet metal. The maximum recommended thickness for the left cut snip is 18 gauge metal.

A larger **scissor type snip,** such as the straight pattern and the duckbill type, are designed for cutting straight and larger radius curves in sheet metal. The maximum recommended thickness for the straight and duckbill snip is 23 gauge metal.

A **drill** is used to bore holes in material such as sheet metal, wood, plastic, dry wall, and concrete. Drills are available in 3 different type drives: the $\frac{1}{4}''$ drive, the $\frac{3}{8}''$ drive and the $\frac{1}{2}''$ drive.

Hex head adapters are used with a drill to install or remove hex head sheet metal screws.

A **countersink** is a type of bit that is used to recess holes in wood and other material, so that screws can be driven flush or slightly below the surface.

Drills and bits are used for drilling into wood, metal, and masonry materials.

A **spade** is a flat bit with two sharp edges and a point. It is used to drill holes into wood.

The **auger bit** has a corkscrew shape and is used to bore holes into wood. Common sizes are $\frac{3}{8}'', \frac{5}{8}'', \frac{1}{2}'', \frac{3}{4}'',$ and 1" in diameter; common lengths are $7\frac{1}{2}''$, 12", and 17".

A **masonry drill** is used to bore holes in brick, block, concrete, stone, or mortar.

A **punch** is used to create an indention, usually in a piece of sheet metal. This indention makes an easy path for a drill or a sheet metal screw to follow.

A **tap** is used for cutting inside threads on metal material.

Dies are used to cut threads on the outside of galvanized and black iron pipe. Before using a die, always ream the inside of the pipe to remove all of the burrs.

A **claw hammer** has a slightly rounded face on the front and a claw on the back that is used to remove nails. A claw hammer ranges from 13 oz. to 28 oz. in weight.

A **ball peen hammer** is designed for metal work. It has a flat back and a round front, which is used to form bends and creases in sheet metal.

A **masonry hammer** has a larger head that is designed for pounding block or concrete. The masonry hammer has a sharp pick on the back for splitting masonry material.

All **hand tools** need to be cleaned regularly with some type of machine oil to prevent corrosion

BASIC CONSTRUCTION

On many occasions in the construction industry, tradesman will be working along side each other to complete a project. Each trade must be familiar with basic construction terms and knowledge. The following will help the technician understand some of the terms used in basic construction.

A residential **roof construction** can be truss built, rafter built, or a flat roof design.

The two most common **residential roof designs** are the truss and rafter built roofs.

A common design of **truss built roof** is pre-manufactured with braces between the top and the base of the truss. Trusses are commonly constructed of 2" × 4", 2" × 6", 2" × 8", or 2" × 10" lumber or steel.

A **rafter built roof** uses 2" × 6", 2" × 8", 2" × 10", or 2" × 12" boards. They connect the ridge of the roof (the peak) to the top of an exterior wall or knee wall (a smaller exterior wall).

Section 1   System Components and Tools

A **truss and rafter built roof** uses 4' × 8' × $\frac{1}{2}$" plywood or oriented strand board (OSB) sheathing as a base for the shingles. This also strengthens the roof by adjoining the rafters or trusses together as one.

There are **various truss and roof designs** such as the W-web truss, the King Post truss, the Scissor truss, Gambrel roof, and Hip roof.

Organic asphalt, fiberglass asphalt, tile, and clay **shingles** are used to seal the roof and protect it from the weather.

A common residential **ceiling** is made of plaster, drywall, fiberglass inserts, or celotex tile that cover 2" × 6" to 2" × 12" wood or steel ceiling joists.

A **ceiling** is generally an 8 foot high flat type, or a cathedral type that rises diagonally to a peak.

A **drywall ceiling** uses 4' × 8' × $\frac{1}{2}$" or 4' × 8' × $\frac{5}{8}$" sheets of drywall that are connected to the ceiling joists with drywall screws or nails.

An external ceiling is **insulated** with fiberglass or cellulose fiber to reduce the heat transfer of the house. Common fiberglass insulation thicknesses are as follows: R-19 is 6 inches, R-25 and R-30 are 9 inches, and R-38 is 12 inches. Increasing the R-value of insulation will lower the heat transfer of the home.

A common residential **wall** uses 2" × 4" or 2" × 6" wood or steel vertical studs placed 16 inches apart. At the top and bottom of the vertical studs are horizontal studs. They are made of the same material as the vertical studs. Horizontal studs are known as top plates and bottom plates.

An **exterior wall** uses 4' × 8' × $\frac{1}{2}$" plywood, $\frac{1}{2}$" OSB sheathing, $\frac{1}{2}$" celotex, or $\frac{1}{2}$" to 1" dense foam sheathing to adjoin the vertical studs. It is required that the corner walls are constructed of plywood or OSB sheathing.

An **exterior wall** has the weight of the roof bearing down upon it. This is why they are considered load bearing walls.

A **house wrap** is used on exterior walls to decrease air infiltration into the home.

The common **exterior finishes** for homes are vinyl siding, wood siding, aluminum siding, and brick.

An external wall is **insulated** with fiberglass or cellulose fiber to reduce the heat transfer of the house. The common fiberglass insulation thicknesses are as follows: R-11 and R-13 are $3\frac{1}{2}$ inches and R-19 is 6 inches.

A **vapor barrier** is placed on the inside of an exterior wall, behind the drywall, to reduce the quantity of moisture entering or exiting the house.

A wall that uses **drywall** as its interior finish, is constructed of 4' × 8' × $\frac{1}{2}$" or 4' × 8' × $\frac{5}{8}$" sheets of drywall. These are connected to the wall studs with drywall screws or nails.

Drywall that is used in damp areas, such as bathrooms, is moisture resistant. This drywall is also known as green board.

An interior wall that connects the living area of a house to a garage must use a $\frac{5}{8}$" **fire code type drywall.**

A residential **floor** can be constructed of 2" × 8" to 2" × 12" wood or steel floor joists, engineered truss joists, or engineered "I" joists.

Common floor joist **bridging** is made of 1" × 2" wood slats or metal brackets. They are placed between the floor joists, at the center or every one third of the floor span to stiffen a floor.

A **sub-floor** is the floor that is directly above the floor joists. It is made of $\frac{3}{4}$" tongue and groove plywood, and OSB sheathing.

An **underlayment floor** is the floor that is directly above the sub-floor. It is made of $\frac{1}{4}$" finished plywood.

A **finish floor covering** is placed above the underlayment and the sub-floor. It can be made of sheet vinyl, vinyl tile, ceramic tile, parquet wood, $\frac{3}{4}$" wood strip, or carpeting.

REFRIGERATION THEORY

A technician must have a good understanding of science, chemistry, heat transfer, and refrigeration theory. He should be capable of applying this knowledge to his service work. The following is information that will aid the service technician.

Refrigeration is transferring heat from one place to another.

Temperature is the intensity of molecular motion.

There are four common **temperature scales.** They are Fahrenheit, Rankine, Celsius, and Kelvin.

Fahrenheit is the U.S. conventional unit, and its absolute zero scale is Rankine.

On the **Fahrenheit scale,** water freezes at 32° F and boils at 212° F.

To convert **Fahrenheit** to **Rankine,** add 460° F.

Example: 60° F + 460° F = 520° Rankine.

Celsius is the SI unit and its absolute zero scale is Kelvin.

SI stands for Systems International. This is also known as the metric system.

On the **Celsius scale,** water freezes at 0° and boils at 100° C.

To convert **Celsius** to **Kelvin,** add 273° C.

Example: 20° C + 273° C = 293° K

To convert **Celsius** to **Fahrenheit,** use the following calculation:

(1.8 × C) + 32 = F

Example: (1.8 × 40° C) + 32 = 104° F

To convert **Fahrenheit** to **Celsius,** use the following calculation:

.56 × (F - 32) = C

Example: .56 × (104° F - 32) = 40° C

Specific heat is the amount of heat required to raise one pound of a substance one degree Fahrenheit.

Sensible heat is the heat that can be measured or sensed by a thermometer.

Latent heat is hidden heat that changes the state of a substance. As the substance changes state, its temperature remains constant even though a large amount of heat energy is added or removed.

Heat is a form of energy and it is measured in Btus (British Thermal Unit).

A **Btu** is the amount of heat required to raise one pound of water one degree Fahrenheit.

To measure the amount of **Btus** of a substance without changing the substance's state, use the following formula:

Specific heat × temperature difference × weight in pounds = **Btus**

Example: If 2 pounds of water is raised from 40° F to 60° F, how many Btus are added to the water?

Btu = SP. HT. × T.D. × WT.

Specific Heat of water (SP. HT.) = 1

Temperature Difference (T.D.) = 20° F

Weight (W.T.) = 2 lbs

1 × 20 × 2 = 40 Btus

The amount of heat that it takes to change 1 pound of 212° F water to 1 pound of 212° F steam is **970 Btus**

The amount of heat required to change 1 pound of 32° F ice to 1 pound of 32° water is **144 Btus.**

The three **states of matter** are solid, liquid and gas. A **solid** exerts pressure downwards. A **liquid** exerts pressure downward and outwards in a container. A **gas** exerts pressure in all directions in a container.

The **Boyle's Law** states that a volume of a gas varies inversely with the absolute pressure, if the temperature remains constant. It also states that as pressure increases, volume decreases, and as pressure decreases, volume increases.

The **Charles' Law** states that at a constant pressure the volume of a gas varies directly with an absolute temperature. As a gas heats, it expands. As a gas cools, it contracts.

The **Dalton's Law** states that the total pressure of a confined mixture of gases is the sum of the pressures of each of the gases in the mixture.

Work = Force × Distance

Example: 500 lbs. × 50 ft. = 25000 ft. lbs. of Work.

Power is the rate of doing work. One horsepower is 33,000 ft. lb of work in 1 minute.

One Horsepower = 746 watts

A **watt** is the measurement of power in an electrical circuit. 1 watt = 3.41 Btus.

Heat is transferred by conduction, convection, or radiation. **Conduction** is the transfer of heat from one molecule to another. **Convection** is the transfer of heat by from one place to another by a fluid, commonly air or water. **Radiation** is the transfer of heat from a heat source to a solid object through space.

One ton of refrigeration effect is a specific quantity of heat that is transferred. One ton of refrigeration is equivalent to 12,000 Btus per hour or 200 Btus per minute.

Latent heat is the heat that is absorbed or extracted when a substance changes its state. When a substance changes from a liquid to a gas, it is called **latent heat of vaporization.** When a substance changes from a gas to a liquid, it is called **latent heat of condensation.** When a substance changes from a solid to a gas, it is called **sublimation.**

A refrigeration system absorbs heat by absorbing the latent heat of vaporization in the **evaporator.**

A refrigeration system extracts heat by removing the latent heat of condensation from the **condenser.**

Section 1 System Components and Tools

Figure 1-1: *Air Conditioning System Components*

AIR CONDITIONING COMPONENTS

An air conditioning system consists of various components. Each component helps keep the system operating properly. The technician is responsible for knowing about every system component, in order to correctly diagnose malfunctions.

The following are common **components** that are used in air conditioning systems:

1. The **compressor** changes the refrigerant from a low-pressure/low-temperature gas into a high-pressure/high-temperature gas.

2. An **oil separator** removes up to 98% of the oil that circulates in a refrigeration system. It is located in the discharge line.

3. The **condenser** changes a high-pressure/high-temperature gas into a high-pressure/high-temperature liquid to remove heat from the system.

4. The **receiver** stores liquid refrigerant until it is needed in the evaporator. It is located directly down stream from the condenser.

5. The **filter dryer** traps particles and small amounts of moisture in the system. It is located in the liquid line.

6a. The **sight glass** aids in charging the refrigeration system, and is located in the liquid line.

6b. A **moisture indicator** is commonly located in the sight glass. It senses moisture that may be present in the systems refrigerant.

7. The **metering device** changes a high-pressure/high-temperature liquid into a low-pressure/low-temperature liquid.

8. The **evaporator** changes a low-pressure/low-temperature liquid into a low-pressure/low-temperature gas. It absorbs heat into the system.

9. An **accumulator** traps liquid refrigerant and prevents it from entering the compressor. It is located in the suction line.

10. A **heat exchanger** subcools the liquid refrigerant in the liquid line and superheats the refrigerant gas in the suction line. It connects the liquid line to the suction line to increase the efficiency of the system.

As the temperature in a **sealed system** increases, the pressure also increases; and as the temperature decreases, the pressure decreases.

A **split air conditioning system** seperates the compressor, known as the condensing unit, and the condenser from the metering device and evaporator coil. The condensing unit is connected to the evaporator coil with refrigerant lines.

In a **residential split system,** the evaporator coil and metering device are located in the plenum of the furnace, while the compressor and condenser are located on the outside of the home. The liquid and suction line connect the outdoor and indoor units.

Outdoor condensing units should be mounted on a level concrete slab. Being level will allow any accumulation of water to run evenly out of the cabinet.

1. Compressor
2. Discharge Line
3. Condenser
4. Condenser Fan
5. Liquid Line
6. Filter Dryer
7. Metering Device (Restrictor)
8. Evaporator
9. Suction Line
10. Indoor Blower

Figure 1–2: *Split System*

A fill of gravel under the **concrete slab** is recommended to prevent settling of the ground. Settling can cause stress and even break the refrigerant lines.

There are several factors to consider for the **location of the condensing unit,** including available space, length of line set, visual appearance, and noise factor (near bedroom windows, patios, etc. . . .).

Manufacturer's **clearance recommendations** must always be followed when installing a condensing unit.

Residential split system outdoor **condensing units** are wired with a 208/240V single phase power supply.

A **fused disconnect** is located near the system to break the power source supply to the unit during service.

Technicians must perform a **visual check** to ensure all electrical connections are secure and tight before starting a newly installed system.

A **residential furnace** has 115V line voltage supplied to the system through a disconnect switch. This switch is used to energize the primary side of the control transformer and blower motor.

The **control transformer** of a residential furnace supplies 24V to the heating and cooling thermostat. This thermostat controls a fan relay coil, a contactor coil, and the heating control circuit.

A **room thermostat** should be installed on an inside wall away from any heat source. It should be 5 feet from the floor, and in an area with air motion but away from a supply air register.

On a heating and cooling **thermostat,** the "R" terminal is the power source from the 24V transformer.

On the heating and cooling **thermostat,** the "G" terminal energizes the blower relay coil.

On the heating and cooling **thermostat,** the "Y" terminal energizes the outdoor units contactor coil.

On the heating and cooling **thermostat,** the "W" terminal energizes the heating control circuit on the furnace.

There are **four mechanical connections** that are used to connect the refrigeration line set to the evaporator and the condensing unit. The four connections are quick connect fittings, compression fittings, flare fittings, and sweat connections.

There are two lines to a **refrigeration line set:** the liquid line and the suction line. The liquid line is the smaller line and the suction line is the larger line. The **suction line** must be insulated to decrease refrigerant temperature rise and exterior moisture condensation.

Refrigeration tubing must be protected by PVC if a line set is installed through a masonry wall. This prevents corrosion and possible refrigerant leaks.

Section 1 System Components and Tools

A **suction line** must be pitched toward the flow of vapor to ensure that oil returns to the compressor.

If **refrigerant lines** are coiled, they must be installed horizontally. The vapor must flow from top to bottom to ensure that oil returns to the compressor.

Quick connect fittings are used on precharged refrigeration line sets. These fittings seal the refrigerant in the tubing until the connections are tightened.

When using **precharged line sets,** excess refrigeration line must be coiled horizontally to prevent the trapping of the circulated oil.

If any alteration is made to the length of a **refrigeration line set,** the system must be recharged to match the new length.

Compression fittings use a brass coupling type nut with an O-ring and a threaded male adapter. The nut with the O-ring is tightened onto the male adapter, compressing the O-ring for a leakless joint.

Flare fittings use a flare and a flare nut connected to the male flare fitting to create a leakless joint.

Sweat (soldered) connections use couplings into which the refrigeration tubing fits. Silver braze is used to seal the tubing together.

When **brazing** a joint, capillary action causes the brazing material to flow throughout the entire joint area.

Non-precharged line sets and evaporator coils are charged with nitrogen and plugged until installation.

Condensing units are pre-charged with refrigerant for a specific length of refrigeration line set. If additional line is needed, the refrigerant charge must be increased according to manufacturer specifications.

Soft copper is used on refrigeration line sets up to 1" OD. Refrigeration line sets 1" OD and above are commonly made of **hard drawn copper** tubing.

OD stands for the outer diameter of copper tubing. When working with refrigeration tubing, the OD or outer diameter of the tubing is always taken into consideration.

The common **metering devices** used for split systems below a $7\frac{1}{2}$ ton capacity are the capillary and fixed orifice types.

TXV metering devices are used with split systems that have a capitity of $7\frac{1}{2}$ tons or higher and that have increased to a 12 SEER (Seasonal Energy Efficiency Ratio) or greater.

Attic-space mounted **split systems** use either a horizontal type furnace or a horizontal air handling system.

All **ductwork** must be insulated when used in an unconditioned space, such as an attic or crawl space.

An **auxiliary drain pan** must be installed on air conditioning systems that are located in an attic space. The drain pan must be $1\frac{1}{2}$ inches deep and not less than 3 inches larger than the air handling unit.

A normally closed **float switch** controls the cooling circuit and is commonly used to prevent damage to ceiling drywall if an overflow occurs.

A **secondary drain line** should extend from the auxiliary drain pan to a conspicuous place. This serves as an alarm which indicates that the primary drain is restricted.

Condensate drains must be **trapped,** as required by the equipment manufacturer.

REFRIGERANTS

There exist many types of refrigerants in the air conditioning and refrigeration industry. Refrigerants are classified into types based on their toxicity, flammability, and total safety. The technician must be informed of refrigerant properties and methods of usage.

The most common **HCFC refrigerant** in residential and commercial air conditioning systems is R-22.

The chemical name for **R-22** is Monochlorodifluoromethane. The molecular formula for **R-22** is $CHCLF^2$. **R-22** uses alkylbenzene oil and boils at -40° F at atmospheric pressure.

Refrigerant tanks must never be stored in an area above 125° F.

A **single component refrigerant** has one refrigerant, such as R-22 or R-12, that can be charged into a system as a gas.

The following are **single component refrigerants:** R-12, R-22, R-134A, R-123, and R-236fa.

The single component refrigerant that is used as a replacement for R-12 is (HFC) **R-134A.**

An **azeotropic refrigerant** is a mixture of two or more refrigerants that performs the same as a single component refrigerant. Azeotropic refrigerants will not separate when charged as a gas.

The following are **azeotropic refrigerants:** R-500, R-502, R-503, R-507, and R-410A.

The azeotropic (blended) refrigerant that is used as a replacement for R-22 is **R-410A.**

A **near-azeotropic refrigerant** is a mixture of two or more refrigerants that have different boiling points. A near-azeotropic refrigerant must be treated as a zeotropic blend. It must never be charged into a system as a gas, or the refrigerant composition will change.

The following are near-azeotropic refrigerants: R-401A, R-401B, R-402A, R-402B, R-403B, R-404A, R-408A, and R-409A.

The near-azeotropic (blended) refrigerant that is used as a replacement for R-12 is **R-409A**.

The near-azeotropic (blended) refrigerant that is used as a replacement for R-502 is **R-408A**.

A **zeotropic refrigerant** is a mixture of two or more refrigerants that have different boiling points. A zeotropic blend must never be charged into a system as a gas or the refrigerant composition will change.

The following are zeotropic refrigerants: R-407A, R-407B, and R-407C.

A **ternary refrigerant** is a blend of three refrigerants that act as a zeotropic refrigerant.

The following are ternary refrigerants: R-401A, R-401B, R-402A, R-402B, R-403B, R-404A, R-407A, R-407B, R-407C R-408A, R-409A, and R-410A.

Zeotropic and **near-azeotropic refrigerants** must be charged into a system as a liquid or the refrigerant will change its composition.

When working with blended refrigerants, pressure temperature charts use **"dewpoint line" and "bubble line" values**. The **"dewpoint line"** value is used to measure superheat. The **"bubble line"** value is used to measure subcooling.

Temperature glide is the temperature difference between the vapor and liquid state of a near-azeotropic, a zeotropic, and a ternary blend refrigerant. This change occurs at a constant pressure during evaporation and condensation.

An azeotropic refrigerant has a **temperature glide** of zero.

A near-azeotropic refrigerant has a **temperature glide** above zero up to 10° F.

A zeotropic refrigerant has a **temperature glide** of above 10° F.

Fractionation occurs when a zeotropic or a ternary refrigerant is added to a system as a gas. The refrigerants separate unevenly in the system, and this causes the refrigerant to change its composition.

The chemical name for **R-410A** is Difluoromethane/Pentafluoroethane. The molecular formula for **R-410A** is CH_2F_2/CF_2CF_3.

R-410A uses polyol ester oil and boils at -62.9° at atmospheric pressure.

Pressure temperature charts are used to determine certain pressures and temperatures of different refrigerants.

LUBRICANTS

Technicians must understand lubricants and their properties in order to use them correctly with proper refrigerants. The following are common terms and definitions of lubricants and their properties.

Mineral oils are often referred to as petroleum oils. These oils originate from the earth.

Synthetic oils are man-made chemicals produced to function as lubricants.

The three most common types of **synthetic oils** are glycol, alkylbenzene, and polyol ester.

R-134a is not miscible with **mineral oil**.

Mineral oils are most commonly used with CFC refrigerants.

Mineral oils are not used with HFC alternative refrigerants.

Mineral oils are not miscible with HFC refrigerants.

Polyol ester oil is the only lubricant suitable to use with HFC-134A.

Alkylbenzene oil is a synthetic oil made by mixing benzene and propylene.

Alkylbenzene oil is commonly used with HCFC refrigerants and can be used with CFC refrigerants.

PAG (polyalkylene glycol) oil refers to lubricants used in automotive air conditioning systems.

A **hygroscopic oil** has a high ability to absorb moisture.

If **moisture levels** are more than 100 PPM (parts per million) in an air conditioning or refrigeration system, corrosion will result.

Good characteristics of oil are proper viscosity, low pour point, non-corrosive, low carbon-forming tendencies, stability, low floc point, and low hygroscopic ability.

Floc point is the temperature at which oil starts to produce a waxy substance that may interfere with the refrigeration system.

Section 1   System Components and Tools

TABLE 1–1:   Pressure Temperature Chart

Examples:

R-22 will be at 40° F when the pressure is 68 PSIG.

R-12 will be at 37 PSIG when its temperature is 40° F.

PSIG	Temperature in Degrees F.		
	R-12	R-22	R-502
25	25	1	-7
26	27	2	-6
27	28	4	-5
28	29	5	-3
29	31	6	-2
30	32	7	-1
31	33	8	0
32	34	9	1
33	35	10	2
34	37	11	3
35	38	12	4
36	39	13	5
37	**40**	14	6
38	41	15	7
39	42	16	8
40	43	17	9
42	45	19	11
44	47	21	13
46	49	23	15
48	51	24	16
50	53	26	18
52	55	28	20
54	57	29	21
56	58	31	23
58	60	32	24
60	62	34	26
62	64	35	27
64	65	37	29
66	67	38	30
68	68	**40**	32
70	70	41	33
72	72	42	34
74	73	44	36

Pour point is the ability of oil to flow with refrigerants at very low temperatures. A lower pour point is better.

Miscibility is the ability of a fluid to readily mix with another fluid.

Lubricants should remain miscible at all temperatures and pressures.

A **synthetic lubricant** absorbs more moisture than a mineral oil.

A **PAG oil** absorbs more moisture than a polyol ester oil.

Polyol ester oil tolerates 3% to 5% mineral oil.

It is harder to **dehydrate** a system using polyol ester oil than a system using mineral oil.

Polyol ester lubricants should be stored in their own container because moisture can permeate through certain materials.

The **viscosity** of a lubricant oil is measured by the Saybolt Universal Viscosity Test.

The **higher the viscosity,** the thicker the oil.

Flash point refers to the temperature at which vapors from the oil surface flashes will ignite with a flame, but not persistently burn.

The main **function of a lubricant** is to decrease friction and minimize wear of internal components.

An oil lubricant helps to form a **liquid seal** from the high and low side of the system.

When retrofitting a system with alternative refrigerants, it is the **technician's responsibility** to recover the oil/refrigerant mixture before charging the system with the new refrigerant and oil.

According to the EPA, any **oil removed** from a system is the property of the service contractor.

COMPRESSORS

A compressor is the most vital component of an air conditioning and refrigeration system. There are many types of compressors. Each is designed to be used for a specific application. Compressor manufacturers are always working on methods to improve the efficiency of their compressors. The following are related terms for air conditioning and refrigeration compressors.

A **compressor** takes a low-pressure/low-temperature refrigerant gas from the evaporator and raises it to a high-pressure/high-temperature gas. Then it pumps the refrigerant gas into the condenser.

The most common compressors used for residential and light commercial air conditioning are the **reciprocating** and the **scroll** designs.

The **positive displacement compressor** designs are reciprocating, rotary, screw, and scroll designs.

Reciprocating compressors use pistons to compress the refrigerant gas.

Screw compressors use two special helical rotors that induce suction vapors then compress them. They are used on large tonnage systems.

Scroll compressors are 10% higher in efficiency then reciprocating types. They use two spiral portions that are meshed together to compress refrigerant gas. Scroll compressors handle liquid floodback better than reciprocating compressors.

A **non-positive displacement compressor** design is the centrifugal compressor.

Reciprocating compressors use pistons, cylinders, and valves to compress the refrigerant gas.

Compressor Ratio = Discharge PSIA/Suction PSIA. An increase in compression ratio will overheat a compressor.

The **three classifications** of compressors are hermetic, semi-hermetic, and open.

The **hermetic compressor** is a welded, non-servicing type that is driven by an internal motor.

The **semi-hermetic compressor** is bolted together and is serviceable. It is driven by an internal motor.

The **open compressor** is bolted together and is serviceable. It is driven by an external motor.

The open compressor can use **two types of drives.** They are the belt drive and the direct drive.

The two types of **compressor lubrication** systems are splash lubrication and forced lubrication.

Compressors are cooled by the low-pressure/low-temperature refrigerant gas from the evaporator and by the condenser fan moving air over its outer shell.

Multi-stage compressors commonly use a two stage thermostat to cycle a two speed compressor. They also use two compressors connected together to operate one system.

Crankcase heaters are used to heat the oil of the compressor during the off cycle. They aid in keeping the liquid refrigerant from mixing with the oil, which could cause a flooded start.

AIR COOLED CONDENSERS

Air cooled condensers are commonly used in split and package type air conditioning systems. The technician must have a good understanding of condensers and their related terms.

A **condenser** removes heat from an air conditioning system by changing the refrigerant from a high-pressure/high-temperature gas to a high-pressure/high-temperature liquid. It also de-superheats and subcools the refrigerant.

Additional superheat is found at the end of the evaporator and the heat from the motor windings. This heat, along with heat of compression, must be removed from the refrigerant before it can be condensed.

The tubes of a **condenser** are made out of copper or aluminum.

The **fins** of a condenser are made out of aluminum.

Condensers must have properly designed clearances on all sides of the unit to prevent re-circulation of air over the coil.

Split system air cooled **condensing units** contain the compressor, the condenser coil, the condenser fan, service valves, and various electrical controls.

Packaged units contain the compressor, the condenser coil, the evaporator coil, the condenser and evaporator fans, service valves, the heating portion of the unit, and various electrical controls.

The **temperature difference** between the ambient air and the refrigerant in the condenser should be between 20° and 30° F.

The **air flow** of a condenser coil is either a side discharge or a top discharge type.

There are two types of **side discharge** condensers. One type pulls air across the coil, through the fan, and out to the ambient air. The other type pulls air across the fan and pushes it through the condenser coil. If the **air** is pushed over the condenser coil, the coil must be cleaned directly in front of the condenser fan blades. If the air is pulled over the condenser coil, the coil must be cleaned at the exposed section of the coil. If **air** is trapped inside a condenser coil the high-side pressure will increase. This causes a decrease in cooling capacity and increase in amperage, and the excessive heat could damage the system.

Section 1   System Components and Tools

Top discharge condensers pull air from the bottom or sides of the condenser coil and push it out the top of the condenser.

WATER COOLED CONDENSERS

Many light commercial air conditioning and refrigeration systems utilize water cooled condensers. It is important for the technician to have a complete understanding of their operation.

The efficiency of a **water cooled condenser** is effected by the temperature difference between the refrigerant and the water in the condenser and the quantity of water flowing through the system.

The **capacity of the condenser increases** as the temperature difference between the refrigerant and the water increases.

In a water cooled condenser, an **increase of water quantity** decreases the power consumption of the compressor.

In a water cooled condenser, a **decrease of water** saves water but increases the power consumption of the compressor.

The **fouling factor** is the measurement of heat transfer characteristics of water. The higher the fouling factor, the poorer the heat transfer.

The **higher the fouling factor** of the water is, the higher the flow rate of water needs to be.

Whenever possible, a **condenser's water** should be discharged to storm drains and storm sewers if local codes permit.

A **water regulating valve** controls the flow of water through a water cooled condenser.

A **water regulating valve** can be located at the inlet or outlet of a water cooled condenser.

A **water regulating valve** operates on spring pressure and acts as the closing force in water cooled condensers. In refrigeration systems, high-side pressure acts as the opening force.

The condensing temperature is commonly **10 degrees higher** than the temperture of the water leaving the condenser.

Mineral deposits and **scale** build up on the inside of a water cooled condenser. This build up helps insulate the water from the refrigerant tube. It also decreases the efficiency. The decrease in efficiency causes an increase in high-side pressure and water consumption.

The most common water cooled condensers are the **tube within a tube type, shell and coil type,** and **shell and tube type.**

The **tube within a tube** type water cooled condenser uses the inner tube for refrigerant and the outer tube for water.

The tube within a tube uses the **counterflow principle** where the cooler water is in contact with the cooler refrigerant liquid and the warmer water is in contact with the hot gas.

The **tube within a tube** condenser is commonly constructed of steel and copper.

There are **two common types** of tube within a tube condensers. One type is the **sealed coil,** which must be chemically cleaned. The other type has **horizontal tubes,** one on top of another. These tubes are flanged on each end. The flanged ends make it possible to manually clean the condenser.

The **shell and coil** type water cooled condenser has the refrigerant in the shell. The water flows through the coil located inside the shell.

The shell of a **shell and coil** condenser acts as a liquid receiver. It holds the refrigerant charge until it is needed in the evaporator.

The **shell and coil** condenser can not be manually cleaned. It must be chemically cleaned.

The **shell and tube** type water cooled condenser has the refrigerant in the shell. The water flows through the tubes located inside of the shell.

The shell of a **shell and tube** condenser acts as a liquid receiver. It holds the refrigerant charge until it is needed in the evaporator.

The **water box** is the end cap of a shell and tube condenser.

A **shell and tube** condenser can be cleaned manually by removing the water box.

INDOOR EVAPORATOR COILS

There are various types and styles of evaporator coils that are used on air conditioning and refrigeration equipment. It is important for the technician to have thorough knowledge of evaporator coil types and understand their function.

An **evaporator** absorbs heat into the refrigerant by the latent heat of vaporization.

On an upflow furnace, the indoor evaporator coil must be **installed at least 2 to 3 inches** above the heat exchanger to prevent the furnace from overheating.

Direct expansion evaporators have refrigerant inside the tubes of the evaporator, and the controlled medium is directly in contact with the coil.

The **tubes** of an **evaporator coil** are made of copper.

The **fins** of an **evaporator coil** are made of aluminum.

A **distributor** is a device that feeds refrigerant evenly through the circuits of an evaporator coil.

Split system evaporator coils are located in the furnace or the air handling unit.

When an **uncased coil** is installed into an existing furnace, metal plates are installed around the outside of the coil to force air through it. If there is any air bypass, this will decrease the coil efficiency.

When an **evaporator coil** is installed, it is important to make sure the coil is level for proper condensate drainage.

An **"A" coil** is two coil sections connected together in the shape of an "A". This is the most common coil used in upflow residential systems today.

An **"N" coil** is three coil sections connected together in the shape of an "N". This is used when a narrow width of a coil is needed in an upflow application.

A **slant coil** is used on upflow or downflow systems where a low coil height is critical.

Upflow system evaporator coils are commonly located on the positive pressure side of the blower after the heat exchanger.

Downflow system evaporator coils are commonly located on the positive pressure side of the blower after the heat exchanger.

An **"H" coil** is used on horizontal flow furnaces and packaged AC units. The coil is designed so air flows over it from one side of the system to the other.

In a residential split system, using a **horizontal flow system,** air is pushed over the coil. In commercial package unit, using a horzontal flow system, air is pulled over the coil.

When an evaporator coil is installed on the negative side of the blower, air is pulled over the coil. A **"P" trap** is mandatory on the drain line to prevent condensation from entering the duct system.

METERING DEVICES

One of the main components of a refrigeration and air conditioning system is the metering device. Although there are several types of metering devices each used for a specific application, they all work to feed refrigerant to the evaporator. The following are types of metering devices.

A **metering device** takes a high-pressure/high-temperature subcooled liquid from the condenser and changes it to a low-pressure/low-temperature saturated liquid. It then directs the liquid into the evaporator.

Flash gas is the instantaneous evaporation of liquid refrigerant as it enters the evaporator.

A **heat exchanger** is commonly used to decrease flash gas.

There are three common types of **metering devices** that are used in residential and light commercial air conditioning. They are the **capillary tube,** the **TXV,** and the **piston** or **fixed bore.**

The pressure drop of a **capillary tube** type metering device is determined by the tubing length and inner diameter.

TXV stands for Thermostatic Expansion Valve.

A **Thermostatic Expansion Valve, TXV,** has three operating pressures: the sensing bulb pressure (opening force), the evaporator pressure (closing force), and the spring pressure (closing force).

Turning the **TXV valve stem** clockwise increases superheat. Turning it counterclockwise decreases the superheat.

Superheat (entering the compressor) = suction line temperature – evaporator's boiling point temperature.

Example:

Suction line temperature = 50° F

Evaporator boiling point temperature = 35° F

Superheat = 50° F – 35° F

Superheat = 15° F

An **external equalizer line** is used on a TXV when there is an excessive pressure drop through an evaporator. The equalizer line compensates for the drop in pressure through the evaporator. The line is located in the suction line 6 to 8 inches downstream from the TXV sensing bulb.

TXV sensing bulb needs to be placed at the 4 o'clock and 8 o'clock positions on the suction line if it is larger than $\frac{7}{8}$ inches in diameter. If the suction line is smaller than $\frac{7}{8}$ inches in diameter, the sensing bulb must be located at the 12 o'clock position.

The pressure drop of a **piston** or **fixed bore** type metering device is determined by its orifice size and the high side pressure.

Section 1   System Components and Tools

SERVICE VALVES

Service valves are used to open and close off the system's flow of refrigerant. They are also used to provide access for charging evacuation and recovering refrigerant. The following are various types of service valves.

Service valves are usually located on the outdoor condensing unit.

A **discharge service valve** is located on the outlet of the compressor. NEVER FRONT SEAT A DISCHARGE SERVICE VALVE.

A **liquid line service valve** is located in the liquid line.

Front seating the liquid line service valve blocks the flow of refrigerant to the liquid line. If the compressor is started, the system pumps down, removing the refrigerant from the liquid line and the evaporator coil. The refrigerant is then stored in the condensing unit.

A **suction service valve** is located on the inlet of the compressor.

Front seating the suction service valve blocks the flow of refrigerant to the compressor and draws a vacuum on the low pressure gauge to test for operation of compressor valves and rings.

Front seating the suction service valve blocks the flow of refrigerant to the compressor.

A **king valve** is a three position service valve that is located on the outlet of a receiver.

A **queen valve** is a two or a three position service valve located on the inlet of the receiver.

A **two position service valve,** located on split system condensing units, uses the back seated position to open the refrigeration lines to the rest of the system for refrigerant flow and to obtain gauge readings.

A **two position liquid line service valve** uses the front seated position to pump a system down or test compressor valves.

A **Schrader valve** is used to check system pressures. It is permanently connected in an air conditioning condensing unit. The valve core seals the refrigerant in the system until a manifold gauge hose is connected. When a manifold gauge hose is connected, the hose will depress the valve core and pressure will be supplied to the gauge.

Three position service valves are used on older split systems and larger packaged air conditioning systems. Three position service valves are located on the suction line and the liquid line.

Back seating a three position service valve blocks the gauge port and allows the system to operate.

Mid-seating a three position service valve (gauge position) allows the system to operate and direct refrigerant pressure to the gauge port.

Front seating a three position liquid line service valve blocks the flow of refrigerant to the liquid line. If the compressor is started, the systems pumps down. This removes the refrigerant from the liquid line and the evaporator coil. The refrigerant is then stored it in the condensing unit.

LEAK DETECTION

There are many different types of leak detection methods and equipment. The following are examples of leak detection methods.

An **oil trace** around copper tube fittings may be an indication of a refrigerant leak.

When testing for a **leak,** the system must be pressurized and electrically de-energized.

Air has a tendency to stay in the condenser and increase the operating high-side pressure.

If there is no pressure in the system, pressure can be increased for **leak detection** by adding nitrogen and a trace amount (usually an ounce or two) of R-22.

A pressure regulator and an in-line pressure relief valve should be used when **pressurizing a system** with nitrogen.

Oxygen must NEVER be used to purge or pressurize a refrigeration or air conditioning system. Doing this could cause an explosion.

Pressures should never exceed the low side **test pressure rating** on the data plate.

A **soap bubble** solution can be used to test for a suspected leak. If a leak is present, bubbles will start to form.

If manufacturer's specifications allow, **refrigerant dye** can be added to a refrigeration system and circulated for 24-hours to detect a leak.

A **halide leak detector** is used with systems containing halocarbon refrigerants. It must never be used near flammable material.

When using a **halide leak detector,** the flame turns green because of the results of chlorine in the refrigerant coming in contact with a hot copper element located in the halide detector. The halide leak detector will not detect a Hydroflurocarbon (HFC) refrigerant.

An **electronic leak detector** is designed to test for extremely small refrigerant leaks. Since refrigerant is heavier than air, the detector should always be held under the tubing.

When a leak is located using an **electronic leak detector,** the detector's sound becomes high pitched and its flashing light becomes steady.

An **ultrasonic leak detector** senses a leak when the high pitched sound waves, created by a leak, enter the detector.

An **ultrasonic detector** can sense a leak in a system that is under positive or negative pressure.

An **ultra violet light,** used with a fluorescent system additive, can detect extremely small refrigerant leaks.

RECOVERY EQUIPMENT

When it becomes necessary to open a sealed refrigeration system, the refrigerants need to be recovered. The following are terms and information related to recovery equipment. Refer to Table 1–2 on page 17 for examples.

Recovery means to remove refrigerant from an appliance and store it in an external container.

The **three methods of recovery** are charge migration, use of the appliance compressor, and use of a refrigerant recovery unit.

The **three types of recovery machines** are vapor recovery units, liquid recovery units, and vapor/liquid recovery units.

There are **three classifications of equipment:** very high pressure, high pressure, and low pressure.

Very high pressure equipment uses refrigerant that boils below -58° F at atmospheric pressure.

High pressure equipment uses refrigerant that boils between -58° and 50° F at atmospheric pressure.

Low pressure equipment uses refrigerant that boils above 50° F at atmospheric pressure.

The **requirements of recovery equipment** are divided into size and usage.

Small appliance equipment must reduce the pressure in a system to 4" Hg.

Recovery cylinders must be approved by the U.S. Department of Transportation (DOT). A **recovery cylinder** must not be filled to more than 80% of its capacity.

To **recycle** means to remove and clean the refrigerant from an appliance without meeting the requirements for reclamation.

To **reclaim** is to process refrigerant to a level equal to new refrigerant specifications by chemical analysis. The reclaimed refrigerant must meet the standards of ARI 700 before it can be resold.

Recycling removes moisture, air, and acids from a refrigerant.

The filters on **recycling machines** must be changed as required. The oil separators also need the oil removed regularly.

SOLDERING AND BRAZING

A large part of installations and some repairs require a technician to perform soldering and brazing techniques. The following terms and information are related to soldering and brazing.

Soldering is accomplished when a temperature below 800° F is used to bond metals.

Brazing is accomplished when a temperature of 800° F or above is used to bond metals.

Never attempt to **solder or braze** a joint if a system is under pressure. This could cause the rapid escape of refrigerant that could result in an injury.

Stay Brite #8 solder is 94% tin and 6% silver. It melts at 535° F.

95-5 solder is 95% tin and 5% antinomy. It melts at 450° F.

Flux is used to keep material clean while soldering.

Braze and **silver solder** melting point temperatures range from 1100° to 1480° F.

Brazing material creates a stronger joint than soft solder or silver solder.

Leaks in coils made from aluminum are commonly repaired with epoxy or aluminum solder.

Never overheat a **filter dryer** when brazing it into the system. This will damage the filter.

When brazing a joint, **capillary action** causes the brazing material to flow throughout the entire joint area.

Section 1   System Components and Tools

TABLE 1–2 *Inches of Mercury Vacuum in Appliances*

	Inches of Mercury Vacuum Equipment Manufactured	
High pressure appliance	*Before 11-15-93*	*After 11-15-93*
HCFC-22 charge (under 200 lbs.)	0	0
HCFC-22 charge (200 lbs and over)	4	10
Other high pressure appliances using CFC-12, 500, 502, 114 (under 200 lbs)	4	0
Other high pressure appliances using CFC-12, 500, 502, 114 (200 lbs and over)	4	15
Very high pressure appliances (Using CFC-13 & 503)	0	0
Low pressure appliance:		
CFC-11, HCFC-123	25Hg.	25mm

Regulators have two gauges. One gauge measures pressure to the torch tip in PSIG. The other measures tank pressure in PSIG.

The operating pressures of an **acetylene regulator** ranges from 0–15 PSIG.

The operating pressures of an **oxygen regulator** ranges from 0–100 PSIG.

An **oxygen** and **acetylene regulator** must be turned clockwise to increase pressure and counterclockwise to decrease pressure.

An **oxyacetylene torch** uses a set of hoses to connect the torch handle to the oxygen and acetylene pressure regulators.

The **oxygen and acetylene** pressure gauges display the remaining tank pressure and the regulated pressure to the torch tip.

A tank of dry nitrogen and a pressure regulator are used to **purge** the refrigeration tubing. This helps prevent oxidation of the tubing during the brazing process.

Oxygen must NEVER be used to purge or pressurize a refrigeration or air conditioning system. Doing this could cause an explosion.

A **tube cutter** is used to cut hard or soft tubing. A **reaming tool** generally accompanies the tube cutter. It is used to remove small burrs from inside the tubing after it has been cut.

Small tube cutters, sometimes called "Imps", are used to cut small diameter tubing. The **reaming tool** is used to remove the burrs of the copper tubing after it has been cut.

Swaging of soft copper tubing is often done to connect the same size tubing without the use of a coupling. When swaging, the inside diameter of one tube is expanded to fit over the outside diameter of the other tube. The depth of the swage is determined by the type of swaging tool that is used.

When **bending** soft copper tubing, the minimum bending radius on smaller tubing is no less than 5 times the diameter of the tubing.

Example: $\frac{1}{2}"$ diameter tubing $\times 5 = 2\frac{1}{2}"$ minimum radius bend

A **tube bender** is used to prevent kinks when bending soft copper tubing. When bending copper tubing, the tubing diameter must never be decreased.

When assisting in a bend, a **spring type tube bender** can fit the inside or the outside of the tubing.

A **lever bender** can make a precision bend up to 180° on soft copper tubing.

A **flare nut** and a **flare fitting** are used in the refrigeration and air conditioning field to create a leak free mechanical joint.

A **flaring block** holds copper tubing in place when using the flaring tool.

A **flaring tool** is the tool used to create a flare by beveling the copper tubing with the flare tool. To prevent splits in the flare, tighten and loosen the tool as the flare is being made.

MANIFOLD GAUGES

A manifold gauge set is one of the most important instruments a service technician can possess. These gauges provide the technician with needed information about the operation and status of the system.

A **manifold gauge set** is used to check operating pressures on a refrigeration and air conditioning system in PSIG (pounds per square inch gauge). The temperature is taken on the inside color-coded scale.

A **manifold gauge set** is adjusted by opening the gauges to the atmosphere and adjusting the pressure to zero PSIG.

Manifold gauge sets are available in analog and digital readout.

The **center yellow hose** of a manifold gauge set is used to charge, evacuate, add oil, and recover refrigerant from a refrigeration and air conditioning system.

A **manifold gauge** set contains two gauges. One is a high-side gauge and the other is a low-side gauge. Each gauge reads positive pressure in PSIG. The low-side gauge can also read negitive pressures in inches of Hg. (mercury).

In an air conditioning system, the **high-side gauge** is commonly connected to the smaller line and the **low-side gauge** is commonly connected to the larger line.

The **blue gauge** on a manifold gauge set is used for testing the low-side pressure. It is called a **compound gauge,** because it measures positive pressures up to 350 PSIG and negative pressures down to 30" Hg.

The **red gauge** on a manifold gauge set is used for testing high-side pressure. It measures positive pressure from 0 to 500 PSIG.

Both high and low **pressure gauges** measure the pressure and temperature of different refrigerants. On a gauge, the pressure measurements are the outer black numbers measured in PSIG. The temperature measurements are the inner colored numbers measured in degrees Fahrenheit.

Four way manifold gauges use front and side valves to evacuate and charge a system without having to switch hoses.

Low loss fittings are used on manifold gauges to decrease the amount of refrigerant that is vented to the atmosphere.

Excessively **high discharge pressure** can be caused by air (non-condensables) in the system, an overcharge of refrigerant, a dirty condenser, re-circulating ambient air, or insufficient air flow over the condenser.

A **decrease in suction pressure** can be caused by a refrigerant leak, a restricted or defective metering device, a dirty evaporator coil, or insufficient air flow over the evaporator coil.

To convert **PSIG to PSIA** (pounds per square inch absolute), add 14.7 to the PSIG.

To convert **PSIA to PSIG,** subtract 14.7 from the PSIA.

SYSTEM EVACUATION

The evacuation process removes moisture and non-condensables (air) from the sealed system. Both moisture and air will hinder satisfactory operation of the system. The following are terms related to evacuation.

Any **system opened** to the atmosphere must be completely evacuated.

A minimum evacuation level for an air conditioning system is **500 microns.**

Dehydration is the removal of moisture in a refrigeration or air conditioning system.

A **larger diameter center hose** decreases the evacuation time of a system.

Higher ambient temperatures decrease the evacuation time of a system.

If a system is **evacuated** and the pressure increases slightly then stops, moisture is present in the system.

If a system is **evacuated** and pressure continually increases, a leak is present in the system.

A **triple evacuation** is accomplished by evacuating a system to 28" Hg. and charging the system to 0 PSIG three times.

A **single** or **deep evacuation** is accomplished by evacuating a system from 50–100 microns. The system should hold the vacuum after the pump is turned off.

A **vacuum pump** is connected to the center hose of the gauge port and removes air and moisture from a refrigeration or air conditioning system.

A **vacuum pump's oil** should be changed after every evacuation.

Vacuum pump oil tends to become full of moisture and acid after an evacuation. This effects the vacuum pump's ability to pull a deep vacuum.

Section 1 System Components and Tools

Always allow air to enter a **vacuum pump** before storing, or the pump could become oil locked.

A **micron gauge** is an instrument that accurately measures the depth of a vacuum.

Both **manifold gauge valves** should be open during an evacuation.

A **thermistor vacuum gauge** is an instrument that accurately measures the depth of a vacuum in microns.

A **valve core tool** is used to remove the valve core on a schrader valve. This decreases the evacuation time of a system.

When a system incorporates a three way **service valve,** it must be in the mid-position during the testing of the systems operating pressure, recovery, evacuation, and recharging.

REFRIGERANT CHARGING

There are various methods used to add refrigerant to a system. The method may depend on the type of application and the refrigerant being used. The following are terms and related information on charging.

When the **weight method** of charging is used, an electronic charging scale weighs the proper refrigerant charge into a system.

A **charging scale** is commonly preset for the proper charge of a system in pounds and ounces, or the scale is set to zero and it measures the amount that is removed.

When the **volume method** of charging is used, a charging cylinder measures the proper refrigerant quantity into a system.

A **sight glass** is a device located in the liquid line that allows a technician to see the refrigerant. It is used to aid in properly charging a system. When a system is fully charged, the sight glass will appear clear or free of bubbles. A **sight glass** in the liquid line assists in a systems refrigerant charge.

When **liquid charging** a system, the system must be de-energized and have refrigerant added into the high side of the system through the high side manifold gauge.

When **gas charging** a system, the system must be energized and have refrigerant added to the low side of the system through the low side manifold gauge.

A **capillary tube system** must have an accurate refrigerant charge. Too much refrigerant will flood the coil, and too little will starve the coil.

To charge a system using the **superheat method,** the technician must first measure the indoor wet-bulb temperature, the outdoor dry-bulb temperature, and possess a superheat cross reference chart.

The **superheat charging method** is used on a system that uses a capillary tube or piston type metering device.

To **measure superheat,** subtract the boiling point temperature of the refrigerant (found on the low side gauge) from the suction line temperature (within 6" from the compressor).

To charge a system using the **subcool method,** the technician must know the high side temperature, the liquid line temperature, and the manufacturers subcool recommendations.

To **measure subcooling,** subtract the liquid line temperature from the high side condensing temperature.

The **subcool charging method** is used on a system that uses a TXV type metering device.

If an air conditioning system has an **excessive leak,** oil must be added.

DEFROST SYSTEMS

There are various types of defrost systems used for specific refrigeration applications. The following are defrost types and related terms.

On a **manual shutdown defrost system,** the evaporator coil is designed to be exposed to the surrounding cabinet air.

A plate type evaporator is most commonly used on a **manual shutdown defrost system.**

On a **manual shutdown defrost system,** the evaporator coil usually operates below 32° F.

On a **pressure operated defrost system,** a low-pressure control is commonly used to shut the system down until the coil is free of frost.

On a **pressure operated defrost system,** the low-pressure control is set to cut-in when the evaporator coil is free of frost and has a minimum of 33° F saturation temperature.

On a **pressure operated defrost system,** the storage temperatures are above 35° F.

On a **pressure operated defrost system,** the evaporator coil is defrosted by de-energizing the condensing unit and keeping the evaporator fans in operation to move the 35°F air over the coil to remove frost build up.

On a **pressure operated defrost system,** the defrost period depends on the cut-in and cut-out settings of the low pressure control.

A **time shutdown** type of defrost timer uses trip pins to set the defrost timer into a defrost mode and determine its length of defrost. The trip pins are located on the defrost timer.

A **time shutdown** type of defrost timer uses a set of "normally closed" contacts to control the condensing unit.

On a **time shutdown defrost system,** the defrost period is initiated by a cam operated switch that is controlled by a time clock.

On a **time shutdown defrost system,** the timer is adjusted to stop and start the compressor at selected times of the day or night.

On a **time shutdown defrost system,** the cut-in and cut-out times must be determined by the manufacturer's recommendation.

When **supplementary heat** such as electric heaters, hot gas solenoids, and warm water is used to aid frost removal during a defrost cycle, products are commonly kept below 29° F.

A **24-hour defrost timer** is used on a system incorporating supplementary heat control with "normally open" and "normally closed" sets of contacts. The "normally open" contacts control the supplementary heat circuit and the "normally closed" contacts control the condensing unit and the evaporator fan motor.

In a **hot gas defrost cycle,** the defrost system must keep the compressor energized during defrost.

On a **24-hour defrost timer** used with supplementary heat, the trip pins are used to start the defrost and an adjustable setting is used to end the defrost.

The maximum amount of **trip pins** that can be installed at one time in a 24-hour defrost timer is 12, one every two hours.

The **length of the defrost cycle** on a 24-hour timer ranges from 2 to 110 minutes.

On an **electric heat defrost system** a supplemental electric heater is used to remove the frost build up on the evaporator coil.

In an **automatic water defrost system,** water may be used directly as a defrosting medium. The system's water is electrically heated. Then the water is sprayed by headers directly over the evaporator's surface, removing frost from the coil.

In an **automatic water defrost system,** excess water and melted frost flow to a drain line. During water defrost, the compressor and evaporator fans are stopped.

In a **hot gas defrost system,** hot gas exits the discharge line of the compressor and enters the inlet of the evaporator located after the **metering device.** A "normally closed" hot gas solenoid controls the refrigerant flow during the defrost cycle.

Hot gas is fed into the evaporator, raising the temperature and pressure of the coil. Since it is likely that the refrigerant may condense, a potential for **liquid slugging** of the compressor is possible and must be taken into consideration during installation. An **accumulator** is commonly added to the suction line to prevent the compressor from developing a liquid slug.

In a **hot gas defrost system,** the hot gas is more effective when initially entering the evaporator.

In a **time initiated temperature terminated defrost system,** the defrost period is terminated by a termination thermostat.

In a **time initiated temperature terminated defrost system,** the termination thermostat may be adjustable or preset.

The **termination thermostat** is used to terminate the defrost cycle on a time initiated temperature terminated defrost system and is commonly controlled by the evaporator coil temperature.

In a **time-initiated, temperature-terminated defrost system,** the termination thermostat can be a single-pole single-throw, or a single-pole double-throw set of contacts.

A time-initiated, temperature-terminated defrost timer commonly uses a **"timer release solenoid"** to change the timer switch contacts back to the cooling mode after the coil has been defrosted.

In a single-pole, double-throw **termination thermostat,** the switch contacts that control the "timer release solenoid" are commonly a "close on rise" type. The contacts that "open on rise" usually control the evaporator fan motor.

In a single-pole, single-throw **termination thermostat,** the switch contacts control a "timer release solenoid." The "timer release solenoid" is energized when the coil is defrosted and the position of the contacts are closed.

The **"timer release solenoid,"** in conjunction with the termination thermostat, will stop the defrost cycle prior to defrost timer settings if the evaporator is clear of frost.

Section 1 System Components and Tools

Example: If the **defrost timer** is set for a 35 minute defrost and the evaporator coil is defrosted after 20 minutes, the termination switch energizes the timer release solenoid causing the system to switch back into the cooling mode.

In a **time-initiated, temperature-terminated defrost system,** the action of the "timer release solenoid" causes the defrost timer switches to change back to the cooling mode. This starts the condensing unit. At the same time, it de-energizes the defrost circuit and the termination thermostat circuit.

FUNDAMENTALS OF GAS COMBUSTION

It is of vital importance that technicians have a knowledge of the fundamentals of gas combustion. The following are basic related terms.

Two common **types of gases** used today are Natural (Methane) and L.P.G. (Liquified Petroleum Gas).

Air has a **specific gravity** of 1.0.

Natural gas has a **specific gravity** of 0.6. It is lighter than air.

L.P.G. has a **specific gravity** of 1.5. It is heavier than air.

Natural gas has a heating value of approximately 1000 Btus per cubic foot.

L.P.G. has a heating value of approximately 2500 Btus per cubic foot.

Combustion is a chemical reaction, or the rapid oxidation of fuel that results in the release of heat and light.

The requirements for proper combustion are **fuel, oxygen,** and **ignition.**

An **oxygen molecule** is 2 oxygen atoms (O_2.)

For **complete combustion,** one part methane (CH_4) is combined with 2 parts oxygen ($2O_2$).

A **carbon dioxide molecule** is 1 carbon atom combined with 2 oxygen atoms (CO_2).

A **water vapor** molecule is 2 hydrogen atoms combined with 1 oxygen atom (H_2O).

During **complete combustion,** (CH_4) + ($2O_2$) will change to (CO_2) + (H_2O) + HEAT.

Air is approximately 21% oxygen and 79% nitrogen.

The proper air and fuel mixture for **complete combustion** is 1 part fuel and 10 parts air. An additional 5 parts of **excess air** is added to the combustion air to insure complete combustion.

Carbon monoxide is a deadly gas caused by incomplete combustion and flame impingement on a cold surface of a heat exchanger.

A **natural gas** (methane) molecule is 1 carbon atom combined with 4 hydrogen atoms (CH_4).

Heat transfer is accomplished in three ways: conduction, convention, and radiation.

Thermodynamics is a science that deals with the relationship of heat and mechanical energy.

GAS FURNACE CONTROLS AND COMPONENTS

Heating service technicians must know the function of numerous controls and components. Technicians must also understand the ignition systems that contain many of these components.

A **standing pilot furnace** uses a continuous pilot designed to ignite the gas at the main burners of the furnace.

A **standing pilot furnace** uses a thermocouple or a thermopile to prove the existence of a pilot.

A **standing pilot ignition system** consists of a main gas valve, a pressure regulator, a pilot assembly, and a thermocouple.

A **thermocouple** is made of two dissimilar metals and is part of the pilot assembly.

The **thermopile** is a pilot generator rated at a higher millivolt rating than a thermocouple.

The **pilot flame** should engulf the thermocouple tip by $\frac{3}{8}$ to $\frac{1}{2}$ inch.

On a standing pilot furnace, the pilot is proven by a **30 millivolt** signal from the thermocouple to the gas valve.

A **Thermopile** millivolt system will range from 550 to 750 millivolts.

An **intermittent pilot furnace** uses an ignition system that ignites the pilot on every cycle of the furnace. The pilot ignition system is commonly a spark or a hot surface type.

A **spark type ignition system** uses an ignition transformer, a cable, and an electrode to produce the spark for the pilot.

A **hot surface type ignition system** controls current through a silicon carbide ignitor which increases the temperature above the ignition point of the fuel. Its operation is much quieter than a spark ignition system.

In a **hot surface ignition system,** the components consist of an ignition control module, a combination gas valve, a flame sensor, and a hot surface ignitor.

Most **hot surface igniters** operate on 120 volts. Igniters in some ignition systems have recently started operating on 24 volts.

In a **direct ignition system,** the components consist of the following: a combination gas valve, ignition control module, and a ignitor/sensor.

A **direct ignition furnace** uses an ignition system that ignites the main burners on every cycle of the furnace. This furnace has no pilot.

The **direct ignition system** is commonly a spark or a hot surface type.

The **voltage** created by an ignition control module commonly ranges from 7,000 to 15,000 volts.

A **spark gap,** used on a spark ignition system, is commonly $\frac{1}{8}$".

The **pilot or main burner flame** should engulf the tip of the rod on an ignitor/sensor by $\frac{3}{8}$ to $\frac{1}{2}$ inch.

All electrical connections to any heating appliance should be done in compliance with the **National Electrical Code.**

All furnaces must have there own separate **120 volt circuit.**

Four **furnace designs** are downflow/counterflow, upflow, horizontal, and multi-positional.

The **combination gas valve** consists of a pilot solenoid, a main valve solenoid, a pressure regulator, a main shut-off, and a pilot adjustment.

The **pilot solenoid** controls the gas flow to the pilot.

The **main gas valve solenoid** controls gas flow to the main burners.

The **pressure regulator** is adjustable and controls the amount of gas pressure to the main burners. It commonly reduces the gas pressure of 7" WC to 3.5" WC for natural gas. If the regulator is turned clockwise the pressure will increase.

WC refers to water column. It is an instrument in which water is placed inside it. The column has increments in inches. It is used for measuring low pressure.

A **burner orifice** is a drilled, threaded port located in the gas manifold. It directs a stream of gas to the burners.

Primary air is air that is needed for complete combustion. It is introduced and mixed with the gas before the point of ignition.

Secondary air is air that ensures complete combustion. It surrounds the flame at the point of ignition.

A **burner** is used to mix combustion air with fuel. Then it directs the mixture to the combustion chamber or heat exchanger for ignition. The burner must be installed properly or a noisy ignition could result.

The **venturi** is part of the burner in which air and fuel is mixed.

A **draft diverter** allows the air within the furnace location to blend with the products of combustion.

Ported, slotted, ribbon, or inshot are the most common types of **gas burner designs.**

A **fan control** starts and stops the blower during the furnace operation. It commonly delays on start and stop.

There are **three common types of fan controls,** a thermal activated type, a 24-volt timed delay type, and a solid state circuit board type.

The **thermal activated fan control** uses a bimetal to sense the temperature of the heat exchanger.

The **timed delay fan control** uses an electric heater and a bimetal fan switch to cycle the blower motor.

The **circuit board fan control** uses a timed delay feature that is built into the solid-state circuit board.

A **limit switch** de-energizes the gas valve if the heat exchanger exceeds a predetermined temperature, commonly 180–200° F.

A **pressure switch** proves the operation of the induced draft motor.

A **door interlock switch** is used to disconnect all electrical power supply to the furnace heating and cooling circuits when the blower door is moved.

A normally open **float switch** senses the level of water in a condensate tank and starts a condensate pump.

A **flame roll-out switch,** or fusible link, de-energizes the gas valve if the flame from the burners is pushed out of the heat exchanger. This is commonly caused by a plugged heat exchanger.

Section 1 System Components and Tools

A **heat exchanger** safely transfers the heat created by combustion to air. The air is moved over the heat exchanger by the blower.

A primary heat exchanger is commonly constructed of **cold rolled** or **stainless steel.**

The joints of a heat exchanger are usually **welded** or **mechanically crimped.**

In a condensing furnace, the products of combustion from a primary heat exchanger are directed to a **secondary heat exchanger.** The gases are condensed to increase the efficiency of the furnace. The secondary heat exchanger is located directly in front of the blower outlet.

An **induced draft blower** is used to move products of combustion out of a heat exchanger and into a vent. It is located at the outlet of a heat exchanger.

An **induced draft blower** creates a positive pressure in the venting system of a condensing furnace and a negative pressure in the venting system of a non-condensing furnace.

An **auxiliary limit control** is used as a backup to the main limit control. It is most commonly found on downflow and multi-positional furnaces.

A **vent safety control** is used to de-energize the gas valve if the vent temperature increases above a predetermined temperature.

A **flame proving device** proves the presence of the pilot or main burner flame.

A **thermocouple** proves a pilot flame by changing heat energy to electrical energy. It operates at 30 millivolts.

A **flame rod sensor** uses a flame rectification system to prove the presence of the pilot or main burner flame. Common micro amp signals range from .2 to 6 uA DC.

A **bimetal flame sensor** proves the presence of a pilot flame. It does so by using the heat from the pilot to switch a set of contacts and energize the main gas valve.

A **mercury flame sensor** proves the presence of a pilot flame. It does so by using the heat from the pilot to switch a set of contacts and energize the main gas valve.

A **gas valve** controls the flow of gas to the main burners.

The operational sequence of a high efficiency indirect ignition furnace is as follows:

On a call for heat, the thermostat closes between the R and W terminals. The induced draft motor is energized, and this causes the normally open pressure switch to close. If the limit control is in the closed position, the ignition module is energized. The ignition system either sparks or energizes a hot surface ignitor. The pilot valve is energized and the flame is proved by the pilot safety circuit. The main gas valve is then energized and the main burners are ignited. After approximately 30 to 60 seconds, the indoor blower is energized by the fan control circuit. The system will operate until the thermostat is satisfied and the R and W contacts open, which terminates the need for heat. The induced draft motor, main burner, and pilot circuits are de-energized. The indoor blower will remain in operation until a predetermined time or temperature. After this period, the indoor blower will de-energize.

COMBUSTION AIR

All fossil fuel equipment needs air for the combustion process to occur. The quantity of combustion air varies with the furnace size and type. The following are terms and information related to combustion air.

Combustion air is the air that is used for the combustion, dilution, and ventilation process of a furnace.

Dilution air is the air surrounding a furnace location. When this air is induced into the diverter, it becomes diluted with the products of combustion (exhaust gasses) leaving the chimney.

Infiltration is outdoor air that enters into a home around loose fitting doors and windows.

Infiltration is measured in air changes per hour.

Infiltration rates:

Tight construction = .3–.6 air changes per hour

Average construction = .6–1.0 air changes per hour

Loose construction = 1.0–2.5 air changes per hour

Gas fired appliances use a **15 to 1 ratio** of gas and combustion air mixture. Every 1 cu. ft. of gas requires 15 cu. ft. of combustion air.

Natural draft appliances use a **14 to 1 ratio** of gas and dilution air mixture. Every 1 cu. ft. of gas requires 14 cu. ft. of dilution air.

Fan assisted draft furnaces do not use dilution air.

Combustion air needs to be free from household chemicals, or corrosion of the heat exchanger and vent system could occur.

Gas appliances can be located in unconfined or confined spaces.

An **unconfined space** is a space with a volume that is *greater than 50 cu. ft./1000 Btu/h* of the total input of all gas fired appliances.

Example: If a 150,000 Btu/h furnace is located in a basement with a 40,000 Btu/h water heater, the total input is 150,000 + 40,000 = 190,000 Btu/h. To determine if the appliances are in an unconfined area use this formula: 190,000 Btu/h/1000 Btu/h × 50 cu. ft. = 9500 cu. ft. If the volume of the basement is greater than 9500 cu. ft., then it is considered an unconfined space.

Combustion air can be taken from the unconfined space if the construction of the home is loose.

Combustion air must be provided from the outdoors to the unconfined space in tight constructed homes.

When introducing **outdoor combustion air** into an unconfined space, two equal openings must be provided. One at or within 12 inches of the ceiling and the other at or within 12 inches from the floor.

Each **opening** used to introduce combustion air into an *unconfined space* must have the total free area of at least *1 sq. in. per 4000 Btu/h input* and *not less than 100 sq. in.*

Example: An 80,000 Btu/h furnace and a 40,000 Btuh water heater are installed in an unconfined space of a loose constructed home. The total input is 120,000 Btu/h/4000 Btu/h = 30 sq. in. per opening. Since the minimum opening is 100 sq. in., each opening must be at least 100 sq. in.

When **horizontal ducts** are used to introduce outdoor air to the unconfined space, the openings must be *1 sq. in. for every 2000 Btu/h input* and *not less than 100 sq. in.*

Louvers and grills reduce the percentage of free area in a combustion air opening. This must be taken into consideration when introducing combustion air into a confined or unconfined space.

Example: A 10" × 10" grill with the free area of 70% is used for combustion air. 10" × 10" = (100 sq. in. × .7). The total free area of the opening is reduced to 70 sq. in.

A **confined space** is a space with a volume *less than 50 cu. ft./1000 Btu/h* of the total input of all gas fired appliances.

Combustion air must be provided from the outdoors to the confined space in tight constructed homes.

When introducing **outdoor combustion** air into a confined space, two equal openings must be provided. One at or within 12 inches of the ceiling and the other at or within 12 inches from the floor.

Each **opening** used to introduce combustion air into a *confined space* must have the total free area of at least *1 sq. in. per 1000 Btu/h* total input and *not less than 100 sq. in.*

Example: An 80,000 Btu/h furnace and a 40,000 Btu/h water heater are installed in an confined space of a loose constructed home. The total input is 120,000 Btu/h/1000 Btu/h = 120 sq. in. Each opening must be at least 120 sq. in.

When **horizontal ducts** are used to introduce outdoor air to the *confined space*, the openings must be *1 sq. in. for every 2000 Btu/h* total input and *not less than 100 sq. in.*

The minimum dimension of a **horizontal duct** must not be less than three inches.

When a **vertical duct** is used to introduce outdoor combustion air into a *confined space*, the opening must be *1 sq. in. for every 4000 Btu/h* total input and *not less than 100 sq. in.*

When a **vertical duct** is used to introduce combustion air into a confined space, it must terminate 1 foot above the floor.

It is extremely important to completely separate the furnace **supply and return air** from the **combustion air,** or the negative pressure from the blower could cause products of combustion to mix within the living space of the home.

Direct vent, high-efficiency condensing furnaces must be installed to manufacturers recommendation.

Many **direct vent,** high-efficiency condensing furnaces allow for a vertical or horizontal venting system. The vent pipe is commonly schedule 40 PVC.

When installing combustion air for a **direct vent,** high-efficiency, condensing furnace, support the pipe every 5 feet, insulate the pipe if necessary, and have an air tight seal.

The **combustion air and vent pipes,** used on a direct vent high efficiency condensing furnace, must be located near the same area to prevent the system from operating under different air pressure zones. If they are too close to each other, it could cause the furnace operation to cycle on the low-pressure switch.

VENTS AND CHIMNEYS

Vents and chimneys are the passage ways for combustion gases. It is important for the service technician and installer to be knowledgeable of the types of vents, chimneys, and their related terms.

Section 1 System Components and Tools

Venting is the removal of combustion gases through a vent system.

A **vent** is the passageway that transports combustion gases to the atmosphere.

A **vent connector** is the horizontal or lateral section of pipe that connects the appliance to a vertical vent or chimney.

Combustion gases are commonly comprised of the following: carbon dioxide (CO_2), hydrogen (H), oxygen (O_2), and nitrogen (N).

Incomplete combustion and flame impingement on a cold surface creates **carbon monoxide** as a product of combustion.

Buoyancy is the moving of gases through a vent system due to high temperatures.

Buoyancy increases as combustion gas temperature increases.

The temperature difference between the ambient air and the flue gases creates a **density difference.** This difference helps move flue gases through the vent system on a natural draft furnace.

The vent height and the temperature difference between the flue gases and the ambient air creates a negative **static pressure** in a class 1 vent system.

Heat loss through a vent minimizes the **static pressure** required for proper venting.

Category 1 operates at a negative pressure, at least 140° F above flue gas dewpoint temperature, and depends on a natural draft to remove the flue gasses.

Category 2 operates at a negative pressure, less than 140° F above flue gas dewpoint temperature, and depends on a natural draft to remove the flue gasses.

Category 3 operates at a positive pressure, less than 140° F above flue gas dewpoint temperature, and requires a fan powered airtight vent to remove the flue gasses.

Category 4 operates at a positive pressure, below 140° F, and requires a fan powered airtight vent to remove the flue gasses.

When induced draft non-condensing and natural draft type furnaces are used, a **category 1** type venting system is most common.

The **vent system** that is used for an induced draft condensing furnace is a category 4 type vent system. Schedule 40 PVC is a common material used for high efficiency furnace venting. Always use manufacturer's recommendation.

"**Wet Time**" is a period at which vent gases may reach a dew point temperature and condense within the vent system.

"**Wet Time**" refers to the time that it takes a vent system to warm up. This happens at the beginning of a furnace cycle. This time should be less than two minutes.

In a natural draft furnace, **dilution air** decreases the dewpoint temperature of the flue gases, which decreases the possibility of condensation.

Because there is no **dilution air** used with an induced draft non-condensing furnace, it has more problems with condensation.

Double wall **type "B" vent** reduces the heat loss through a vent system and aids in keeping flue gases from reaching their dewpoint temperature.

To resist corrosion, the inner wall of a **type "B" vent** is made of aluminum.

The outer wall of a **type "B" vent** is made of galvanized steel.

An air space located between the inner and outer wall on a **type "B" vent** will decrease heat loss through the vent by 50% when compared to standard single wall vent pipe.

A **type "B" vent** must be used when passing through floors, ceilings, or an unconditioned space.

A **fire stop,** made of non-combustible material, must be used when vent pipe passes through floors.

A **type "B" vent** must have a 1 inch to 3 inches clearance from combustibles.

Standard **single wall vent** pipes must have a 6 inch clearance from combustibles.

A **vent connector** with a diameter of 5 inches or less must be at least 26 gauge galvanized metal.

A **vent connector** with a diameter from 6 to 9 inches must be at least 24 gauge galvanized metal.

A **vent connector** with a diameter from 10 to 16 inches must be at least 22 gauge galvanized metal.

A **vent connector** with a diameter that is greater than 16 inches must be at least 16 gauge galvanized metal.

All **vent connectors** entering a masonry chimney must be cemented to the chimney with an approved cement.

A **vent connector** must have a vertical rise of at least $\frac{1}{4}$ inch per 1 foot of horizontal run.

Example: 10 feet of pipe must have a 2.5 inches vertical pitch. (.25" × 10' = 2.5")

A single wall uninsulated **horizontal vent connector** must not exceed $1\frac{1}{2}$ feet per inch diameter of vent pipe.

One **90 degree elbow** has the resistance of 20 diameters in length of the vent being used.

Example: One 6 inch elbow has the same resistance as

1. 6" × 20 = 120"
2. 120"/12" = 10'

The following problems cause **condensation** in an induced draft non-condensing furnace:

1. Too low of a temperature rise across the heat exchanger caused by gas pressure adjustment set too low or blower speed adjustment set too high.
2. Too large of a venting system.
3. An oversized furnace or heat anticipator set too low which causes short cycling.
4. Too long of a single wall vent horizontal run.
5. Too many 90 degree elbows.

The following problems cause **pressurization** in an induced draft non-condensing furnace:

1. Too small of a vent pipe.
2. Too high of a gas pressure.
3. Too long of a single wall vent horizontal run.
4. Too many 90 degree elbows.

When installing a **main vent,** a straight pipe or two 45 degree elbows are acceptable, but never use two 90 degree elbows.

A **main vent** must terminate a minimum of 5 feet above the nearest connected appliance vent hood.

The minimum height of a **main vent** is 5 feet from the nearest connected appliance vent hood.

The height of the **main vent** depends on the pitch of the roof.

If a **main vent** is located near a vertical wall, the vent height should be not less than 2 feet above the wall within 10 feet away from the wall.

When installing the **vent pipe** of a direct vent high efficiency condensing furnace, the pipe should be pitched towards the furnace a $\frac{1}{4}$ of an inch per linear foot of vent pipe used.

The **vent pipe** of a direct vent high efficiency condensing furnace should have an air tight seal and be supported every 5 feet of pipe length.

All chimneys or vents must have a **minimum area** equal to the area of the furnace connection and a **maximum area** of no more than two sizes greater than the vent diameter.

A **masonry chimney** must be built to building code standards.

A **masonry chimney** commonly uses clay tile liners and brick with a refractory mortar.

A **masonry chimney** that is designed for a gas appliance must use a clay tile liner.

A **masonry chimney** must extend 2 feet minimum height and not less than 2 feet above any part of a building. It must be within 10 feet of a chimney.

A **chimney liner** must be used when installing an induced draft non-condensing furnace.

The space surrounding a **chimney liner** in a masonry chimney can not be used for venting.

Vent tables are used to properly vent a category 1 heating appliance. Category 2, 3, and 4 appliances must be vented to manufacturers specifications.

A **vent table** will help a technician to determine

1. the type of vent connector that can be used, single wall or double wall.
2. the size of vent pipe from the appliance to the main vent.
3. the size that the main vent should be to properly vent the appliances that are being used.

Single wall vent connectors are commonly used with lower efficient furnaces that use dilution air.

Double wall type "B" vent pipe is used as the main vent between ceilings, floors, and unconditioned spaces.

Double wall type "B" vent pipe is commonly used as the vent connector in induced draft non-condensing furnaces.

The following are some typical venting applications:

A type "B" double wall vent with a type "B" double wall vent connector serving one category 1 appliance.

A type "B" double wall vent with a single wall vent connector serving one category 1 appliance.

Section 1   System Components and Tools   27

Figure 1–3:   *Minimum Vent Height Chart*

A type "B" double wall vent with a type "B" double wall vent connector serving two or more category 1 appliances.

A type "B" double wall vent with a single wall vent connector serving two or more category one appliances.

To use a **vent table,** determine the table that matches the type of vent system that is being installed. When vent sizing a single appliance to a single vent system, the procedures are different from common venting.

Use the following procudures to size a vent system using a single appliance with a single vent system:

1. Determine the table that is to be used. (Is the vent connector going to be single wall or double wall? Is a vent or a chimney going to be used?)
2. Determine the height of the vent or chimney being used.
3. Determine the lateral distance the appliance will be from the vent.
4. Using the proper vent table, determine the vent connector type and size.

Example: An induced draft non-condensing 80,000 Btu/h furnace is to be installed in a home that uses a 15 foot double wall type "B" vent. The furnace will be 5 feet from the vent.

1. Use Table 1–3. This table shows the capacity of Type "B" Double Wall Vents with Single Wall Metal Connectors serving a single catagory 1 appliance.
2. Follow the chart and find the *15 foot vent height* that is being used.
3. Since the furnace is going to be *5 feet from the vent,* find the 5' column for the lateral vent connector.
4. In a fan assisted draft furnace, follow the *5 foot lateral* to the point where the 80 (80,000 Btu/h falls between the Min/Max numbers.
5. Follow this column up to the top of the chart to determine the proper size for the vent.
6. Since a *4 inch single wall vent* is found on the chart, it can be used.

Common venting is a term that is used when connecting two or more appliances to one vent system.

When **common venting** two appliances:

1. The smaller appliance with lower Btu rating should connect as high as possible above the larger appliance with greater Btu rating.

TABLE 1–3: CAPACITY OF TYPE B DOUBLE WALL VENTS WITH SINGLE WALL METAL CONNECTORS SERVING A SINGLE CATEGORY 1 APPLIANCE

VENT AND CONNECTOR DIAMETER IN INCHES

Height	Lateral	3"			4"			5"			6"		
H (ft)	L (ft)	Fan Min	Fan Max	Nat Max	Fan Min	Fan Max	Nat Max	Fan Min	Fan Max	Nat Max	Fan Min	Fan Max	Nat Max
6	0	38	77	45	59	151	85	85	249	140	126	373	204
	2	39	51	36	60	96	66	85	156	104	123	231	156
	4	NR	NR	33	74	92	63	102	152	102	146	225	152
	6	NR	NR	31	83	89	60	114	147	99	163	220	148
8	0	37	83	50	58	164	93	83	273	154	123	412	234
	2	39	56	39	59	108	75	83	176	119	121	261	179
	5	NR	NR	37	77	102	69	107	168	114	151	252	171
	8	NR	NR	33	90	95	64	122	161	107	175	243	163
10	0	37	87	53	57	174	99	82	293	165	120	444	254
	2	39	61	41	59	117	80	82	193	128	119	287	194
	5	52	56	39	76	111	76	105	185	122	148	277	186
	10	NR	NR	34	97	100	68	132	171	112	188	261	171
15	0	36	93	57	56	190	111	80	325	186	116	499	283
	2	38	69	47	57	136	93	80	225	149	115	337	224
	5	51	63	44	75	128	86	102	216	140	144	326	217
	10	NR	NR	39	95	116	79	128	201	131	182	308	203
	15	NR	NR	NR	NR	NR	72	158	186	124	220	290	192

Appliance Input Rating in Thousands of Btu/h

Terms for Table

NAT = Natural Draft

FAN = Fan Assisted

NR = Not Recommended

H   = Height

L   = Lateral

Section 1   System Components and Tools

2. The smaller appliance should have a shorter horizontal run than the larger appliance.

3. The smaller appliance must never have more elbows than the larger appliance.

Use the following procedures to size a vent system using two or more appliances with a single vent:

1. Determine the table that is to be used. (Is the vent connector single wall or double wall? Will a vent or a chimney be used?)

2. Determine the height of the vent or chimney being used.

3. Determine the type of venting that each common vented appliance will use. The options are natural or forced draft.

NAT = Natural Draft

FAN = Fan Assisted

There are three different combinations: FAN + FAN, FAN + NAT, NAT + NAT

4. Determine the height of the vent connector rise for each individual appliance. (Common vent tables assume that the connector length is no more than 1 and $\frac{1}{2}$ feet per inch of vent diameter.)

5. Using the vent connector capacity chart, determine the vent connector type and size for each appliance.

6. Using the common vent capacity chart, determine the main vent size.

Example: An induced draft non-condensing 85,000 Btu/h furnace with a 3 foot vent connector rise is to be common vented with a 40,000 Btu/h natural draft water heater with a 2 foot vent connector rise. The home has an existing 5 inch double wall type "B" main vent that is 20 feet high.

1. Use Table 1–4 on the next page. This table shows the capacity of a Type "B" double wall vent with a Type "B" double wall vent connector serving two or more category 1 appliances.

2. To size the vent connector for the 40,000 Btu/h water heater, use the *vent connector capacity chart* on the next page and follow the column on the chart that uses a *20 foot vent height*. Using a *2 foot connector rise,* follow the column over until it reads 66 or 66,000 Btu/h. By following the column up, one sees that a natural draft appliance that is less than 66,000 Btu/h can use a *4 inch vent connector.*

3. To size the vent connector for the 85,000 Btu/h induced draft non-condensing furnace, use the *vent connector capacity chart* on the next page and find the column on the chart that uses a 20 foot vent height. Using a *3 foot connector rise,* follow the column over until it reads a MIN of 35 or 35,000 Btu/h and a MAX of 110 or 110,000 Btu/h. By following the column up, one sees that an induced draft appliance that is 85,000 Btu/h can use a *4 inch vent connector.*

4. To size the main vent, first add the total input of all appliances that are to be common vented. In this case, the total input is 125,000 Btu/h. Use the *common vent capacity chart* on the next page and find the column on the chart that uses a *20 foot vent height*. Since one appliance is a fan assisted induced draft and the other appliance is a natural draft appliance, the *FAN + NAT column* will be selected. Follow the column over until 183 or 183,000 Btu/h is found. This is the maximum input that the main vent can handle. Follow this column up and one sees that the existing 5 inch double wall "B" type vent can be used.

TABLE 1–4: CAPACITY OF TYPE "B" DOUBLE WALL VENTS WITH TYPE "B" DOUBLE WALL CONNECTORS SERVING TWO OR MORE CATEGORY 1 APPLIANCES

* Vent Connector Capacity Chart   VENT CONNECTOR DIAMETER IN INCHES

Height H (ft)	Connector Rise L (ft)	3" Fan Min	3" Fan Max	3" Nat Max	4" Fan Min	4" Fan Max	4" Nat Max	5" Fan Min	5" Fan Max	5" Nat Max	6" Fan Min	6" Fan Max	6" Nat Max
8	1	22	40	27	35z	72	48	49	114	76	64	176	109
	2	23	44	32	36	80	57	51	128	90	66	195	129
	3	24	47	36	37	87	64	53	139	101	67	210	145
10	1	22	43	28	34	78	50	49	123	78	65	189	113
	2	23	47	33	36	86	59	51	136	93	67	206	134
	3	24	50	37	37	92	67	52	146	104	69	220	150
15	1	21	50	30	33	89	53	47	142	83	64	220	120
	2	22	53	35	35	96	63	49	153	99	66	235	142
	3	24	55	40	36	102	71	51	163	111	68	248	160
20	1	21	54	31	33	99	56	46	157	87	62	246	125
	2	22	57	37	34	105	66	48	167	104	64	259	149
	3	23	60	42	35	110	74	50	176	116	66	271	168

Appliance Input Rating in Thousands of Btu/h

TABLE 1–5: COMMON VENT DIAMETER IN INCHES

* Common Vent Capacity Chart

Vent Height (ft)	4" FAN +FAN	4" FAN +NAT	4" NAT +NAT	5" FAN +FAN	5" FAN +NAT	5" NAT +NAT	6" FAN +FAN	6" FAN +NAT	6" NAT +NAT
8	101	90	73	155	129	114	224	178	163
10	110	97	79	169	141	124	243	194	178
15	125	112	91	195	164	144	283	228	206
20	136	123	102	215	183	160	314	255	229

Section 1   System Components and Tools   31

GAS PIPING

Gas piping carries fuel gases to an appliance. It is very important for the installer to know basic terms and methods involved in gas piping. The following will help the technician/installer to better understand gas piping.

A **drip leg** should be installed on a vertical gas line to catch contaminants that may flow with the gas.

A permit is required by most municipalities to install or alter **fuel gas piping.**

A **manual gas cock valve** shut-off should be installed externally to the furnace before the union, the drip leg, and the appliance.

All **manual gas cock valves** must be approved for gas fuel only.

Gas shut-off valves should be installed within three feet of the appliance.

When gas piping supplies gas to more than one building in one area, a **separate shut-off valve** is needed for each building.

A **system shut-off valve** should be installed near the liquefied petroleum gas tank point of connection.

A **gas union** should be installed between the manual gas cock valve, the combination gas valve, and the gas valve control.

The gas supply line to a **mobile home** should be a minimum of $\frac{3}{4}$ of an inch.

The gas piping used to supply **mobile homes** should be installed at least 18 inches below grade (below ground level).

Gas pipe installations should be kept at least 6 inches above the grade or the structure. Examples of structures are porches, steps, breezeways, etc.

After an installation, **gas piping** must be pressure tested to ensure that there are no leaks present.

Most **gas piping** is constructed of standard weight wrought iron steel. It may be galvanized or black.

Gas piping that is not protected can be installed above grade in recesses or channels that are approved by a mechanical inspector.

Plastic gas piping should be at least 18 inches below grade.

Leaks in gas piping can be located with a soapy bubble solution.

Never use an **open flame** to locate a gas leak.

When **liquefied petroleum gas (LPG)** is used, it must maintain a pressure of 11" WC from the regulator at the tank to the appliance.

When using natural gas, a **gas pressure** of 7" WC is required from the meter to the appliance.

When using natural gas, a **gas pressure** of 3.5" WC is commonly required from the furnace to the burner.

Specific gravity refers to the relative weight of an amount of gas as compared to the weight of an equal amount of air.

The **specific gravity** of natural gas is approximately 0.65 sp. gr.

Supports for hanging gas piping should be constructed of metal straps or hooks. They should be placed at approved intervals.

Underground gas piping that is ferrous and metallic should be at least 12 inches below grade.

Underground piping should be supported by compacted soil or sand.

To properly size a gas pipe system, follow the following procedures:

1. Sketch a diagram of the gas piping layout. Include the following:
   A. The appliance Btu/h input ratings.
   B. The distance of each appliance from the main supply line.
   C. The length of the main line and the measurements between each tee.
2. Convert the Btu/h input rating to Btu/h / cu. ft. If it is natural gas, divide the Btu/h input rating by 1000 Btu/h / cu. ft.

   Example: 35,000 Btu/h input / 1,000 Btu/h / cu. ft. = 35 cu. ft.
3. Determine the distance between the gas meter and the furthest appliance. Use this distance for sizing all of the sections on the chart.
4. Starting at the furthest appliance, determine the amount of gas each section of pipe will have to carry, and size accordingly.

   Example:

   To solve this problem, use Figure 1–4 on page 32 and Table 1–6 on page 33.
   1. The distance between the meter and the furthest appliance (the water heater) is 55 feet. Since there is no entry for 55 feet on the table, 60 foot length will be used.

Figure 1–4: *Gas Piping Layout*

2. Total the Btu/h input ratings on all appliances in the system.

   150,000 Btu/h + 75,000 Btu/h + 50,000 Btu/h = 275,000 Btu/h.

3. Convert the Btu/h input rating to Btu/h / cu. ft. by dividing 275,000 Btu/h by 1,000 Btu/h / cu. ft. = 275 cu. ft.

4. On the chart provided, find the column that uses 60 feet and follow it down until it reads 275 cu. ft., or the column succeeding 275 cu. ft. This shows that the main gas line from points "A" to "B" will be $1\frac{1}{4}$ inches. The function of the line is to carry all the gas necessary to supply all appliances that will branch off it. This line has to be larger than the other pipes in the system, because it is carrying all the initial gas from the meter.

5. Going to the furthest appliance (the water heater at 50 cu. ft.) and staying at the 60 foot length, follow the column to 50 cu. ft. The table shows that the gas pipe section from "D" to "E" and "C" to "D" will be $\frac{1}{2}$ inches in size. This occurs because only 50 cu. ft. will be carried by this section of pipe.

6. The next appliance is the furnace that uses 150 cu. ft. of gas. Staying on the 60 foot length, follow the table until it reads 150 cu. ft. (or the column that succeeds 150 cu. ft.). This column is 195 cu. ft. The section of pipe from "C" to "F" will be 1" in size.

7. Since section "B" to "C" carries gas to the furnace and the water heater, the quanity of gas from both appiances will have to be totaled to size the section. 150 cu. ft. + 50 cu. ft. = 200 cu. ft. Staying on the 60 foot length, follow the table down until it reads 200 cu. ft. (or the column that succeeds 200 cu. ft.). This column is 400 cu. ft. The section from "B" to "C" will be $1\frac{1}{4}$ inches in size.

8. The last appliance is the gas range. It uses 75 cu. ft. of gas. Staying on the 60 foot length, follow the table down until it reads 75 cu. ft. (or the column that succeeds 75 cu. ft.). This column is 105 cu. ft. The section of pipe from "B" to "G" will be $\frac{3}{4}$ inches in size.

The following is the list of gas pipe sections and their sizes.

"A" to "B" is 1¼"

"B" to "C" is 1¼"

"C" to "D" is ½"

"D" to "E" is ½"

"C" to "F" is 1"

"B" to "G" is $\frac{3}{4}$

Section 1   System Components and Tools

TABLE 1–6: PIPE EQUIVALENT CHART

Length of Pipe in Feet

Nominal Iron Pipe Size in Inches	10'	20'	30'	40'	50'	60'
$\frac{1}{4}"$	32	22	18	15	14	12
$\frac{3}{8}"$	72	49	40	34	30	27
$\frac{1}{2}"$	132	92	73	63	56	50
$\frac{3}{4}"$	278	190	152	130	115	105
1"	520	350	285	245	215	195
$1\frac{1}{4}"$	1,050	730	590	500	440	400
$1\frac{1}{2}"$	1,600	1,100	890	760	670	610

Maximum Capacity of Pipe in Cubic Feet Per Hour

OIL FURNACES

An oil heating service technician needs to understand fuel fundamentals as well as the controls and components the system utilizes. The following is information needed when servicing oil furnaces.

No. 2 fuel oil has a heating value of approximately 140,000 Btus per gallon.

High pressure gun burners commonly operate at 100 PSIG. (Some models operate between 140 to 200 PSIG).

The primary voltage of an **ignition transformer** is 120 volts. The secondary voltage is 10,000 volts. An ignition transformer is used to create the spark for the ignition of the oil on a high-pressure gun burner.

Electrodes are metal rods wrapped in a ceramic insulator that carry the high voltage from the ignition transformer to the point of ignition.

A common **spark gap** between the electrode tips is an $\frac{1}{8}$ of an inch.

A **bus bar** is a brass or copper plate that connects electrodes to the ignition transformer.

An **oil nozzle** is used to atomize the fuel oil before ignition. Spray angles range from 30 to 90 degrees.

Nozzles have three different spray patterns: solid, hollow, and semi-hollow.

Nozzles are rated in gallons per hour (GPH).

A **squirrel cage combustion blower** is used to supply air for combustion.

The **burner motor** is a split phase motor used to turn the oil pump and the combustion air blower.

The speed of a **burner motor** can be 1725 RPM or 3450 RPM.

There are **three locations** for oil storage tanks: indoor, aboveground, and underground.

An **indoor tank** must be installed within 7 feet of the burner. The fill pipe must be a 2 inch iron pipe size and the air vent must be a $1\frac{1}{4}$ inch iron pipe size. An indoor tank is commonly installed as a single pipe system.

Indoor tanks commonly have a 275 gallon capacity.

An **indoor tank globe valve** must be of the fuse type to prevent oil leakage during a fire.

A 275 gallon fuel tank is **constructed** of #14 gauge steel.

Fuel oil tanks must be approved by Underwriter Laboratories (UL).

A **below ground tank** commonly has a 500, 1000 or 1500 gallon capacity. The fill pipe must be a 2 inch iron pipe size, the air vent must be a $1\frac{1}{4}$ inch iron pipe size, and both must be installed as a two pipe system.

A **decrease of primary air** increases CO_2 percentage and increases combustion gas temperatures, decreases stack temperatures, and increases efficiency.

An **increase of primary air** decreases CO_2 percentage, decreases combustion gas temperatures, increases stack temperatures, and lowers efficiency.

Viscosity is the oil's resistance to flow. The higher the viscosity, the thicker the oil.

The **pour point** is the temperature at which oil will start to flow.

Cloud point refers to the temperature at which wax crystals begin to form in oil.

Single-stage fuel pumps are used with aboveground fuel tank applications.

Two-stage fuel pumps are used with underground fuel tank applications.

The **cad-cell** is a separate component from the cad-cell relay primary control. The cad-cell is wired to the primary control. It receives and sends messages to the cad-cell's primary control relay.

A **cad-cell** is a photo type cell that has a high resistance to darkness. It is made of cadmium sulfide.

The **cad-cell** should be 1,500 ohms or less when sensing light.

The **cad-cell** should be around 100,000 ohms or more when sensing darkness.

A **cad-cell relay** is a control device used to sequence and control the burner motor and ignition transformer while sensing the **light** of the flame.

The **cad-cell relay** uses a black wire to supply line voltage to the relay, a white wire to supply the common wire to the relay, and an orange wire to feed line voltage to the burner motor and ignition transformer.

A **stack mount relay** is a control device used to sequence and control the burner motor and ignition transformer while sensing the **heat** of the flame.

The **stack mount relay** uses terminal number 1 to supply line voltage to the relay, terminal number 2 to supply the common wire to the relay, and terminal number 3 to feed line voltage to the burner motor and ignition transformer.

A **coupler** is a device that is used to connect the shaft of the fuel pump and the shaft of the burner motor.

An **air shutter** is used to adjust the primary air on a high-pressure gun burner.

A **Barometric damper** is a device that is installed in the flue of an oil fired furnace. It is used to control draft through the chimney.

There are three **fuel filters** in the fuel oil system: one on the inlet line, one in the fuel pump, and another at the nozzle.

Turbulent flow refers to air flow created by spiral vanes within the air blast tube.

Most residential oil furnaces use an **oil supply line** that is a $\frac{3}{8}$" OD copper tubing.

The **lubrication** fluid used in burner motors is a number 10 weight non-detergent motor oil.

A **delayed oil valve** is used to allow the oil pump to build up pressure on the start up of a high-pressure gun burner. It is wired in parallel with the oil burner motor.

In an oil furnace, the **products of combustion** are:

Carbon dioxide (CO_2)

Carbon monoxide (CO)

Oxygen (O)

Nitrogen (N)

Water vapor (H_2O)

High CO_2 readings indicate a hot fire.

A **flue gas analyzer** is commonly used to measure CO_2.

Common **CO_2 readings** are from 10% to 12%.

Common **CO_2 readings** for natural gas furnaces should be from 8.5% to 9.5%.

A **draft gauge** is used to measure the amount of draft created in a furnace.

A common **over the fire draft** reading for an oil furnace ranges from -.01" WC to -.02". Over the fire draft reading refers to the exhaust draft, a draft reading after a fire (combustion process).

A common **breech draft** reading for an oil furnace ranges from -.03" WC to -.06" WC.

Breech draft reading refers to the draft within the chimney piping.

A draft for an oil fired furnace can be adjusted using the the **barometric damper** located on the stack. Opening the damper decreases the draft and closing the damper increases the draft.

A **smoke tester** is used on an oil furnace to measure the amount of smoke that is created in the combustion process.

Common **smoke readings** for an oil fired furnace are from #1 to #2 as seen on the smoke sample card. The smoke card is part of a combustion analysis kit.

Section 1   System Components and Tools

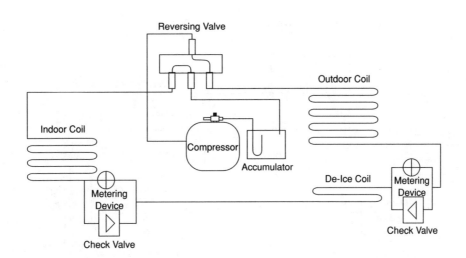

Figure 1–5: *Heat Pump Diagram Cooling Mode*

A **high smoke reading** indicates that there is not enough combustion air.

A **low smoke reading** indicates that there is too much combustion air.

A **stack thermometer** commonly measures temperatures that range from 200° to 1000°F.

To obtain the **net stack temperatures** of flue gases, measure the gases exiting the flue and subtract the ambient air temperature.

The **gross stack temperature** is room air tempeature plus the net stack temperatures, or the reading of the stack thermometer when placed in the furnace flue.

Common **gross stack temperatures** are from 500° to 550° F.

If **combustion air** must be supplied to an oil furnace, there must not be less than 1 sq. in. per 5000 Btus per hour.

HEAT PUMPS

In mild winter climates and in the air conditioning season, heat pumps are very popular. It is pertinent that service technicians are knowledgeable of the various types of heat pumps. The following will aid technicians in building on their knowledge.

Cooling Mode

During the cooling mode of the heat pump, high-pressure/high-temperature refrigerant gas flows from the compressor, through the reversing valve, to the outdoor coil. Sensible and latent heat is removed and the refrigerant is changed into a high-pressure/high-temperature liquid. The liquid flows from the outdoor coil, through the check valve, to the de-ice subcooler coil where the refrigerant temperature is lowered to increase the system's capacity. The high-pressure/high-temperature liquid refrigerant flows from the de-ice coil to the metering device where the refrigerant is changed into a low-pressure/low-temperature saturated liquid. The refrigerant flows through the indoor coil, absorbing heat, causing the refrigerant to change from a saturated liquid into a superheated gas. From the indoor coil, the refrigerant flows through the reversing valve, to the accumulator, and back into the compressor. For an example see Figure 1–5.

Heating Mode

During the heating mode of the heat pump, high-pressure/high-temperature refrigerant gas flows from the compressor, through the reversing valve, to the indoor coil. Sensible and latent heat is removed and the refrigerant is changed to a high-pressure/high-temperature liquid. The liquid flows from the indoor coil, through the check valve, to the de-ice coil (subcooler), where the refrigerant temperature is lowered to increase the system's capacity. This de-ice coil also prevents the formation of ice build up on the bottom of the outdoor coil during the heating mode. The high

Figure 1-6: *Heat Pump Diagram Heating Mode*

pressure/high-temperature liquid refrigerant flows from the de-ice coil to the metering device where the refrigerant is changed to a low-pressure/low-temperature saturated liquid. The refrigerant flows through the outdoor coil. As it flows, it absorbs heat, causing the refrigerant to change from a saturated liquid to a superheated gas. From the outdoor coil, the refrigerant flows through the reversing valve to the accumulator, and back into the compressor. For an example see Figure 1-6.

The **reversing valve** is used in a heat pump to change the refrigerant flow. During the cooling mode it causes the outdoor coil to be a condenser and the indoor coil to be an evaporator. During the heating mode, the reversing valve causes the outdoor coil to be an evaporator and the indoor coil to be a condenser. For an example see Figure 1-7 on page 37.

The **reversing valve solenoid coil** is designed to be de-energized during the winter in northern climates and during the summer in southern climates. This will extend the life of the valve and conserve electrical energy.

The **reversing valve** has four tubes connected to the system. The top tube is connected to the discharge of the compressor. The bottom center tube is connected to the suction side of the compressor.

The bottom left and right tubes can connect to either the outdoor or indoor coil. In warmer climates, the **left tube** connects to the indoor coil and the **right tube** connects to the outdoor coil. In colder climates, the **left tube** connects to the outdoor coil and the **right tube** connects to the indoor coil.

A heat pump system must have a pressure difference of 75 PSIG or higher between the high and low side of the system for the **reversing valve** to operate properly.

The **vapor line** is the larger line that connects the outdoor coil to the indoor coil.

During the cooling cycle, the **vapor line** carries a low-pressure/low-temperature refrigerant gas from the indoor coil to the outdoor compressor.

During the heating cycle, the **vapor line** carries a high-pressure/high-temperature gas from the compressor to the indoor coil.

The **liquid line** is the smaller line that connects the outdoor coil to the indoor coil.

During the cooling cycle, the **liquid line** carries a high-pressure/high-temperature liquid refrigerant from the outdoor coil to the indoor coil.

During the heating cycle, the **liquid line** carries a high-pressure/high-temperature liquid refrigerant from the indoor coil to the outdoor coil.

The indoor and outdoor coil have one **check valve** and one metering device. The check valves are connected in parallel with one metering device and in series with the other.

A **ball type check valve** must be installed vertically with its seat located on the bottom. The "seat" refers to where the ball covers the valve opening. When flow stops, the ball will rest over the opening, therefore closing the valve. When soldering a **ball type check valve** into the system, use a heat sink to prevent warping.

Section 1   System Components and Tools

Figure 1-7: *Reversing Valve*

An **accumulator** is commonly located between the bottom center tube on the reversing valve and the inlet to the compressor. The accumulator will prevent liquid from entering the compressor.

When the system is returned to the heating mode, the **liquid refrigerant** is typically found in the vapor line immediately after the defrost cycle.

A **defrost relay** is used in conjunction with an electrical defrost system to achieve the defrost function.

A proper **defrost function** must have the following qualifications:

1. The **reversing valve** must be positioned for the cooling mode.
2. The outdoor coil's **fan** must be de-energized.
3. The system must remain in defrost until the **outdoor coil** has reached a pre-determined temperature.
4. **Auxiliary strip heaters** must be energized to increase supply air temperature.

The heat pump commonly develops frost during the heating cycle. **Frost builds up** on the outdoor coil.

A **defrost termination control** is used to de-energize the defrost cycle when the outdoor coil is free of frost.

The **defrost termination thermostat** is commonly located on the outdoor coil and opens at a preset temperature of 55° F.

Four common types of **defrost systems** are temperature initiation - temperature termination; time initiation - temperature termination; static pressure initiation - temperature termination; and static pressure / time initiation-temperature / time termination.

A **temperature initiation - temperature termination** defrost system senses the temperature difference between the outdoor ambient air and the outdoor coil.

A **time initiation - temperature termination** defrost system uses a defrost timer to initiate the defrost cycle and a temperature sensor located on the outdoor coil to terminate the defrost cycle.

A **pressure initiation - temperature termination** defrost system uses a pressure switch that measures a pressure difference across the outdoor coil to initiate the defrost system and a temperature sensor located on the outdoor coil to terminate the defrost cycle.

The **pressure switch** commonly closes at a preset pressure difference of .5" WC to .65" WC.

A **pressure/time initiated - temperature terminated** defrost system uses a defrost timer and a pressure switch to initiate the defrost system and a temperature sensor located on the outdoor coil to terminate the defrost cycle.

Temperature split is the temperature difference between the air entering a coil and the temperature of the coil.

An **outdoor ambient thermostat**, OAT, is used to energize the auxiliary heat operation if the outdoor ambient temperature drops below a predetermined temperature.

A **hold back thermostat** is located on the outdoor coil of a heat pump.

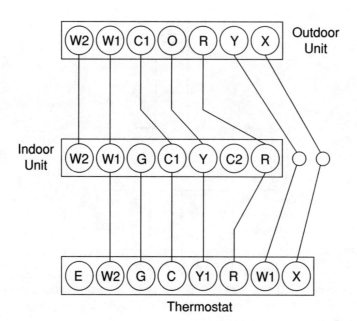

A **hold back thermostat** is used to prevent ice build up on an outdoor coil during the heating cycle.

A **hold back thermostat** senses the outdoor ambient temperature. It is used to keep the system energized during a defrost cycle if the room thermostat becomes satisfied.

A **discharge temperature thermostat** is located on the indoor coil of an air to air heat pump. It is used on a system that incorporates fossil fuel as a means of auxiliary heat.

A **discharge temperature thermostat** de-energizes the fossil fuel system when the discharge temperature reaches 95° F during the defrost cycle.

A heat pump must always be **sized** for the cooling load of the building.

The proper amount of **auxiliary heat** must be installed to handle the total heating requirement of the building.

Auxiliary strip heaters are found in heat pumps to aid in the heating process during low outdoor ambient temperatures.

Auxiliary strip heaters are used during the defrost cycle to increase the temperature of the supply air.

The **auxiliary strip heaters** are located at the discharge side of the blower and down stream from the indoor coil on an air to air heat pump. They are electrically controlled by the second stage of a two stage thermostat.

The **de-ice coil** is the lower section of the outdoor coil. It is used during the cooling mode to increase the subcooling of the refrigerant.

The **de-ice coil** prevents the formation of ice build up on the bottom of the outdoor coil during the heating mode.

The **COP,** coefficient of performance, is a heat pump's rating in the heating mode. It is determined by comparing the total amount of heat produced (heat output) to the amount of electrical energy used (electrical input).

When installing an **outdoor coil** on an air-to-air heat pump, the unit should be located as close as possible to the indoor coil. The length of the vapor line is critical when the system is in the heating mode. Longer lines are less efficient, while shorter lines are more efficient.

An **outdoor unit** must not be located under an overhang. This could cause re-circulation of air and decrease the efficiency of the system.

When installing the **outdoor coil,** a gravel foundation should be placed below the pad, or base, the unit will rest on.

Refrigeration lines must be kept as short as possible to prevent pressure drop, heat loss, or heat gain through the line set.

The **vapor line** must be insulated to decrease refrigerant temperature rise and exterior moisture condensation during the cooling mode. It will also prevent refrigerant temperature drop during the heating mode.

Common **static pressures** in the design of a heat pump duct system are .15" WC for the supply duct and .05" WC for the return duct.

Section 1   System Components and Tools

A common **temperature rise** over the indoor coil during the heating mode is 35° F.

For service accessibility, a minimum of a **20 inch clearance** is commonly required for the indoor unit.

A **1 inch air space** is required between the supply air duct and any combustible material.

When installing an indoor downflow type unit, a **non-combustible floor base** is required between the floor and the system.

On horizontal heat pump installations, an **auxiliary drain pan** must be located under the entire unit.

When an indoor coil is installed on the negative side of the blower, air is pulled over the coil. A **"P" trap** is mandatory on the drain line to prevent condensation from entering the duct system.

A **secondary drain line** should extend from the auxiliary drain pan to a conspicuous place. This serves as an alarm, indicating that the primary drain is restricted.

An **electrical disconnect** must be installed at the indoor unit if circuit breakers are not a part of the unit.

A standard heating and cooling **thermostat** cannot be used for a heat pump system.

An air to air heat pump uses an **8 strand, 18 gauge** wire for the control circuit.

The **"Y1" terminal** on a heat pump thermostat controls the reversing valve.

The **"W1" terminal** on a heat pump thermostat controls the compressor motor operation.

The **"W2" terminal** on a heat pump thermostat controls the auxiliary heat circuit.

The **"R" terminal** on the heat pump thermostat is the power supply from the transformer.

The **"C" terminal** on the heat pump thermostat is the 24 volt common from the transformer.

The **"G" terminal** on the heat pump thermostat controls the fan relay coil.

The **"X" terminal** on the heat pump thermostat controls the trouble light at the thermostat.

The **"E" terminal** on the heat pump thermostat controls the emergency heat light at the thermostat

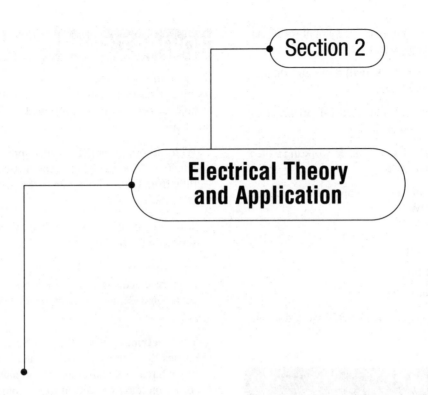

Section 2: Electrical Theory and Application

HVAC/R technicians must have a thorough understanding of electricity and electrical controls. Theory, application, meters usage, and troubleshooting are covered in this section.

ELECTRICAL THEORY

Electrical theory is the basis for understanding electricity and electrical controls. Understanding the following electrical terms will help the technician with certification exams as well as troubleshooting electrical circuits.

An **atom** consists of electrons, protons, and neutrons.

An **electron** has a negative charge.

A **proton** has a positive charge.

A **neutron** has a neutral charge.

An atom that has a **negative electrical charge** is one that has an excess of electrons.

An atom that has a **positive electrical charge** is one that has an excess of protons.

An atom that has a **neutral charge** has the same number of protons as electrons.

Like charged particles repel each other. Unlike charged particles attract each other.

Conductors are materials that easily pass electrons. Some examples of conductors are gold, silver, and copper.

Insulators are materials that do not easily pass electrons. Some examples of insulators are rubber, plastic, and glass.

A power supply is created by moving a conductor through a magnetic field. This induces a **current** in the conductor.

Electrical pressure is measured in **voltage.** The symbol for voltage is "E".

Electrical pressure, also known as **voltage** (or electromotive force), is the potential difference that pushes the electron through an electrical circuit.

A **Millivolt** is a thousandth (1/1,000) of a volt. Millivolts should be measured with a millivoltmeter.

A **coulomb** is a quantity measurement of electrons. One coulomb contains 6.25×10^{18} electrons.

Electrons that flow in an electrical circuit are known as **current.** The symbol for current is "I".

Current is measured in **amperes** or **amps.**

One amp is defined as one coulomb per second.

A **milliamp** is a thousandth (1/1,000) of an amp, and should be measured with a Milli-ammeter.

A **microamp** is a millionth (1/1,000,000) of an amp, and should be measured with a micro-ammeter.

The opposition to current flow in an electrical circuit is known as **resistance.** The symbol for resistance is "R".

Resistance is measured in **ohms.** The symbol for ohms is the Greek symbol Omega (Ω).

A **megaohm** is a million (1,000,000) ohms and should be measured with an ohmmeter.

When electrons flow through an electrical circuit, electrical **work** is being accomplished.

Power is the rate at which electrons do work in an electrical circuit. Its symbol is "P".

A **watt** is the measurement of the power in an electrical circuit. Its symbol is "W".

   1 watt = 3.412 Btus

   1000 watts = 1 kilowatt

   1 kilowatt = 3412 Btus

   1 Horsepower = 746 watts

A **wattmeter** measures the true electrical work done in an electrical circuit.

MAGNETISM

Many HVAC/R electrical controls and loads operate using the principle of magnetism (electromagnetism). The following terms and concepts are used to aid the technician in a better understanding of magnetism theory.

All **magnets** have a north and south pole.

Unlike **poles** attract each other and like **poles** repel each other.

Magnetic lines of force flow between the north and south poles of a magnet.

Magnetic lines of force are known as **flux.**

A **permanent magnet** is a piece of material that holds its magnetic force after passing through a magnetic field.

An **electromagnet** is produced by coiling a wire around an iron core and applying current to the wire.

Magnetism is used in inductive loads such as relay coils and motor windings.

Magnetic flux lines are created when current flows through a conductor. When the conductor is coiled, a magnetic field is created.

To **intensify the magnetic field** in an electromagnet, add more turns to the conductor or increase the current flow through the conductor.

An **iron core** tends to center itself in the coil when there is current flow through the conductor.

DIRECT CURRENT

Direct current flows in one direction, from negative to positive. Understanding direct current will help the service technician become more proficient in diagnosing systems problems.

With **direct current (DC),** electrons flow in one direction. When using electron theory, the electrons flow from the negitive pole to the positive pole of a DC power source.

A simple circuit consists of 3 components, **a power source, a path, and a load.**

The **power source** is the voltage or potential used to push the electrons. Higher voltage creates a greater push. On a DC power source, red is positive and black is negative.

The **electrical path** is the wire (conductor) that carries the electrons through the circut. Larger wires have less resistance and greater capability for an increase of current flow. Longer wires increase the resistance.

The **electrical load** is a current consuming device that converts electron flow into heat, light, or magnetism.

Switches are placed in an electrical circuit to control the electron flow. A closed switch allows current to flow in the circuit. An open switch breaks the circuit, disrupting the current flow.

Switches can be operated manually or automatically. Examples of manually operated switches are push button, toggle, or lever types. Examples of automatically operated switches are thermostats, pressure switches, magnetic relays, and timed delay relays.

Ohm's Law is used for DC (direct current) or AC (alternating current) circuts that are purely resistive.

The following is the formula for Ohms law:

   E = Voltage I = Amperage R = Resistance

   $E = I \times R$ or $I = E/R$ or $R = E/I$

Watt's law is used to find the power in an electrical circuit.

The following is the formula for Watts law:

   E = Voltage I = Amperage R = Resistance P = Power

   $P = E \times I$ or $E = P/I$ or $I = P/E$

   $P = I^2 \times R$

   $P = E^2/R$

Section 2 Electrical Theory and Application

A **power factor** is used to calculate the power rating of an inductive circuit. Power in an inductive circuit is lost due to the out of phase relationship that the current has with the voltage.

The calculation of power in an AC inductive circuit is
P = E × I × PF.

SERIES CIRCUITS

A common series circuit is a switching device that is wired in series with a load for the purpose of control. The following is information on series circuits.

A series circuit is a circuit that has **one path** for current to flow through.

Current stays the same throughout a series circuit.

Voltage drops across loads that are wired in series.

Current decreases as loads are added in series.

Total resistance increases as loads are added in series.

The calculation for resistance in a series circuit is
RT = R1 + R2 + R3 . . .

PARALLEL CIRCUITS

Load devices in HVAC/R electrical circuits are usually wired in parallel. The following is information on parallel circuits.

A parallel circuit is a circuit that has **two or more paths** for current to flow through.

Voltage stays the same at each load that is wired in parallel.

Current divides across the individual branches in a parallel circuit.

Current increases as loads are added in parallel.

Total resistance decreases as loads are added in parallel.

The calculation for resistance in a parallel circuit is
**RT = R1 × R2 / R1 + R2 or
1/RT = 1/R1 + 1/R2 + 1/R3 . . .**

ALTERNATING CURRENT

Most circuits in HVAC/R industry operate on an alternating current. It is important for the service technician to understand the principles of alternating current. The following is information that will be helpful to technicians.

An **alternating current (AC)** is a current that reverses the electron flow in a circuit.

Alternating current is generated by moving a conductor through a magnetic field.

Frequency is the number of complete cycles that occur in an AC circuit in one second.

Frequency is measured in **hertz (Hz).**

In a **60 Hertz** electrical circuit, electrons change direction 120 times in one second.

An **AC sine wave** is a graphic illustration that is used to represent the postion of the windings of a generator. Generators are used to produce alternating current.

One complete cycle **sine wave** represents one complete cycle of the generator.

An **AC sine wave** contains a total of 30 electrical degrees. It reaches its peak positive voltage at 90°, returns to a value of 0 volts at 180°, increases to its maximum negative voltage at 270°, and returns to 0 volts at 360°. Each complete wave is called a cycle.

The resistance to the passing of magnetic flux lines through a substance is called **reluctance.**

Capacitance is the ability to store electrons in an electrical circuit.

Capacitive reactance is the opposition to change in voltage caused by the capacitor in an AC capacitive circuit. Current leads voltage in a capacitive circuit.

The resistance in an AC circuit due to the change of current flow is called **reactance.**

Inductance is the generation of a voltage by the variance of the current flow rate in an AC circuit.

Inductive reactance is the opposition to current flow in an AC inductive circuit caused magnetic lines of flux. Voltage leads current in an inductive circuit.

Impedance is the total resistance of an AC circuit, caused by inductive reactance, resistance, and capacitive reactance.

A **single-phase AC** power supply uses two wires to bring voltage to an electrical load. A 120V circuit uses one leg of power and a common (neutral) wire. A 240V circuit uses two legs of power and does not use a common (neutral) wire in the electrical circuit.

The **hot leg** is the wire used in a power supply that feeds a voltage source to an electrical circuit. In a 120V power supply there is only one leg and it must be fused. In a 240V single-phase power supply both legs of power must be fused.

The hot leg on a 120V power supply and both hot legs on a 240V power supply must be **fused.** The neutral leg does not need to be fused.

A **three-phase system** has three hot legs that are 120 degrees out of phase with each other.

A **three-phase alternating current** uses three wires to bring voltage to a load. All three legs on a three-phase system must be fused.

The **voltage** between any two hot legs on a 240V three-phase power supply is 240V and on a 480V power supply is 480V.

Polarity is the condition that indicates the direction of electron flow. In a 120 volt circuit black is hot, white is neutral, and green or bare copper is ground.

A **ground** wire connects from the frame of an electrical load to the earth ground. It protects against electrocution during a short to ground situation. Its color code is green.

POWER DISTRIBUTION

It is important to understand how utility companies distribute power to customers. The following is information on voltage phases and distribution configurations that utility companies use to distribute electrical power.

Power supplied to substations is **120,000V** or higher. The substation reduces the voltage to **4800V** for residential areas and **34,000V** for commercial areas.

A **step down transformer** will reduce the voltage from 4800V to a common low voltage residential application of 240V single-phase.

Light commercial applications use either a **Delta** or a **Wye** configured transformer. A Delta configuration steps the voltage down to 240V, three-phase (60 Hz). A Wye configuration steps the voltage down to 208V, three-phase (60Hz).

A **208 volt three-phase wye transformer** is considered a balanced system. Across any two hot legs, a reading of 208 volts should be obtained. From any hot leg to ground, the voltage reading should be 120 volts.

Figure 2–1: *Wye Configured Transformer*

A **240 volt three-phase delta transformer** is an unbalanced system. Across any two hot legs, a reading of 240 volts should be obtained. From two hot legs to ground, the voltage reading should be 120 volts. From one hot leg to ground, the voltage reading should be 208 volts. The leg that reads 208 volts to ground is considered the wild leg.

Figure 2–2: *Delta Configured Transformer*

Larger commercial and industrial applications will step voltage down to 460V, three-phase (60 Hz), using a Delta configured transformer. When a Wye configured transformer is used, the voltage will be 480V, three-phase (60 Hz).

Section 2   Electrical Theory and Application

When placing a voltmeter across any two legs on a **460V three-phase Delta system,** a reading of 460V should be obtained.

When placing a voltmeter across any two legs on a **480V three-phase Wye system,** a reading of 480V should be obtained.

When testing a **480 volt three-phase Wye system,** any hot leg to ground should measure 277V.

The voltage that is supplied to an electrical load must be within **10%** of the designed voltage. An electrical load that is designed for 120V can operate on a voltage from 108V to 132V. An electrical load that is designed for 240V can operate on a voltage from 216V to 264V.

AC MOTORS

All service technicians need to understand AC motor principles. The following information covers types of motors, motor controls, capacitors, starting relays, and related terms. This section will help the technician with competency exams as well as in the field.

Induction motors use induced magnetism to convert electrical energy to mechanical energy.

Motors are designed to operate with a supply voltage that is **+ or - 10%** of the voltage that is labled on the nameplate.

The **nameplate** is a plate located on the motor that lists the design characteristics of the motor.

The **rotor** is the rotating iron core that turns in the magnetic field. The rotor is connected to the shaft of the motor.

The **stators** are the stationary electromagnets. They are also known as poles.

The **stators** have opposite magnetic polarities.

Motor speeds depend on the number of stationary poles in a motor.

A **2 pole** motor is 3600 RPM.

A **4 pole** motor is 1800 RPM.

A **6 pole** motor is 1200 RPM.

Motors are rated in **horsepower.** One horsepower is the amount of power required to lift 550 pounds in one second.

746 watts equals one horsepower.

External motors are **cooled** by air motion. Hermetically sealed compressors are cooled by refrigerant gas and air motion.

Single phase motors use a **run winding** for motor operation and a **start winding** that gives a motor the starting torque.

The **run winding** is less resistant than the start winding.

The **run winding** is in the circuit during the full operation of the motor.

The **start winding** has a higher resistance than the run winding and is in the circuit on motor start up.

The **start winding** is taken out of the circuit after the motor has reached 75% of its running speed.

The **start winding** is 90 degrees out of phase with the run winding.

The **run and start windings** are wired in parallel to each other.

When **testing** the resistance of motor windings with an ohmmeter, always use the R × 1 ohms scale.

When **measuring the resistance** of the motor windings, run to start is the highest resistance, common to start is the next lowest resistance, and common to run is the lowest resistance.

When measuring resistance, an **open** run winding will read infinite resistance between run and common. An open start winding will read infinite resistance between start and common.

When measuring resistance, a **shorted** run winding will read zero or a low resistance between run and common. A shorted start winding will read zero or a low resistance between start and common.

When testing for a **short to ground motor winding,** always use the highest ohm scale.

If a **motor winding** is shorted to ground or the internal windings are shorted together, it commonly trips a circuit breaker or blows a fuse.

When testing for a **grounded motor winding,** a measurable resistance from any motor terminal to ground shows a short to ground condition.

The **contacts** of a start relay are wired in series with the start winding.

Start relays used on refrigeration compressors must be external due to the arcing of the contacts. This could cause acids to form in a refrigeration system.

A **centrifugal switch** is a normally closed switch that takes the start winding out of the circuit.

A **centrifugal switch** uses the centrifugal force of the rotor to open a set of normally closed contacts that are in series with the start winding.

A **current relay** uses the operating current of the run winding to control a set of contacts.

A **current relay** has a normally open set of contacts in series with the start winding.

A **potential relay** uses the back electromotive force (back voltage) of the start winding to operate a set of contacts.

A **potential relay** has a set of normally closed contacts in series with the start winding.

On a **potential relay**, terminals 1 and 2 are the normally closed contacts and terminals 2 and 5 are the coil.

On a **potential relay**, terminal 1 connects to the start capacitor, terminal 2 connects to the start winding, and terminal 5 connects to common terminal located on the compressor.

A **solid state starting relay** uses a positive temperature coefficient (PTC) thermistor in series with the start winding.

On a **solid state relay** as the temperature increases the resistance increases.

FLA, or full load amps, is the operating or running amperage of a motor.

LRA, or locked rotor amps, is the starting amperage of the motor.

The **LRA** is commonly five times higher than normal running amperage.

Overload protectors are commonly a bimetal snap disc or a thermistor.

Overload protectors open a set of normally closed contacts by sensing excessive heat, excessive current, or both.

Overload protectors can be external or internal types.

External overloads are located outside a motor.

Internal overloads are located in the windings of a motor.

Thermal Overload (pilot duty)   Overload Protector (line break)

Overload protectors can be used in a pilot duty or line break circuit.

Pilot duty overload protectors use a normally closed set of contacts in series with a contactor or starter coil that controls the power supply to the motor being protected. When testing an open internal overload protector with an ohmmeter, use the R×1 scale.

Line break overload protectors use a normally closed set of contacts in series with the common side of a motor or compressor.

When testing for an **open internal overload protector**, the ohmmeter will read measurable resistance across run and start, infinite resistance across run and common, and infinite resistance across start and common.

When testing for an **open internal overload protector** using an ammeter and a voltmeter, the ammeter will read zero while the voltmeter will read source voltage across the motor.

Capacitor

Capacitors are used to give a motor starting torque or to aid a motor in its running operation.

Capacitors are rated in **microfarads** (mfd or uf).

A capacitor is designed with a specific **VAC (volts AC)** rating. The voltage to a capacitor must never exceed this rating or the capacitor could be damaged.

The **start capacitor** is used on Capacitor Start (CS) and Capacitor Start/Capacitor Run (CSR) motors. It is a dry type capacitor and uses a 20,000 ohm 2 watt bleed resistor to discharge the capacitor during the off cycle.

Start capacitors are designed to energize the start windings of a motor until its rotor reaches about 75% of its running speed.

The **start capacitor** is de-energized by a start relay. The most common start relays are the current relay, the potential relay, the solid state relay, and the centrifugal switch.

The **run capacitor** is used on the Permanent Split Capacitor (PSC) and Capacitor Start/Capacitor Run (CSR) motors. It is an oil filled capacitor, and the oil dissipates heat during operation.

Run capacitors are used to give the motor a small starting torque. They are left in the circuit to increase the running efficiency of the motor.

A **dual capacitor** is two run capacitors set in one container.

Section 2 Electrical Theory and Application

Dual capacitors are usually found on residential and window air conditioning systems. They are connected to a PSC type compressor and condenser fan motor.

Capacitors can be wired in series or in parallel to increase or decrease the capacitance in a circuit. The following are calculations that are used to figure the circuits capacitance.

C1 = Capacitor number one

C2 = Capacitor number two

TC = Total capacitance

If two **capacitors** were wired in parallel, the calculation for the capacitance is

C1 + C2 = TC.

If two **capacitors** were wired in series, the calculation for the capacitance is

$$\frac{C1 \times C2}{C1 + C2} = TC.$$

When replacing a run or start capacitor, the **voltage rating** must always stay the same.

When **testing** a start capacitor with an ohmmeter, the bleed resistor must be temporarily removed.

When **testing** a capacitor with an ohmmeter, a good capacitor will move from infinite resistance to zero resistance then back to infinite. A shorted capacitor will stay at zero or a readable resistance. A capacitor that is internally burned open will stay on infinite resistance.

A **capacitor analyzer** is used to measure the microfarad, the power factor, and the condition (shorted or open) of the capacitor.

SHADED-POLE MOTORS

Shaded-pole motors are used for small condenser and evaporator fans and small, low Btu output furnaces.

Shaded-pole motors have copper rings located on the stator of the shaded-pole. A current is induced in the shaded-pole from the main winding. The shaded-pole produces a magnetic field that is out of phase with the main winding, thus giving the motor its starting torque.

Reversing the direction of a shaded pole motor can be done by disassembling the motor, removing the rotor, turning the rotor 180°, and reinstalling it.

Shaded-pole motors can have many different **speeds**, but most have two or three speeds.

SPLIT PHASE MOTORS

A **split phase type compressor motor** is used on refrigeration systems that incorporate a capillary tube metering device. Split phase motors are also used for oil burners, exhaust fans, and small water pumps.

A **split phase motor** uses an inductive run winding to operate the motor and a resistive start winding to give the motor its starting torque.

A **split phase motor** does not use a run or start capacitor.

A **split phase motor** is commonly a fractional horsepower type motor.

A **split phase motor** commonly uses a centrifugal switch, a current relay, or a solid state relay to de-energize the start winding after the motor has reached 75% of the running speed.

Reversing a split phase motor is done by reversing the leads on the start winding.

CAPACITOR START INDUCTION RUN MOTORS (CSIR)

A **capacitor start induction run motor** produces a high starting torque. The CSIR motor is a split phase motor with a start capacitor added in series with the start winding.

When a CSIR motor reaches **75% of its running speed**, the start winding and the start capacitor are de-energized by a centrifugal switch or a starting relay. The motor run winding stays energized to keep the motor in operation.

A **CSIR motor** is used on compressors up to 5 HP, larger water pumps, large belt drive fans, and with compressors where the refrigeration system uses a TXV type metering device.

If a start capacitor has too low of a **microfarad rating** for the CSIR motor, the motor will not start.

PERMANENT SPLIT CAPACITOR MOTOR (PSC)

The **PSC motor** is designed to have a low staring torque with a high running torque.

A **run capacitor** is wired in series with the start winding. The start and run windings are in the circuit during the full operation of the motor.

The **run capacitor** limits the electron flow through the start winding so it will assist with the running torque of the motor.

A **PSC** motor does not need a start relay or a start capacitor.

PSC compressors are used with systems that equalize during the off cycle.

Multi-speed PSC motors with direct drive blower motors are very common.

If a run capacitor has too low of a **microfarad rating** for the PSC motor, the motor might not start. It could also increase the amperage of the motor during the running cycle.

CAPACITOR START CAPACITOR RUN MOTOR (CSR)

The **CSR motor** produces high staring and running torque. It is actually a CSIR motor with a run capacitor added to the start winding.

When the motor reaches **75% of its running capacity,** the start capacitor is de-energized by a starting relay. The run capacitor controls a small amount of current through the start winding during the running operation of the motor. This causes the motor to operate with an increase of running torque.

A **CSR motor** is used where high starting and running torque is required, such as large hermetic and semi-hermetic compressors.

A **CSR motor** commonly uses a potential type starting relay.

If a start capacitor has too low of a **microfarad rating** for the CSR motor, the motor will not start.

If a run capacitor has too low of a **microfarad rating** for the CSR motor, the motor may not start or it could increase the amperage of the motor during the running cycle.

THREE-PHASE MOTORS

Three-phase motors are large rugged motors. They are much stronger than single phase motors.

Three-phase motors use three windings that are 120 electrical degrees out of phase. This causes a large starting torque and an excellent running efficiency.

Three-phase motors do not have start windings, capacitors, or starting relays like single phase motors.

Common **voltages** for three-phase motors are 208/240 and 440/480 volts.

They have two basic types of **winding configurations,** the Wye and the Delta.

To **reverse the direction** of a three-phase motor, reverse any two power leads connected to the motor.

Three-phase motors are used for large compressor motors, large blower motors, and large pumps normally above 5 H.P.

An **integral controlled motor (ICM)** is a variable speed motor controlled by the output of a circuit board.

ELECTRICAL CONTROLS

There are numerous electrical controls in the HVAC/R industry. The following information will give the technician a better understanding of electrical control operation.

Fuse

Section 2 Electrical Theory and Application

Fuses and circuit breakers are considered over-current protectors. They are used to open a circuit if the current reaches a value that will cause excessive or dangerous temperatures in the conductor or conductor insulator.

A **fuse** is rated for a maximum amperage and a maximum voltage.

To **test a fuse** with an ohmmeter, disconnect the power supply, remove the fuse from the circuit, set the ohmmeter to the Rx1 scale, and place the leads across the fuse. If the fuse is in good condition, there will be zero resistance. If it is burnt open, there will be infinite resistance.

To **test a circuit breaker** with a clamp-on ammeter, clamp the ammeter around one wire that leads to the circuit breaker. Set the circuit breaker to the "ON" position and check the amperage reading of the circuit. If the amperage exceeds the rating of the circuit breaker, test the circuit for an electrical short or a grounded condition. If the amperage does not exceed the rating of the circuit breaker and the circuit breaker trips the circuit off, the breaker is defective.

A heating and cooling **thermostat** is used to sense space temperature. There are many different types of sensing elements that are used in thermostats such as liquid or gas charged sensing bulbs, bimetal elements, and thermistors.

A **bimetal sensing element** is two different metals bonded together, commonly brass and invar, to cause a warping action. This warping action will either open or close a set of contacts.

A **bimetal snap disc** type sensor is used on thermostats, limit controls, and fan controls.

A liquid, gas, or cross charged (gas and liquid) **sensing bulb** is commonly used to operate a bulb type temperature control.

The **liquid charged** sensing bulb is extremely sensitive. It is completely full of liquid and has no vapor present.

A **gas charged** sensing bulb contains all gas. It has no liquid in the bulb at anytime. For proper operation, it must be installed so that the bulb is the coldest part of the control.

A **cross charged** sensing bulb always has liquid and vapor present. The temperature control can be installed where the bulb is at a warmer or colder location than the rest of the control.

A **positive temperature coefficient thermistor** is used on electronic thermostats as the sensing element. Its resistance increases as the temperature increases.

A **negative temperature coefficient thermistor** is used on electronic thermostats as the sensing element. Its resistance increases as the temperature decreases.

Heating Thermostat

A **heating thermostat** opens on a temperature rise and closes on a temperature drop.

The heat anticipator is used to maintain stable **temperatures** within a space. It will prevent system overshoot (too long of an on cycle) and system lag (too short of an off cycle).

The **heat anticipator** must be set to the heating control amperage.

The **heating control amperage** can be measured by placing an in-line ammeter across terminals R and W. It also can be measured by placing a 10 loop multiplier wire across terminals R and W and measuring the amperage with a clamp on ammeter.

Mercury is commonly used in a sealed glass bulb filled with an inert gas to close switch contacts on heating and cooling thermostats.

Single stage heating and cooling thermostats commonly use **four strand wire** for control. The red wire is power from the transformer, the white wire is to the heating control circuit, the green wire is for the fan relay circuit, and the yellow wire is for the cooling circuit.

Two stage thermostats will have two extra wires for the second stage cooling circuit and the second stage heating circuit.

Automatic changeover is commonly incorporated in commercial thermostats. This allows for the system to change from heating to cooling automatically.

Deadband is the point at which an automatic changeover thermostat is balanced between the heating and cooling cycle.

Outdoor thermostats are commonly used on electric heating systems to sense the outdoor temperatures.

When the outdoor temperatures drop below a predetermined set point, the thermostat energizes a bank of electric strip heaters.

Cooling Thermostat

A **cooling thermostat** opens on a temperature drop and closes on a temperature rise.

A **cooling anticipator** is wired in parallel with the cooling circuit. It prevents system lag.

The **cooling anticipator** is commonly a non-adjustable resistor placed across the cooling thermostat.

An **outdoor ambient thermostat** is used on cooling systems to control condenser fan speeds if the temperature drops below a predetermined point. During low ambient conditions, the fan speed is decreased.

An **outdoor ambient thermostat** is used on cooling systems to de-energize the compressor circuit if the temperature drops below a predetermined point. During low ambient conditions, the compressor is locked out to prevent compressor damage.

A **programmable thermostat** is used to control an HVAC system more accuracy than a standard bi-metal thermostat.

A programmable thermostat uses a **thermistor** instead of a bimetal to sense room temperature.

A programmable thermostat's **cooling anticipator** has a fixed setting.

A programmable thermostat's **heat anticipator** can be an adjustable or have a fixed setting.

LED lights are commonly used on programmable thermostats as indicator lights.

An **emergency heat light** is used with a programmable heat pump thermostat to indicate the operation of the emergency heat cycle.

A **24 volt power supply** is commonly brought to a programmable thermostat to energize the LED lights located in the thermostat.

A **two stage programmable** heat pump thermostat uses the first stage to energize the compressor and the second stage to energize the auxiliary strip heaters.

A **battery** is used in a programmable thermostat to save the program and maintain clock time if there is a power failure.

Automatic changeover is commonly used in a programmable thermostat. This allows for the system to change from heating to cooling automatically.

Programmable thermostats are available with a **five day or seven day** programming capability.

Temperature set back is the temperature at which a thermostat is programmed during an unoccupied time for the purpose of energy conservation.

Set point is the setting of the thermostat during an occupied time.

Computed or intelligent recovery is commonly used in programmable thermostats. This feature determines the rate that the space temperature deviates from the systems set point temperature to the systems set back temperature. The thermostat then starts the system to allow the space temperature to increase gradually to the predetermined set point at a pre-set time.

The **override** feature in a programmable thermostat is used to temporarily bypass the system's programmed schedule. This can be accomplished by pressing the override or hold button.

Limit Switch

A **limit switch** is used to de-energize a gas valve if the temperature of the furnace rises above a predetermined temperature.

A **limit switch** commonly uses a bimetal as the sensing element and can be an automatic or a manual reset type.

Section 2 Electrical Theory and Application

Low Pressure Switch

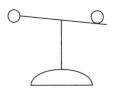

A **low-pressure switch** is used in an air conditioning system to sense the low side pressure on a system. The low-pressure switch opens when there is a drop in pressure and closes when there is a rise in pressure.

The **low-pressure switch** can be used as a safety control to sense a drop in system pressure due to a loss of refrigerant charge. This will protect the compressor from overheating and loss of lubrication.

The **low-pressure switch** can also be used as an operating control. As the space temperature decreases, the system's operating pressure decreases. This fluctuation in system pressure can cause a low-pressure switch to open and close the circuit to the compressor.

High Pressure Switch

A **high-pressure switch** is used in an air conditioning system to sense a rise in high side pressure. If the pressure increases above a predetermined point, the high-pressure switch will open and de-energize the compressor.

The **high-pressure switch** is used as a safety control. It is either an automatic or a manual reset type.

A **reverse-acting high-pressure switch** is used during low ambient conditions to control condenser fans on an air cooled condenser. By cycling the condenser fans, the control of high side pressure is accomplished.

Relay

A **relay** is used to energize or de-energize a load circuit by using the signal of a control circuit. When the coil of a relay is energized, the contacts in the relay change their positions. A normally open contact will close and a normally closed contact will open. Relays commonly energize motors, heaters, lights, contactors, or other relay coils.

The **contacts** of a relay must never exceed 20 amps.

When using an ohmmeter, a relay coil should have a **measurable resistance** across the coil.

When **control voltage** is applied across the coil, the switch contacts should change positions.

The **inrush current** on the start up of a relay is higher than the holding current during relay operation.

If a **relay coil** is shorted or burnt open, the switch contacts will not change positions.

When measuring the resistance of a relay coil and an **infinite resistance** is obtained, the coil is burnt open.

When measuring the resistance of a relay coil and a **zero resistance** is obtained, the coil is shorted.

A **shorted relay coil** commonly damages the control transformer.

When **replacing a relay,** it is important to know the relay's switch configuration and the coil voltage.

Contactor

A **contactor** is used to energize or de-energize high voltage loads. When the coil of a contactor is energized, the contacts in the contactor change their positions. A normally open contact will close and a normally closed contact will open. Contactors commonly energize motors and electric heaters. The coil voltage of a contactor can be 24, 120, 240 and 480 volts.

When using an ohmmeter, a contactor coil should have a **measurable resistance** across the coil.

When **control voltage** is applied across the coil, the switch contacts should change positions.

If a **contactor coil** is shorted or burnt open, the switch contacts will not change positions.

When measuring the resistance of a contactor coil and an **infinite resistance** is obtained, the coil is burnt open.

Figure 2-3: *A Motor Starter*

When measuring the resistance of a contactor coil and a **zero resistance** is obtained, the coil is shorted.

A **shorted contractor coil** commonly damages the control transformer.

When **replacing a contactor,** it is important to know the contactor's switch configuration, coil voltage, and contact rating.

The difference between a **relay and a contactor** is the amount of current that the contacts can handle. A relay can handle under 20 amps and a contactor can handle over 20 amps.

A **starter** is comprised of contacts, a solenoid coil, and a set of overloads. When control voltage is applied across the coil of the starter, the contacts change positions. A normally open contact will close and a normally closed contact will open. Starters commonly control the operation of motors. Overload protectors are built into the starter to de-energize the coil if an overload occurs.

When using an ohmmeter, a starter coil should have a **measurable resistance** across the coil.

If a **starter coil** is shorted or burnt open, the switch contacts will not change positions.

When measuring the resistance of a starter coil and an **infinite resistance** is obtained, the coil is burnt open.

When measuring the resistance of a starter coil and a **zero resistance** is obtained, the coil is shorted.

A **shorted starter coil** commonly damages the control transformer or main line fuse.

When **replacing a starter,** it is important to know the starters switch configuration, coil voltage, and contact rating.

The difference between a **contactor** and a **starter** is the overload protectors that are built into the starter.

There are **three types of overload protectors** used in starters. They are the bimetal relay, the thermal relay, and the molten-alloy relay.

The **bimetal relay** uses two dissimilar metals bonded together. When current increases, more heat is produced. This causes the bimetal to warp, breaking the circuit to the starter coil.

The **thermal overload** uses a wire that is in series with the load being controlled. On an overload condition, the wire overheats and opens a set of contacts that are in series with the starter coil. **Overload condition** refers to excessive electrical current in the form of heat flowing through the wire.

The **molten-metal relay** places the molten-alloy in series with the load being controlled. If an overload condition occurs, the metal melts. This will cause a ratchet wheel to open a set of contacts that are in series with the starter coil.

Section 2   Electrical Theory and Application

Time Delay Relay

A **time delay relay** (TDR) is used to delay the starting or stopping of a load. TDRs are used on furnaces to delay the blower operation, or on air conditioning systems as an anti-short cycle device for the compressor. The length of the delay can be preset or adjustable.

A **time delay relay** is a control switch used to transfer an action in the control circuit to the operation of a higher voltage load. The time delay relay has a heater or a solid state circuit controlling a set of normally open or normally closed contacts.

The time delay relays can either be a **delay on make** or **delay on break** type. The delay circuit is energized by the control circuit.

A **delay on make relay** delays the opening or closing of contacts for a predetermined amount of time.

The **delay on make relay** is commonly used in air conditioning circuits to delay the starting of the compressor circuit.

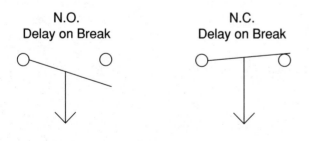

A **delay on break relay** delays the opening or closing of contacts for a predetermined amount of time.

The **delay on break relay** is commonly used in heating circuits to delay the stopping of the indoor blower after the heating cycle.

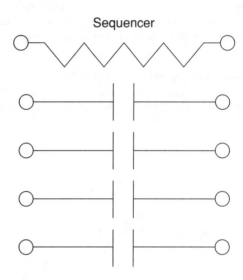

Figure 2–4: *A Sequencer*

A **sequencer** is a series of switches that sequentially control loads in an HVAC system.

There are many different methods of **sequential operating controls,** such as thermal operated timed delay relays, solid state circuit boards, and cam operated timers.

A **sequencer** is commonly used in an electric heating system. The sequencer energizes the heaters, delaying current flow to each heater element. This will decrease the inrush current on the start up of the system.

Microprocessors are commonly used as sequencers. They are pre-programmed circuit boards that have many different switching combinations.

Figure 2–5: *Defrost Timer*

A **defrost cycle** is commonly controlled by an electric defrost timer or a solid state circuit board.

An electric **defrost timer** (Figure 2–5) uses a synchronous motor to control a cam which in turn controls switch contacts. Defrost timers are commonly used to control a defrost cycle located in a refrigeration or heat pump system.

Microprocessors can be used as defrost timers. They can be programmed to energize a defrost circuit at a specified period in time or when the need for a defrost is detected.

Solenoid Coil

A **solenoid valve** is used to control the flow of refrigerant in an air conditioning or refrigeration system. When energized, a normally closed solenoid valve will open and a normally open solenoid valve will close.

On an **automatic pump down system,** a normally closed solenoid valve is placed in the liquid line. A thermostat directly controls the solenoid valve and a low pressure switch controls the compressor. When the thermostat closes, the solenoid valve opens. This will allow refrigerant to flow to the evaporator coil. When refrigerant enters the evaporator, the pressure increases. This will close the pressure switch and energize the compressor. When the thermostat is satisfied, it will open and de-energize the solenoid valve. This will cause the valve to close and it will block the flow of refrigerant to the evaporator coil. The compressor will continue operating until the refrigerant is pumped out of the evaporator. This will cause the pressure switch to open and de-energize the compressor.

During a defrost on refrigeration systems, a **hot gas solenoid valve** is used to direct hot gas from the discharge line to the inlet of the evaporator.

A **reversing valve solenoid** is used to reverse the flow of refrigerant in a heat pump system.

Hot gas bypass solenoid valves are used for capacity control on commercial air conditioning systems.

Crankcase heaters are placed in or around the shell of a compressor. They are used to prevent migration of refrigerant into the crankcase during low ambient temperatures.

Crankcase heaters are commonly energized during the system's off cycle.

Electric Heater

Electric heater elements are used in electric furnaces as the primary heat source.

Electric heaters are rated in watts. One watt is equal to 3.412 Btus.

Auxiliary strip heaters are used in heat pumps to aid in the heating and defrost cycle during low outdoor ambient temperatures.

The **auxiliary strip heaters** are located down stream of the indoor coil on a heat pump. They are electrically controlled by the second stage of a two stage thermostat.

Fusible Link

Fusible links are safety controls used to break a circuit on an increase of temperature.

On auxiliary strip heaters, **fusible links** are wired in series with the limit control and the electric heater. This secondary safety control is used to prevent overheating of the heating element.

Fusible links are also used as flame rollout sensors on many gas fired furnaces.

Transformer

The **step up transformer** takes a lower voltage and increases it to a higher voltage. They are used to apply a spark across an electrode for the ignition of a heating cycle. An ignition transformer steps voltage up from 240V, 120V, or 24V to 7000V to 10000V.

The **step down transformer** takes a higher voltage and decreases it to a lower voltage. They are commonly used as the control voltage for relays, contactors, and other low voltage controls in an HVAC/R system. A step down transformer commonly steps voltage down from 120V or 240V to 24V.

The **turn ratio** on the secondary winding compared to the primary winding of a transformer determines if it is a step up or a step down type. If there are more turns on the secondary winding than on the primary winding, it is a step up transformer. If there are more turns on the primary winding than on the secondary winding, it is a step down transformer.

If a **transformer** uses a 240V primary winding that has 400 turns and the secondary has 200 turns, the output voltage will be 120V.

When replacing a step down transformer, the **VA rating** is important. This is the power or wattage that is being consumed in a control circuit. If the VA rating of the transformer is lower than the power rating of the control circuit, the transformer will burn out.

To calculate the **VA rating,** measure the amperage of the control circuit and multiply it by the voltage of the control circuit.

Example: The amperage of a systems control circuit is 1.6 amps and the voltage being used is 24 volts. **E × I = W:** 24V × 1.6 A = 38.4 W. The transformer that can be used on this circuit must have a minimum VA rating of 38.4. The transformer that would be used for this circuit would commonly be a 40 VA rated transformer.

When testing a transformer **using a voltmeter,** source voltage at the primary and secondary side of the transformer shows that the transformer is functional.

If there is **voltage** at the primary side but not at the secondary side of the transformer, the problem could be a bad primary or secondary winding.

When testing a transformer using an **ohmmeter,** measurable resistance should be across the secondary and the primary windings.

An infinite resistance across the **secondary winding** of a transformer shows an open secondary winding.

A zero or low resistance across the **secondary winding** of a transformer shows a shorted secondary winding.

An infinite resistance across the **primary winding** of a transformer shows an open primary winding.

A zero or low resistance across the **primary winding** of a transformer shows a shorted primary winding.

SOLID-STATE ELECTRONICS

New HVAC/R equipment uses solid-state components. The service technician must be able to understand the components of solid-state controls in order to be capable of troubleshooting system malfunctions. The following are terms related to solid-state electronics.

Solid-state circuit boards are becoming more common in residential and light commercial HVAC/R. They are being used to replace electrical and electromechanical controls that are used for safety and operation.

Never use an ohmmeter when testing a circuit board. The voltage from the battery located in the meter can damage the solid-state circuitry.

Semiconductors conduct electricity in a controlled manner. There are N-type, negatively charged and P-type, positively charged semiconductors.

Diodes allow current to flow in one direction. One end of a diode is an anode and the other end is a cathode.

Transistors are made of N-type and P-type semiconductor material. They have three terminals, a base, a collector, and an emitter. Transistors can be used as a solid-state relay or an amplifier.

Thermistors are semiconductors that are used to sense temperature. There are two types of thermistors, a PTC or positive temperature coefficient type and an NTC or negative temperature coefficient type.

A **PTC thermistor's** resistance increases when the temperature increases and an NTC thermistor's resistance decreases when the temperature increases.

A **capacitor** is used to store electrons. In a DC circuit a capacitor blocks current flow and in an AC circuit a capacitor aids current flow.

A **resistor** is used to drop the voltage in a solid-state circuit.

A **rectifier** is used to change AC to DC.

WIRING LAYOUTS AND ELECTRICAL DIAGRAMS

It is imperative that the technician be well versed in interpreting electrical diagrams and layouts. Frequently, the technician has to refer to schematic diagrams to familiarize himself with a system's components and system operations. The following are terms and information on wiring layouts and electrical diagrams.

A residential **furnace electrical** circuit is sized to provide 120V, single phase, 60 Hz through a properly sized circuit breaker.

An electrical **disconnect switch** must be located at the furnace to disconnect the power supply during service.

An electrical **disconnect** must be installed near an air conditioning condensing unit to disconnect the power supply during service.

Common 120V **loads** that operate in an HVAC system are the indoor blower motor, the induced draft blower, the transformer's primary winding, an electronic air cleaner (if used), and a humidifier transformer (if used).

Any electrical **pigtail** wires that are twisted together must be connected with the proper sized wirenut.

A 240V, 60Hz single phase **power supply** is used for the condensing unit on residential packaged or split systems.

The **loads** that are operated by the 240V circuit on a package HVAC unit are the compressor, condenser fan, transformer's primary winding, the supply air blower, the induced draft blower, and crank case heater.

The input of a **control transformer** for a residential split air conditioning system is 120V, 60Hz, single phase and the output is 24V.

The input of a **control transformer** for a packaged HVAC system is 208/240V, 60Hz, single phase and the output is 24V.

The **24V control circuit** directly energizes relay coils, contactor coils, gas valves, and control modules in an HVAC system.

A **schematic** is a group of lines and electrical symbols arranged in a ladder form to show the operational sequence of an electrical system. It is used when troubleshooting an HVAC/R system.

In a **schematic,** the vertical lines represent the power supply and the horizontal lines represent the individual circuits.

A **legend** is used with a schematic diagram to identify the label of electrical circuit components.

The **top portion** of a schematic is usually the line voltage side of the electrical circuit.

The **bottom portion** of a schematic is usually the low voltage side of the electrical circuit.

A **transformer** is commonly located between the top portion and the bottom portion of the schematic diagram.

A **pictorial diagram** shows the actual factory installed wiring and location of system components.

A **pictorial diagram** is also known as a label or line diagram.

A **factual diagram** consists of an electrical schematic and a pictorial diagram.

An **installation diagram** is an aid for the system installer. It displays the system's fuse size, wire size, and its color code.

An **installation diagram** commonly shows the placement of the thermostat and power supply wiring for the system.

Factory installed wiring is the electrical wiring that is color coded, sized, and installed in a system at the factory.

Field installed wiring is the wiring that is brought to an HVAC system by an installation technician as the system is being installed.

A **heavy dashed line** on an installation wiring diagram indicates field installed line voltage wire.

A **heavy solid line** on an installation wiring diagram indicates factory installed line voltage wire.

A **light dashed line** on a wiring diagram indicates field installed low voltage wire.

A **light solid line** on a wiring diagram indicates factory installed low voltage wire.

A **connection diagram** shows the actual wiring of electrical components along with their proper color coded wire.

Section 2   Electrical Theory and Application   57

Figure 2–6: *Schematic of a Split Air Conditioning System*

Figure 2–6 is a schematic of a split air conditioning system. The operational sequence is as follows:

On a call for cooling from the cooling thermostat, (**CTH**) terminals **R** and **Y** close, energizing the contactor coil (**C**), and the fan relay coil (**FR**) in the 24 volt circuit. The **C1** and **C2** contacts change position and energize the compressor, **Comp.**, and condenser fan, **CF**, in the 240 volt circuit. The **FR1** and **FR2** contacts change positions and energize the indoor blower motor (**IBM**) in the 120 volt circuit. If the low pressure switch (**LP**) opens, the contactor coil (**C**) will de-energize and contacts **C1** and **C2** will open. This will break the circuit to the compressor (**Comp.**) and condenser fan (**CF**).

Figure 2–7: *Schematic of a Forced Air Gas Furnace*

Figure 2–7 is a schematic of a forced air gas furnace. The operational sequence is as follows:

With the door switch (**DS**) in the closed position, the transformer (**TR**) is energized by 120 volts. 24 volts will then be available on the control side of the transformer. On a call for heat, the heating thermostat closes terminals **R** and **W,** and energize the induced draft relay (**IDR**) with 24 volts. The normally opened **IDR** contacts close, energizing the induced draft motor (**IDM**) with 120 volts. After the motor is energized, the pressure switch (**PS**) closes proving the operation of the induced draft motor. The **PS** directs current to the high limit (**HL**) and secondary limit (**SL**). If both limit controls are in the closed position, the ignition control module is energized across terminals **TH** and **TR.** This starts a sequence of operation by energizing the pilot valve (**PV**) and the spark ignition system. When the pilot lights, the flame is proven by the flame sensor. Then the main valve (**MV**) is energized and the main burners are ignited. The heat from the heat exchanger causes the fan switch (**FS**) to close, energizing the indoor blower motor (**IBM**) with 120 volts. If the thermostat's "auto/on" switch is set to the "on" position, the thermostat will close from terminals **R** to **G.** This will energize the fan relay (**FR**) coil causing **FR1** and **FR2** contacts to change positions. The normally closed **FR2** contacts will open, de-energizing the low speed circuit, and the normally open **FR1** contacts will close, energizing the high speed on the indoor blower motor (**IBM**) with 120 volts.

Section 2　Electrical Theory and Application

Figure 2–8: *Schematic of a Furnace Circuit Board*

Figure 2–8 is a schematic of a circuit board common to many forced air furnaces. The operational sequence is as follows:

On a call for heat, the heating thermostat will close from **R** to **W** and bring 24 volts to the **W** terminal on the circuit board. Once the **W** terminal has been powered by 24 volts, this will start the sequence of operation by energizing the **GAS1** and **GAS2** terminals with 24 volts. This will energize the pick coil on the gas valve through the normally closed set of contacts on the bimetal pilot sensor. At the same time, it will energize the pilot igniter and the hold coil on the gas valve. With the pick coil and pilot igniter energized, gas will flow to the pilot and be ignited by a spark from the pilot igniter. The pilot flame will heat the bimetal pilot sensor and a set of normally closed contacts will change to the open position after approximately 45 seconds. This will de-energize the pick coil and the pilot igniter. The hold coil will stay energized, since it is directly controlled from the **GAS1** terminal and is not controlled by the bimetal flame sensor. The hold coil will keep the pilot valve open, which will keep the pilot lit and ready for the ignition of the main gas burners. The bimetal pilot sensor also has a set of normally open contacts that will close once the pilot flame is sensed, approximately 45 seconds after pilot ignition. This proves the existence of the pilot and energizes the main gas valve. The main gas valve will feed gas to the main burners, and the gas will be ignited by the pilot. When voltage is sent to terminal **GAS3**, *simultaneously,* voltage will also be sent to the solid-state time delay circuit. It is this voltage that will start the energizing of the solid-state time delay. After 45 seconds, the fan will energize on the heating speed. When the thermostat is satisfied, terminals **R** and **W** will open. This will de-energize the (hold) pilot valve and the main gas valve. The blower will operate for 45 seconds, and then it will shut down, completing the cycle.

Figure 2–9: *Schematic of an Oil Fired Forced Air Furnace*

Figure 2–9 is a schematic of an oil fired forced air furnace. The operational sequence is as follows:

With the disconnect switch (**Disc**) and the limit switch, (**Lim**) in the closed position, 120 volts is brought to the black and white wires. This will energize the cad cell control. On a call for heat, terminals **R** and **W** close on the thermostat (**TH**). This will cause the orange wire on the control to energize the burner motor (**BM**) and the ignition transformer (**IT**) with 120V. The burner motor drives the fan and the oil pump which supplies the fuel and the air for combustion. The ignition transformer supplies the spark for ignition. The heat from combustion transfers into the heat exchanger and increases the temperature causing the fan switch (**FS**) to close and energize the indoor fan motor (**IBM**) with 120 volts. The cad cell proves the flame by sensing its light. This will keep the system energized until the thermostat is satisfied. If the cad cell does not sense the light that is produced by the flame, the cad cell controller will de-energize the burner motor and the ignition transformer after approximately 45 seconds. If the cad cell senses light before the system is energized, the system will not start.

Figure 2–10 is a schematic of an air to air heat pump. The operational sequence is follows:

COOLING CYCLE

When the thermostat is set to the cooling cycle and the fan switch is set in the auto position, the control relay (**CR**), the fan relay (**FR**), and the compressor contactor (**C**) are energized. The control relay (**CR**) energizes the reversing valve solenoid which directs refrigerant so that the outdoor unit is used as the condenser and the indoor unit is used as the evaporator. The fan relay (**FR**) energizes the indoor fan on high speed, and the contactor (**C**) energizes the compressor and condenser fan.

Section 2 Electrical Theory and Application

Figure 2–10: *Schematic of an Air to Air Heat Pump*

Legend for Figure 2–10

ADR	Auxiliary Defrost Relay
APS	Air Pressure Switch
C	Contractor
CCH	Crankcase Heater
CF	Condenser Fan
Comp.	Compressor
CR	Control Relay
CT1	Cooling Thermostat 1st Stage
CT2	Cooling Thermostat 2nd Stage
DR	Defrost Relay
FR	Fan Relay
FU	Fusible Link
H1	Auxiliary Strip Heater #1
H2	Auxiliary Strip Heater #2
H3	Auxiliary Strip Heater #3
HP	High Pressure Switch
HR	Holding Relay
HT1	Heating Thermostst 1st Stage
HT2	Heating Thermostst 2nd Stage
IFM	Indoor Fan Motor
LIM	Limit Switch
ODT	Outdoor Thermostat
RV	Reversing Valve
TD1	Timed Delay Relay #1
TD2	Timed Delay Relay #2
TM	Timer Motor

HEATING CYCLE

When the thermostat is set to the heating cycle and the fan switch is set to the auto position, the compressor contactor (**C**) and the fan relay (**FR**) are energized. The reversing valve solenoid is in the de-energized position. This directs the refrigerant so the outdoor unit is used as the evaporator and the indoor unit is used as the condenser. The fan relay (**FR**) energizes the indoor fan on high speed and the contactor (**C**) energizes the compressor and condenser fan. If the room temperature drops two degrees below the setting of the first stage thermostat, the second stage heating circuit is energized. The second stage thermostat (**H2**) closes, energizing the timed delay relay #1 (**TD1**). This sequentially closes two timed delay relay contacts, which, in turn, energizes two auxiliary strip heaters, H1 and H2. If the outdoor temperature is below the predetermined temperature on the outdoor thermostat (**ODT**), the timed delay relay #2 (**TD2**) is energized. This energizes strip heater #3 (**H3**).

EMERGENCY HEAT CYCLE

During the emergency heat cycle, the normal/emergency switch is placed in the emergency position. The outdoor unit, containing the compressor and condenser fan, is de-energized. The fan relay (**FR**) energizes the fan motor, while the timed delay relay #1 (**TD1**) energizes auxiliary strip heaters **H1** and **H2**. The red light (**R**) at the thermostat indicates that the system is operating on the emergency heat cycle.

DEFROST CYCLE

During the defrost cycle, the system temporarily changes from the heating to the cooling cycle. The initiation of the defrost cycle is accomplished by the air pressure switch (**APS**), the outdoor coil temperature, and defrost time (**TM1**). If the pressure difference across the outdoor coil reaches a predetermined point, the air pressure switch (**APS**) closes, energizing the auxiliary defrost relay (**ADR**). This will close one set of contacts across the pressure switch, forming a holding circuit on the ADR coil. A second set of **ADR** contacts will close, energizing the timer motor. This will cause the timer motor's cam to turn and aid in closing the **TM1** contacts. The **TM1** contacts will close if the temperature of the coil is below a predetermined point and the timer's cam turns to a specific location. This will break the contacts to the condenser fan and make to the defrost relay (**DR**) coil. Energizing the defrost relay coil will close a set of contacts to the reversing valve, close a set of contacts to the timer motor, open a set of contacts to the auxiliary defrost relay (**ADR**), and close a set of contacts to the timed delay relay #2 (**TD2**), which will energize the second stage heaters **H3**. After the system is defrosted, the outdoor coil temperature increases, reaching a predetermined temperature and the timer's cam turns to a specific location. The system returns back to the heating cycle.

ELECTRICAL METERS

Understanding how to apply electrical meters when troubleshooting is extremely important to the HVAC/R Technician. This section covers the use of voltmeters, ohm meters, and ammeters. It will also help technicians with competency exams on electrical meter usage.

An **analog meter** uses the magnetic field in an electrical circuit to cause a deflection of a needle. The needle deflection used with a scale provides an analog read out. The most accurate reading of an **analog meter** is from the middle to the upper two thirds of the scale.

Digital meters use solid-state circuitry to create a digital read out. They have no moving parts, and are more rugged and accurate then analog meters.

Many times digital meters have **autoranging.** This will automatically set the volts, amps, or ohms scale.

A **voltmeter** measures the operating and source voltage to an electrical load.

A **voltmeter** should always be set to the highest scale when working with an unknown voltage.

A **voltmeter** should be placed in parallel with the load being tested.

When **testing for voltage** across a set of opened contacts, a voltmeter should record source voltage.

When **testing for voltage** across a set of closed contacts, a voltmeter should read zero volts.

A **millivolt** is a thousandth of a volt. When measuring millivolts, a millivoltmeter must be used.

Analog ammeters use a needle type of readout and usually have a hold button to lock in the reading.

Digital ammeters have a digital readout and are more accurate than analog meters.

When using an **ammeter,** the circuit that is being tested must be energized.

An **ammeter** measures the start up and operating amperage of a load.

An **ammeter** should always be set to the highest scale when working with an unknown amperage.

An **in-line ammeter** should be placed in series with the load that is being tested.

Section 2   Electrical Theory and Application

Always clamp around one wire of a load when checking amperage using a **clamp-on ammeter.**

A **milliamp** is a thousandth (1/1000) of an amp. When measuring milliamps, a milliamp meter must be used.

A **ten loop multiplier** and a clamp-on ammeter can be used to measure milliamps. If a wire is coiled 10 times, the amp reading will be multiplied by 10.

VOMS are used to measure volts, ohms, and milliamps of an electrical circuit.

An **ohmmeter** measures the resistance of a load and the continuity of a switch or circuit.

An **ohmmeter** uses a battery for its power source.

The source voltage should always be disconnected from the circuit when using an **ohmmeter**.

When testing a component with an **ohmmeter,** the component must be isolated from the circuit.

When testing across a set of normally closed contacts on a relay, an **ohmmeter** should read zero ohms of resistance.

When testing across a set of normally open contacts on a relay, an **ohmmeter** should read infinite ohms of resistance.

When using an **ohmmeter,** an open circuit will read infinite resistance and a short circuit will read zero resistance.

A **wattmeter** is used to read the power output rating of a load. When measuring the output rating of a motor, the power factor must be taken into consideration.

Example: ($P = E \times I \times PF$) If a motor has a power factor (PF) of 85% or .85 and the voltage (E) of 240V with the amperage (I) at 5 amps the power rating (P) would be: ($P = 240V \times 5 A \times .85$) $P = 1020$ watts.

ELECTRICAL TROUBLESHOOTING

Troubleshooting is an everyday occurrence in the work of a service technician. The following are terms and information that will help the technician be proficient in his work and on certification exams.

Using a voltmeter:

A **voltmeter** should always be placed in parallel with the power supply being tested.

A **voltage test** is one of the first tests that should be made on inoperable an electrical circuit.

Energizing a 24 volt relay or contactor coil should change the positions of contacts. If a contact is normally open, it should close and if a contact is normally closed, it should open.

Closed contacts in an operating system should read zero volts. If a measurable amount of voltage is read, the contacts may be burned.

Open contacts in an operating system should read "source voltage."

If **source voltage** is read to a load such as a motor, a heater, or a relay coil, the load should operate. If the load does not operate, it is usually burnt open.

Hopscotch is a method of using a voltmeter to test an electrical circuit. To use this method, one of the voltmeter leads is placed on the common side of a load. The other lead is placed on the line side of the load to test for voltage. If there is no voltage to the load, the lead located on the line side is moved toward the power supply, across wires and switches until source voltage is found.

Using an ohmmeter:

Always disconnect the power supply to the system when using an **ohmmeter.**

Always isolate the component that is being tested when using an **ohmmeter.**

Fuses are generally tested using the ohmmeter. A good fuse will measure zero ohms and a bad fuse will measure infinite ohms.

Closed contacts in a non-operating system should read zero ohms. If a measurable amount of resistance is obtained, the contacts may be burned.

Open contacts in a non-operating system should read "infinite resistance." If zero ohms are found, the contacts are stuck in the closed position.

When **measuring the resistance** of a load, always start on the lowest (R×1) scale.

Measurable resistance should be obtained when using an ohmeter to test an electrical load such as a motor, a heater, or a relay coil.

Continuity means a continuous or complete circuit. It can either be zero ohms or a measurable resistance.

A **shorted** electrical load will read zero or very low ohms of resistance.

An open electrical load will read "infinite" ohms of resistance.

To test a load for a **short to ground condition,** isolate the load from the circuit, use the highest ohm scale, and place one of the meter leads to the line side of the load. Place the other lead to the uninsulated shell or ground terminal of the load.

Using an ammeter:

The system must be in operation to obtain amperage when **using an ammeter.**

To read amperage on a 24V control circuit using a clamp-on ammeter, a multiplier must be used to amplify the current reading.

A **multiplier** is a wire that is coiled 10 times and placed in series from the power supply. This will amplify the current by 10 to obtain the proper current. The reading must be divided by 10.

Example: Using a 10 loop multiplier an amperage of 8 amps is obtained. 8 / 10 = .8 amps total.

An in-line milli-ammeter can be used to test for proper amperage on a 24V control circuit.

A **short circuit** will have an excessive current draw. This will blow fuses or trip breakers in a 120V or 220V circuit. In a 24V control circuit a short will overheat and burn out a control transformer when using the cooling circuit, and will burn out the heat anticipator when using the heating circuit.

An **open circuit** does not allow current to flow and will read zero amps.

The **VA rating** is the power output of a transformer. To test for an underrated transformer, divide the VA rating of the transformer by the control voltage. This will give the maximum allowable amperage of the circuit.
VA/V = A

Example: 40VA transformer / 24V control voltage = 1.66 amps maximum.

Section 3

Air Flow Components and Duct Fabrication

This section consists of air flow components and types of duct fabrication. It is very important for a technician to have a good understanding of air flow test equipment, fans/blowers, types of duct systems, materials used to fabricate duct work, types of air side components, air filters, humidifiers, and temperature/humidity measurements. The following will aid the technician when studying for competency exams.

AIR FLOW TOOLS AND MEASUREMENTS

There exist many types of air flow test equipment that are used daily. The technician needs to be knowledgeable in the use of air flow tools and where and when to obtain measurements. The following is related information that will assist the technician.

A **Tachometer** is used to measure the speed of a rotating object such as a motor or a blower. A common type of tachometer uses a strobe light and measures RPM's. This is known as a strobe tachometer.

A **Pitot-tube** is a tool that is used to test air velocity pressure, static pressure, and total pressure on a duct system.

The small hole in the end of a **Pitot-tube** is used to measure total pressure. The holes on the side of the tube are used to measure the static pressure in the duct system.

Total pressure minus static pressure = **velocity pressure.**

The velocity and static pressures are added and the sum is the **total pressure.**

The **speed of the air** traveling in a duct is refereed to as velocity. It is measured in inches of water column (WC) or feet per minute (FPM).

Extremely small amounts of pressure are measured using a water column gauge. **Water column (WC)** is the pressure that it takes to push a colunm of water up vertically. A half-inch of water column is the amount of pressure it would take to push a column of water in a tube half an inch.

A **Pitot-tube Traverse** is an air flow testing method used to take an average air velocity pressure reading in an air distribution system. This is important when determining air volume within the system.

A **Pitot-tube Traverse** can be done by taking air velocity pressure readings in the main supply and return duct and/or at individual branch runs.

A **Velometer** measures the speed of the air in a duct system or at a diffuser. Meter readouts can be in analog or digital form.

A **diffuser probe** used with a Velometer will measure air speed at the outlet of a diffuser.

A **Pitot-tube** used with a Velometer will measure air speed in a duct system.

Vane anemometers can have a digital or an analog readout. The digital is more accurate than the analog. The anenometer measures the velocity of air in feet per minute (FPM).

A **rotating vane anemometer** uses a propeller type wheel that measures air velocity in feet per minute (FPM).

Thermal anemometers use a hot wire that senses the velocity of the air in FPM when placed in an air stream. They are designed to sense low velocities.

The **static pressure** of a duct system is the outward air pressure that pushes on the duct. It is measured in inches of WC.

A **manometer** used with a Pitot-tube will test a system's velocity pressure, static pressure, and total pressure.

Static Pressure + Velocity Pressure = Total Pressure (SP + VP = TP).

A **U-tube manometer** uses a U-shaped tube and a liquid, usually water, to measure air velocity pressure, static pressure, or total pressure in a duct system.

A **U-tube manometer** is used to test pressures above 1.0" WC.

An **incline manometer** is a transparent plastic measuring device, filled with a liquid, that measures pressure in a duct system.

The **incline manometer's** scale is on a slope, and measures pressures below 1.0" WC.

A **Calibrated or Differential Pressure Gauge** uses a high and a low pressure port to measure the pressure difference between two different pressure locations.

Volume of air in a duct system is the quantity that is flowing through the system. It is measured in cubic feet per minute (CFM).

An **airflow hood** is a tool that is placed over a diffuser to determine the volume of air in CFM.

Air Flow Calculations

Air Flow Through a Duct System

V = velocity in FPM (Feet Per Minute)
A = area of a duct in sq. ft. (Square Feet)
Q = quantity of air in CFM (Cubic Feet Per Minute)

Q = AV   or   (CFM = sq. ft. × FPM)
V = Q/A   or   (FPM = CFM / sq. ft.)
A = Q/V   or   (sq. ft. = CFM / FPM)

Calculation for Furnace Blower Output

CFM = Output (Btu/h) / 1.08 × temperature difference

Calculation for Weight of Air

Weight of air = W   Volume = V   Density = D
W = DV   or   W/D = V   or   W/V = D

Duct sizing calculators are used to help design duct systems. An example of a duct sizing calculator is on page 67.

To use the duct calculator, first find the CFM of the equipment, then determine the velocity or allowable friction loss per 100 feet.

Example:

1. Assume the required volume of air is 1600 CFM and the velocity is 1100 FPM.
2. Set the 1600 CFM opposite to the velocity 1100 FPM. The friction scale will be set at .10 of an inch per 100 feet of duct.
3. The round duct diameter scale shows that a 16 inch round duct must be used.
4. The equivalent rectangular duct chart shows that the area of a 16 inch duct is the same as 8 inches × 30 inches.

Section 3   Air Flow Components and Duct Fabrication

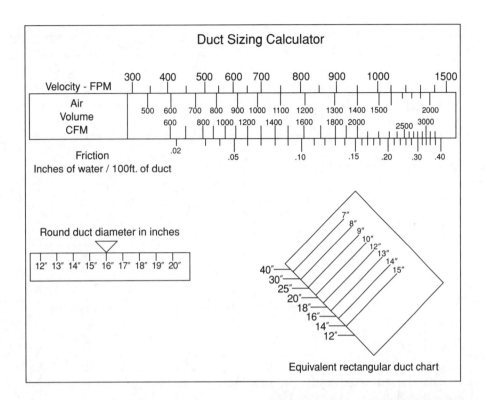

A **friction loss chart** is used to size duct systems if a duct calculator is not used. A friction loss chart has the same features as a duct calculator except in the way the round duct is converted to rectangular duct. There is an example of a friction loss chart on page 68.

The **horizontal lines** on the chart show the cubic feet per minute (CFM) of air.

The **vertical lines** show the friction loss of water column per 100 feet.

The **diagonal lines** from right to left show the air velocity in FPM.

The **diagonal lines** from left to right show the round duct diameter sizes.

Two variables are needed to use the **friction loss chart.**

Follow the chart using these examples.

Example 1

If the air quantity of a system is 70 CFM and the round duct size is 5 inches, the friction loss of the duct would be .1" WC per 100 feet.

The velocity would be 500 FPM.

Example 2

If the velocity of the air is 900 FPM and the friction loss per 100 feet is .15" WC, the quantity of the air is 400 CFM.

The duct size of the system is a 9 inches round.

BLOWERS AND FANS

A large part of troubleshooting HVAC/R equipment is directly related to blowers and fans. The technician must know about air volume, bearing types, lubrication, motor amperage, motor speeds, fan blades, etc. The following is information that will aid the technician in this area.

Dry bearings cause excessive amp draw on a motor.

A **bushing** or **sleeve bearing** is a bronze sleeve around a motor shaft.

A **sleeve bearing** can be permanently oiled and sealed at the factory, or can require field oiling.

Oil type motors have oil ports and must be oiled with 3 to 6 drops of non-detergent oil every six months. Over oiling can cause an accumulation of dirt to build up in the stator windings.

Ball bearing type motors are more rugged and can handle more of a load than sleeve bearing type motors.

Sealed ball bearing motors should be disassembled and re-greased every three to five years, depending on use.

Larger bearing type motors, commonly over 1 horsepower, use zerk fittings to allow the motor to be greased approximately every three months.

Most outdoor **condenser fans** use an axial or propeller type blade designed to move the proper quantity of air across the condenser coil.

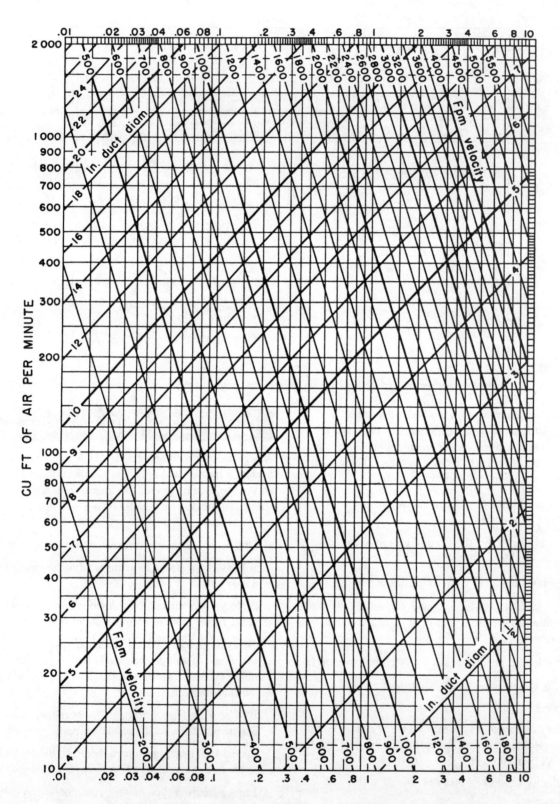

Figure 3–1: *Friction Loss Chart*

Section 3 Air Flow Components and Duct Fabrication

Air volume over a fan will change if any of the following factors are changed:

Speed of motor rotation

Number of fan blades

Blade pitch

Blade length

Blade width and shape

The **motor amperage** will increase on a fan motor if the air volume is increased, and will decrease if the air volume is decreased.

Fan motors are weatherproofed by using rain shields to protect the motor from water.

A **condenser grill** shields the operating fan blade to protect against injury during fan operation.

Direct drive condenser fans lower than five horsepower are commonly a permanent split-capacitor (P.S.C.) type of motor.

During **low ambient conditions,** condenser fans must be cycled to prevent the high-side pressure from dropping too low. When the high-side pressure is too low, it causes the evaporator coil to become starved of refrigerant liquid. This decreases the efficiency of the system.

Condenser fans that have a two speed type of motor often use a reverse acting pressure control to cycle the fan depending on the high-side pressure or outdoor ambient temperature.

A **solid-state motor control** varies the speed of the condenser fan so a constant high-side pressure is maintained. It commonly uses a thermistor to sense the liquid line temperature. It will lower the speed of the condenser fan on a decrease of liquid line temperature and raise the speed on an increase of liquid line temperature.

Indoor blowers use a centrifugal or squirrel cage fan design to move the proper quantity of air over the heat exchanger and evaporator coil.

A **blower door switch** is used to de-energize the blower if the panel near the blower is removed. This will prevent injury during blower operation.

Direct drive indoor blowers lower than five horsepower are commonly use a P.S.C. motor.

Belt drive indoor blowers lower than 5 horsepower commonly use a capacitor start induction run type of motor.

Single phase direct drive blower motors are commonly multi-speed. A three speed direct drive 120 volt blower uses different color coded wire taps for the different speeds.

The following are color coded wires that are commonly used for different **speeds:**

Red- low speed

Black- high speed

White- common speed

Other **speed taps,** such as medium low, medium, and medium high, have a various range of color codes depending on the manufacturer. Color codes that are used for the mid range speeds are blue, yellow, and brown.

When installing a blower motor, the wires controlling the **extra speeds** that are not being used must be capped with a wirenut and taped or tie strapped to prevent an electrical short.

To **change the speeds** on a direct drive blower motor, the colored wire taps must be changed. The white or orange common wire must remain in the circuit.

Fans and blower motors that are five horsepower or above are commonly three-phase motors.

Fans and blowers that have **bent blades** will decrease the air flow performance and cause excessive vibration.

Single phase motors often use run and start capacitors to aid in motor starting torque and operation. Three-phase do not use capacitors.

FLA is the full load amperage of a motor. It can also be stated **RLA,** which stands for running load amperage.

LRA is the locked rotor amperage, or starting amperage of the motor.

The **fan speed calculation** for a belt drive blower is:

Fan RPM = diameter of motor pulley / diameter of fan pulley × RPM

Example: A motor has a 3" diameter pulley and it turns a blower with a 6" diameter pulley. Motor has a 1050 RPM. 3" / 6" × 1050 RPM = 525 RPM.

Air volume is measured in cubic feet per minute (CFM). When using an air conditioning system, a common volume of air per ton of refrigeration effect is **400 CFM.**

A higher **CFM** of air is used during a cooling cycle and a lower CFM is used during a heating cycle.

The common external **static pressure** for residential and light commercial blowers is .5 WC.

As the **static pressure** in a duct system increases, the amperage on the blower motor increases. As the static pressure decreases, the amperage decreases.

Nameplate information including voltage, amperage, rotation, RPM, and horsepower is important when replacing a fan motor.

AIR DUCT SYSTEMS AND FABRICATION

The air duct system contains and directs air from the blower unit to the conditioned space. Air duct systems and duct fabrication are important for a technician in the HVAC/R field to understand. The following are many of the terms that are related to air duct systems and fabrication.

Horizontal rectangular air ducts up to 60 inches wide must be **supported** with 1 inch wide galvanized steel hangers at intervals not exceeding 10 feet.

Joints, seams and fittings in air duct systems must be made air tight by the use of tapes, mastics, gaskets, and other means.

A **round metal duct** crimped joint must have a contact lap of at least $1\frac{1}{2}$ inches.

A **round metal duct** must be connected with at least three screws that are spaced evenly apart.

Vibration isolators that are installed between metal ducts and mechanical equipment must not exceed 10 inches in length.

A **metal duct system** must be installed at least 4" above the earth surface.

When a **metal duct system** is installed under a concrete slab, the duct must be encased in at least 2 inches of concrete.

Circular bands that are used to connect round duct must be at least 1 inch wide and must use the same gauge metal as the duct.

The minimum total area of a **supply and return air duct** used on a blower type warm air furnace must be at least two square inches per 1000 Btu/h approved output rating.

The minimum total area of a **supply and return air duct** used on a heat pump must be at least six square inches per 1000 Btu/h approved output rating.

The following must be printed on the outside of external **duct insulation:** The name of the manufacturer, the "R" value at the installed thickness, and the flame spread index.

An air duct system that is installed in an unconditioned space, such as a crawl space, attic area or on a roof top, must be **insulated and sealed with an approved vapor barrier.**

When a duct system is located on the exterior of a roof, it must be **insulated** with at least 3 inches of a mineral fiber blanket with an approved weatherproof barrier.

When a duct system is located in an attic garage or crawl space, it must be **insulated** with at least 1 inch of a mineral fiber glass blanket with an approved vapor barrier.

In a single dwelling unit, exposed **rectangular ducts or plenums** 14 inches or less must be fabricated with a minimum of 28 gauge galvanized steel.

In a single dwelling unit, exposed **rectangular ducts or plenums** over 14 inches must be fabricated with a minimum of 26 gauge galvanized steel.

A **cross break** on a section of duct increases its rigidity and decreases the noise caused by expansion and contraction.

Low pressure duct systems are designed for air pressure up to 2" WC.

When air ducts are run through floors, a **fire stop** around the duct is required. A common type of fire stop is fiberglass insulation.

When installing **air ducts** in a basement, the supply and return air trunks should run perpendicular to the floor joists.

In a home with a **slab floor construction,** the air duct system is commonly installed in the attic area.

In a home with a **crawl space construction,** the air duct system is either installed in the attic or crawl space area.

An **extended plenum** duct system is one where the furnace is in a central location of the home and the main duct supply is branched from 2 sides of the furnace.

Each main duct supply on an **extended plenum system** should not exceed 24' in length without reducing the duct area.

When using a **reducing extended plenum,** the furnace is located at one end of the house and the trunk size is reduced as it increases in distance from the furnace.

The decrease in size of a **reducing extended plenum** causes an increase in air velocity.

A **duct transition** is a section of duct that connects two different sizes of duct.

Section 3 Air Flow Components and Duct Fabrication

A **crimping tool** is commonly used to decrease the diameter of the male fitting on a round duct.

Branch runs are the supply air ducts that distribute air to each room in the house. They should never be connected to the end of any plenum system, as this will cause imbalanced air flow.

Balancing dampers must be provided in branch runs as close to the supply air plenum as possible.

When adjusting a **balancing damper**, if the adjusting handle is in line with the duct, the damper is 100% open. If the adjusting damper is perpendicular to the duct, the damper is 100% closed.

A **main supply air plenum** must be located at least 1 inch from a combustible surface.

A **register** is adjustable and is used on the supply air side of a system. A **grill** is non-adjustable and is used on the return air side of a system.

Supply air registers used with first floor upflow and attic furnace installations commonly use a ceiling type mount.

Supply air registers used with basement, crawl space, or first floor down flow installations commonly use a floor type mount.

Bathrooms, kitchens, and utility rooms use a **side wall** or **toe kick** type supply air registers. This reason is to prevent water damage to the furnace and duct system.

A **perimeter loop** duct design is used for homes with a slab floor construction. Before the concrete slab is poured, a duct loop is placed around the perimeter of the home. Supply air openings in the duct are made, and four branch feeder ducts are run to feed the supply air from the furnace to the perimeter loop. The concrete slab is then poured around the duct system.

Perimeter radial or **central plenum** duct systems are found when furnaces are installed on the first floor of a home. The duct system will either be run in an attic or a crawl space area. The furnace is centrally located in the home and the plenum is extended above or below the furnace. When it is extended below the furnace, the plenum is located in a crawl space and branch supplies are connected from the plenum to the outside of the perimeter of the home in the floor. If it is extended above the furnace, the plenum is located in the attic and branch supplies are connected from the plenum to the outside of the perimeter of the home in the ceiling.

Return air systems can use an extended plenum, or reducing extended plenum, when the furnace is located in the basement.

Return air systems use floor joists, wall studs, and gypsum board as channels for the return of conditioned space air back to the furnace.

In basement installations, **return air trunks** are mounted tight to the floor joists and are commonly made of 30 gauge sheet metal.

Central return systems are common with attic and first floor furnace installations.

The common locations for **return air grills** are high or low on the inside wall.

Return air grills that are located high on the wall are used in a region where the cooling season is more predominant.

Return air grills that are located low on the wall are used in a region where the heating season is more predominant.

Below are diagrams of common metal duct connectors:

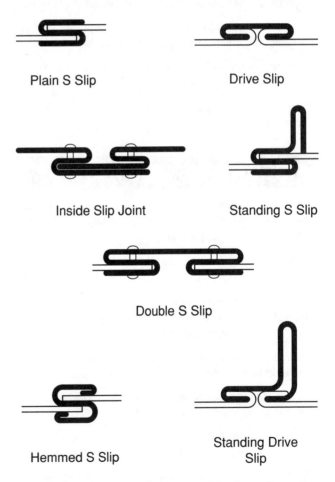

"S" slips are commonly connected horizontally on the duct sections.

Drive slips are commonly connected vertically on the duct sections.

FLEXIBLE (FLEX) DUCT

A flexible air duct is generally used in approved areas where it is more appropriate to use than a standard sheetmetal duct. The following information pertains to flex duct.

When installing **flexible duct,** the section being installed must be fully extended to prevent the reduction of air flow.

The **radius** of a flexible duct must not be less than the diameter of the duct being used.

Sheetmetal collars that are connected to a flexible duct must be a minimum of 2 inches in length.

Sheetmetal sleeves used for joining two sections of flexible duct must be a minimum of 4 inches in length.

A horizontal flexible duct must be **supported** every 4 feet.

A vertical flexible duct must be **supported** every 6 feet.

A **saddle material** must be used when supporting flexible duct to prevent any restriction of the duct diameter.

The **minimum width** of saddle material used to support a flexible duct is $1\frac{1}{2}$ inches.

The **maximum sag** for a flexible duct is $\frac{1}{2}$ inch per foot between the supports.

FIBERGLASS DUCTBOARD

Many HVAC/R companies prefer to use fiberglass ductboard instead of sheetmetal. The following information is related to ductboard and their applications.

Fiberglass ductboard is commonly 1 inch or $1\frac{1}{2}$ inches thick.

The **maximum static pressure** that can be obtained in a fiberglass ductboard system is 2" WC negative pressure and 2" WC positive pressure.

The **maximum air velocity** in a fiberglass ductboard system is 2400 FPM.

The **maximum inside temperature** for a fiberglass ductboard system is 250° Fahrenheit.

The **maximum outside surface temperature** for a fiberglass ductboard system is 150° Fahrenheit.

When **screws** are used to fasten duct board to a metal surface, a #10 sheet metal screw with a $2\frac{1}{2}"$ square washer must be used.

Screws used to fasten ductboard material must be a minimum of a $\frac{1}{2}$ inch longer than the board thickness.

The following materials must be used when **sealing a fiberglass ductboard**: pressure sensitive aluminum foil tape, heat activated aluminum foil tape, mastic, and glass fabric tape.

A shiplap joint flap must be **stapled** every 2 inches on center.

Aluminum foil tape is used over a stapled shiplap joint flap.

Crosstabs are used to hold seams in position when staples are not used.

A **crosstab** is made of aluminum foil tape with a minimum length of 8 inches. It is placed across a taped joint every 12 inches.

The **minimum width** of pressure sensitive aluminum foil tape that is used on fiberglass ductboard is $2\frac{1}{2}$ inches.

Aluminum foil tape must **overlap** each side of a fiberglass ductboard joint by a minimum of 1 inch.

If aluminum foil tape is stored in an area less than 50° F, the tape must be **preheated.**

When using **heat activated tape,** the tape must be heated from 550° to 600° F after being applied to the fiberglass ductboard.

The **minimum width** of heat activated tape used on fiberglass ductboard is 3 inches.

Mastic with a glass fabric system can be used to seal the seams of a fiberglass ductboard system.

A **tie rod** or **channel reinforcement** must be used on larger ductboard systems to increase duct rigidity.

A **tie rod** is commonly placed vertically in a fiberglass ductboard system.

A **tie rod** used for ductboard reinforcement is made of 12 gauge galvanized steel wire.

When a transverse joint is used in a ductboard system, the **tie rod** must be placed 4 inches from the end of the female joint.

When a butt joint is used in a the ductboard system, the **tie rod** must be placed 3 inches from either side of the joint.

There are generally three types of seams used with **fiberglass ductboard.** They are shiplap joints, butt joints, and V-grooves.

Section 3 Air Flow Components and Duct Fabrication

Air must flow from the male to the female end of a shiplap joint.

V-Grooves can be used on single flat board sections to create a rectangular duct.

Square cut or butt joints are not permitted on corners of ductboard sections.

The following are diagrams of common fiberglass duct-board closures.

B = Square Cut
BF = Square Cut with Flap
M = Male Shiplap
S = Female Shiplap
SC = Shiplap Corner Fold
SF = Female Shiplap with Flap
V = V-Groove
VC = V Corner Groove

AIR SIDE COMPONENTS

Technicians must understand component functions and terms that pertain to air side components, which might include types of air dampers and damper systems.

There are two general types of **damper** blades. They are the **parallel blade** and the **opposing blade.**

The blades of the **parallel blade type damper** opens together uniformly. The blades of the **opposing blade type damper** are constructed so that each blade opens and closes in the opposite direction.

The **opposing blade type damper** gives more uniform air flow through the system.

Mixed air control mixes return air with outside air for ventilation and free cooling. This system is controlled by an economizer circuit.

The **economizer circuit** uses a damper motor along with enthalpy sensors, discharge air sensors, and/or mixed air sensors to determine the amount of outside air to bring in for free cooling.

Exhaust or relief dampers are used in systems to prevent the pressurization of a building.

In a constant air volume system, **face and bypass dampers** are used together to control the amount of air over a heating or cooling coil. This type of control is found on larger commercial systems.

A **variable air volume (VAV) system** controls a quantity of air to a conditioned space depending on the load. During cooling, air volume increases as load increases. During heating, air volume decreases as heat load increases.

The air **temperature difference** between conditioned air and ventilation air must not exceed 10° F.

Dampers balance airflow by opening or closing certain ducts (zone control).

Some **dampers** are located in the diffuser, grill, or right in the duct.

Dampers should fit tight for accurate air control and minimum leakage.

Commercial and industrial buildings must have **automatic fire dampers** installed where the duct sections are run between rooms.

Fire dampers are usually held open by a fusible link that will melt with heat and cause the damper to close.

Smoke dampers use a photoelectric device to detect smoke. This causes the damper to close.

Proper ventilation is needed to increase oxygen and decrease carbon dioxide levels in the conditioned space.

An **Energy Recovery Ventilator (ERV)** is used to add fresh ventilation air and remove stale exhaust air in a home.

AIR FILTERS

There are various types of filters and filtering equipment that used to help clean the air we breathe. The following are filter types and terms that relate to air filters.

Air filters should always be placed in the return air system upstream from a heat exchanger or a coil.

Air filters must be located so that they can be readily removed.

Return air boots must have turning vanes to evenly distribute the air across the filter media.

Ionizing means to charge a particle with a negative or positive charge.

Electronic air cleaners (EAC) are used to remove small particles of dust from an air distribution system. There are three common types: an ionized plate type, a charged media non-ionizing type, and a charged media ionizing type.

The **ionized plate type** generates positive ions on wires that come in contact with dust particles. The wires pass the positively charged ions to the dust that is sent through collector plates. The collector plates have an electrical field that attracts and holds the dust particles. This system uses a high DC voltage of 12,000 volts.

The charged media **non-ionizing type** uses a dielectric medium consisting of fiber-glass and cellulose mats supported by an electrically charged grid. The electrical charge attracts the dust particles into the filter media.

The **charged media ionizing type** uses a combination of ionization of dust particles and electrically charging the filter media for an effective filtration system.

Most **electrostatic air cleaners** have a screen to trap large particles.

An **electronic air cleaner** removes particles as small as 0.0001 micron.

An **electronic air cleaner** may be equipped with a pressure drop indicator and an automatic shut off control.

Cells of an **electronic air cleaner** should be cleaned every two or three months using chemicals or detergents.

The framework (casing) of the **electronic air cleaner** is connected with sheetmetal screws to the return air side of the furnace.

Portable electronic air cleaners can be installed in ceiling applications. Most electronic air cleaners use a **pre-filter,** installed before the electronic element, to catch large dust particles.

The **electrostatic filter** puts a static electrical charge on all particles passing through it, which then is drawn to collector plates with an opposite electrical charge.

Because of high voltage, the **electrostatic filter** is designed to shut off automatically when the service door is open.

A **manometer** is used to measure pressure drop across an air filter. Ther pressure drop is measured in inches of WC.

Disposable filters should be changed at least twice a year.

An **adhesive filter** is coated with an adhesive liquid or oil to trap particles in an air distribution system.

Adhesive filters remove up to 90% of dust particles.

Section 3 Air Flow Components and Duct Fabrication

HUMIDIFIERS

There are various models and types of humidifiers installed on heating appliances. A technician needs to know how to troubleshoot and how to install them correctly. The following provides information that will help the technician when dealing with humidifiers.

Three general **types of humidifiers** are evaporative, atomizing, and vaporizing.

Five common **evaporative type humidifiers** are plate type, rotating plate type, rotating drum type, fan powered type, and flow through type.

A **rotating drum type humidifier** commonly uses a bypass duct to circulate supply air through the humidifier, and back into the return air. This increases the efficiency of the humidifier and, with the use of a damper, allows the humidifier to be closed off during the summer months.

The media of a **plate type humidifier** revolves in a pan full of water.

A **fan powered type humidifier** uses a fan to induce air over the humidifier media.

The **flow through type humidifier** uses a stationary media and a solenoid valve for water control. When the solenoid is energized, water flows over the top and through the media allowing the warm air from the furnace to evaporate the water.

A **flow through type humidifier** commonly uses a **solenoid valve** for water control.

A **flow through type humidifier** commonly uses a bypass duct to circulate supply air through the humidifier back into the return air. This increases the efficiency of the humidifier and, with the use of a damper, allows the humidifier to be closed down during the summer months.

An **atomizing type humidifier** commonly uses a rotating disc to atomize the water.

An **atomizing type humidifier** commonly uses a fan to move water vapor.

The **vaporizing type humidifier** uses an electrical heating element at the bottom of a water pan to evaporate the water.

The **vaporizing type humidifier** can be installed in the supply or return air section of the heating system.

The plate type, rotating plate type, rotating drum type, and fan powered humidifiers are most commonly installed on the **supply air duct** of a forced air heating system.

A **float valve** is used to regulate water into the pan of plate type, rotating plate type, rotating drum type, and fan powered humidifiers.

The **humidifier media** is part of the humidifier that holds the water until it is evaporated into a heating system.

A **plate type** of humidifier uses a porous plate type media that is located in a rack inside of the humidifier.

The media of a **rotating plate** type humidifier revolves in a pan full of water.

The media of a **rotating drum** type humidifier revolves on a drum located in a pan full of water.

The average recommended relative humidity for a home during the winter months is **35% to 45%**.

If the proper **humidity level** is used in a home during the winter months, it will decrease the rate of moisture evaporation off of the skin. This will improve the comfort level.

Humidifiers commonly have their own 24 volt 20VA control transformer.

The **humidistat** is a control that senses humidity in a home. When the humidity level decreases, the humidistat closes, energizing the humidifier motor or solenoid valve.

The **humidifier circuit** is commonly energized when the indoor blower is energized during the heating cycle.

A **current sensing relay** is commonly used to energize the primary side of the 120V humidifier control transformer or the 24V humidifier circuit. It senses the flow of current through the heating speed of the blower and closes a set of normally opened contacts.

Many modern furnace controls use terminals marked **H or HUM** for the humidifier circuit. These terminals can energize the 120V primary side of a control transformer or a 24V humidifier circuit.

TEMPERATURE AND HUMIDITY

A technician needs to know various types of temperature sensors. He must further be capable of performing humidity calculations and understand the importance of humidity in a system. The following section covers temperature and humidity.

A liquid column **thermometer** is a glass thermometer that is filled with mercury or red alcohol. Its common range is from -30° F to +120° F.

A **dial thermometer** uses a bimetal sensor, when it incorporates an analog read out and a thermistor, when using a digital read out. Its range is from -40° F to +160° F.

Electronic thermometers use a thermistor type sensor to measure temperatures. A thermistor's resistance changes when its temperature changes. An electronic thermometer is much more accurate than a dial or bimetal thermometer.

Infrared thermometers use an infrared beam to check the surface area temperature of an object. Infrared thermometers are usually used to check temperatures of objects that are difficult to reach with standard thermometers.

Recording thermometers are used to check and record the temperatures of a system over a period of time. Recording thermometers are usually used on a system from one to seven days.

The **dry-bulb delta-T** is the dry-bulb temperature difference between the conditioned (inside space) compared to the unconditioned (outside space).

The **wet-bulb delta-T** is the wet-bulb temperature difference between the conditioned (inside space) compared to the unconditioned (outside space).

A **sling psychrometer** is used to measure the wet-bulb and the dry-bulb of air to determine the relative humidity.

A **sling psychrometer** uses liquid filled glass tube thermometers that are extremely accurate.

The **dry-bulb temperature** of air is the sensible temperature measured with a standard thermometer.

The **wet-bulb temperature** of air is measured by placing a wick that is saturated with water around a thermometer bulb and rapidly moving air over the wick. The dryer the air, the lower the wet-bulb temperature will be.

Electronic **humidity measuring equipment** is used to measure the relative humidity and the dew point of the air. Some electronic meters use remote sensing probes that are placed in an air stream to measure the relative humidity level.

A **Psychrometric chart** is used to determine the properties of air such as dry-bulb temperatures, wet-bulb temperatures, dew point temperatures, moisture content, relative humidity levels, specific volume and more.

The **vertical lines** are used to measure dry-bulb temperatures.

Figure 3–2: *Vertical Line Psychrometric Chart*

The **horizontal lines** are used to measure grains or pounds of moisture and dew point temperatures.

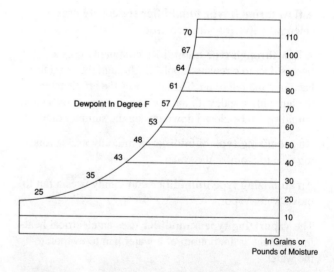

Figure 3–3: *Horizontal Line Psychrometric Chart*

The **45 degree diagonal lines** are used to measure the wet-bulb and the total heat content.

Figure 3–4 *Diagonal Line Psychrometric Chart*

The **diagonal lines** are used to measure the specific volume.

The **curved lines** are used to measure the percentage of relative humidity.

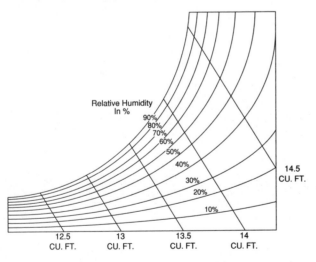

Figure 3–5 *Curved Line Psychrometric Chart*

Complete **air conditioning** means to heat, humidify, cool, dehumidify, ventilate, filter, and circulate conditioned space air.

A typical **comfort goal** for the HVAC industry is maintaining air to a 75° F dry-bulb temperature at 50% relative humidity.

To high of a dry-bulb **temperature** decreases the rate of heat transfer from a person to the surrounding air. To low of a dry-bulb temperature increases the rate of heat transfer from a person.

To high of a **relative humidity** decreases the rate of evaporation of moisture from a person to the surrounding air. To low of a relative humidity increases the rate of evaporation from a person.

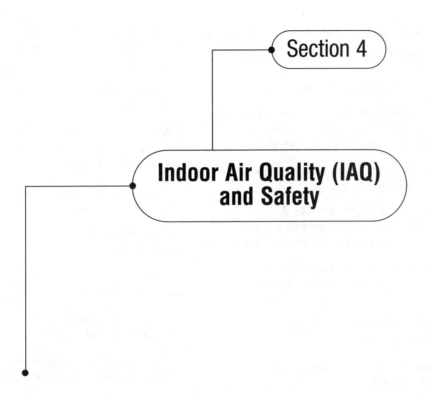

Section 4: Indoor Air Quality (IAQ) and Safety

This section provides information on indoor air quality, personal safety, refrigerant safety, soldering/brazing safety, and electrical safety. It also consists of a study guide for EPA Certification Exams. A technician must always be aware of the conditions and situations he is called to work under. He must always work in a safe manner, so that he will not invite injury to himself or others. The following section will aid the technician in understanding IAQ and safety when working with electrical and mechanical systems.

INDOOR AIR QUALITY (IAQ)

There are may types of contaminants that are carried by air flow through supply registers and return air ducts of a heating and cooling system. Contaminants in the air flow can result in poor health conditions for human life. Mild to severe health conditions may occur, depending on the type of contaminant. The following information will provide the technician with a better understanding of indoor air quality.

The difference between a **contaminant** and a **pollutant** is that a contaminant may or may not be a health risk and a pollutant is always a health risk.

Respirable particles that are breathed deep into the lung are commonly 10 microns or less.

Asbestos is a name used to describe a number of fibers that are made from a mineral called **chrysolite.** If asbestos is breathed into the lungs, it could cause scarring of the lung tissue, lung cancer, and damage to the abdomen.

The first thing a technician should do to **improve air quality** is to change a system's air filters.

An inexpensive **disposable filter** is only about 7% efficient.

A **pleated filter** is commonly about 30% efficient.

Carbon type filters are commonly used to efficiently remove tobacco smoke particles.

Stratification refers to air that settles in different temperature layers. To eliminate stratification, air must be kept in motion.

Air contaminants are classified into four groups: gases/vapors, inert particles, microorganisms, and pollen.

Manmade fibers, dust, and cigarette smoke are considered **inert particles.**

Inert particles, such as dust, are the main cause of poor air distribution in HVAC/R equipment.

Microorganisms include fungi, mold, and bacteria.

Gasses and vapors originate from the human body, hair sprays, paints, adhesives, solvents, radon, and ozone.

Pollen comes from grass, weeds, seasonal plants, and trees.

Sick building syndrome (SBS) commonly causes the following symptoms: eye, nose, and throat irritation; dizziness; headaches; drowsiness; and fatigue.

A high content of **carbon dioxide** depletes the oxygen level in a building. A satisfactory CO_2 measurement in an indoor environment is a maximum of 1000 PPM.

Indoor air contaminants are commonly two to five times higher than outdoor pollutants.

Legionnaires' disease is an illness caused by the Legionella pneumophila bacteria. This bacteria is commonly caused by improperly treating cooling tower water.

A solution of **90% water** and **10% bleach** is commonly used to clean non-porous materials and prevent mold growth.

Some **air leakage** into a building will help decrease the percentage of indoor air contaminants.

A **high relative humidity** within a building can increase the growth of mold.

A well balanced **ventilation** system will exhaust stale indoor air and introduce the same quantity of fresh outside air.

Common types of **ventilation systems** are passive, exhaust only, non-heat recovery, balanced mechanical, and heat recovery types.

For proper **ventilation,** a building must have an outside air supply and an exhaust system.

A **VOC (volatile organic compound)** is a highly evaporative chemical that gives off gases.

VOCs can cause a person to have rashes, nausea, sinus congestion, drowsiness, headaches, and respiratory problems.

Radon is a radioactive gas that comes from the Earth's crust. It enters the home through cracks, joints, non-trapped drains, sump pumps, water supplies, underground service openings, and crawlspaces.

Radon commonly stays near ground level because it's nine times heavier than air.

Radon is not considered harmful to humans until it decays.

Radon levels inside should be considered harmful when the average annual concentration in the normal living areas exceeds 800 becquerels (Bq) per cubic meter. Becquerels is a measurement of radon levels.

Polynuclear aromatic hydrocarbons (PAHs) are produced when materials containing carbon and hydrogen are burned.

PAHs outside the home are commonly caused by wood burning appliances, coal burning equipment, and internal combustion engines.

PAHs inside the home are commonly caused by cigarette smoke, the burning of food, and fireplaces.

Ozone (O^3) is an unstable oxygen molecule made up of three oxygen atoms.

Ozone is produced when air is heated by an electrical arc during an electrical storm.

The levels of **ozone** inside the home is directly correlated with the levels outside the home.

Nitric oxide (NO) is a toxic gas which is produced during high temperature combustion.

Nitric oxide and nitrogen dioxide can be produced inside a home when fuel burning appliances are not vented. Another indoor source is cigarette smoke.

Once nitric oxide is released into the air, it quickly changes to **nitrogen dioxide.**

Common indoor levels of **nitrogen oxide** are 0.25 ppm (parts per million) or less.

Carbon monoxide (CO) is a poisonous gas that is produced during incomplete combustion of fossil fuels.

Common indoor sources of **carbon monoxide** are cracked heat exchangers, cracked chimneys, unvented appliances, and cigarette smoke.

Common indoor levels of **carbon monoxide** range between 0.5 to 5 ppm.

If a home has an excessive negative pressure, the combustion gases of an appliance could be drawn back into the home and produce **carbon monoxide.**

Formaldehyde is classed as a volatile organic compound (VOC). It is in many building materials such as medium density fiberboard (MDF) and particle board.

An increase of temperature and humidity levels in a home is directly proportional to the rate that a **formaldehyde** gas is released.

Section 4 Indoor Air Quality (IAQ) and Safety

PERSONAL SAFETY

Personal safety should always be on a technician's mind. He needs to be aware of all unsafe conditions that he may be asked to work under. The technician must always be conscious of the unknown. The following information on personal safety will help the technician work in a safe manner.

Wear **hard hats** to protect against falling objects.

Wear **safety glasses** to protect against flying objects, soldering materials, and chemical splashes.

Wear **gloves** to protect against scrapes, cuts, heat, or burns.

Wear **steel toe safety boots** to protect feet against falling objects and punctures.

Wear **ear plugs** to protect ears while in a high noise level area.

Use a **respirator** to protect against breathing hazards and suffocation.

Use a **safety harness** to protect against falls when working in an area more than 6 feet above the ground.

Regularly **inspect safety equipment** and care for it according to the manufacturer's specification.

Never alter safety equipment in any way.

Never wear loose clothing, loose gloves, jewelry, or expose long hair around machinery.

Always keep **hand tools** and equipment clean and in good working condition.

To prevent an injury, never use a **broken or defective tool.**

When **using a wrench,** it is safer to pull instead of push. This will prevent an injury.

When **lifting objects,** keep your back straight and use your leg muscles.

When **lifting an object,** always move close to it, squat down, lift slowly, and keep the object centered.

When **lifting an object,** always use the full hand; do not use finger tips.

Wear a **back support belt** to aid in lifting objects.

Always ask for assistance when lifting **heavy objects.**

Keep **floors and work area** free of oil, grease, water, and crowded items which may cause a personal injury.

To **prevent a laceration,** wear hand protection when working with sheetmetal.

To **prevent serious burns,** never use gasoline to clean tools or mechanical components.

Eye protection must be used to prevent an injury when working with hammers and chisels.

To prevent a personal injury, never use a **worn chisel.**

To prevent a personal injury, **eye protection** is required when using power tools such as drills and saws.

Always keep hands and feet at a three point connection when climbing **ladders.**

REFRIGERANT SAFETY

Working with refrigerants that are under pressure in a sealed system could invite injury to the technician if he is not careful. Refrigerants can instantly freeze eyes, hands, and fingers if one is not careful. Also, under certain conditions, refrigerant may become harmful to breathe. The following will help the technician stay knowledgeable and to work safely.

To prevent an injury always **wear gloves and safety glasses** when working with refrigerants.

Refrigerants can cause **frost bite burns and suffocation** if not used properly.

To **prevent a refrigerant burn,** never try and stop a refrigerant leak with your hand.

To **prevent suffocation,** never work with refrigerants in a confined area.

Since air is lighter that refrigerant vapor, **air will be displaced** by the refrigerant.

Equipment rooms using large quantities of A1 refrigerants (refrigerants with low toxicity and low flammablity), must be equipped with **refrigerant or oxygen deprivation alarms and sensors.**

When A1 refrigerants are in contact with an open flame, **phosgene gas** is produced.

To prevent suffocation, asphyxiation, and possible brain damage, use a **breathing apparatus** when working around toxic fumes.

A **self contained breathing apparatus** must be available in equipment rooms where systems use a large amount of refrigerants.

A pressure regulator and an in-line pressure relief valve should be used when **pressurizing a system** with nitrogen.

To prevent an explosion, **never connect a nitrogen cylinder and refrigerant tank** to a refrigeration system at the same time.

Never exceed the **low-side pressure rating** when pressurizing a system.

When pressurizing a system, **150 PSIG** of dry nitrogen is usually satisfactory for leak detection on high-pressure systems.

Never use a **halide torch** to detect leaks if the system is pressurized with nitrogen.

A mixture of **oxygen and oil** is extremely explosive.

To prevent an explosion, **never use oxygen** when pressure testing an air conditioning system for refrigerant leaks.

Technicians should always follow the **manufacturer's recommendations** for charging refrigerant into a system.

To **prevent an explosion,** never front seat (close) the discharge service valve on a compressor while it is in operation.

Care must be taken near the system's discharge line to **prevent a burn** while working on an operating system.

Always wear hand protection when handling air conditioning coils to **prevent scrapes and cuts.**

Always wear eye and hand protection when working with acidic refrigerant oil to **prevent burns.**

Never use an open flame during the leak detection of natural gas or flammable refrigerants.

To prevent the danger of excessive pressures never **overcharge** refrigeration systems.

To **prevent refrigerant burns,** always remove the refrigerant from a system before opening to the atmosphere.

To prevent an injury caused by a liquid entering a compressor (liquid slug), never add **liquid refrigerant** into the low side of a compressor while it is in operation.

To prevent a possible explosion, **never refill** a disposable refrigerant cylinder.

To prevent a possible explosion, **never refill** a rechargeable refrigerant cylinder over 80% of its capacity.

Always take into account the cylinder weight, known as the **"Tare pounds,"** when estimating the net weight of the refrigerant in a rechargeable cylinder.

To prevent an explosion, **never heat refrigerant cylinders** with a torch.

Refrigerant **containers** must always be properly labeled and color coded.

Recovery cylinders are **color coded** with a gray bottom and yellow top.

When heating a refrigerant cylinder with warm water, the temperature of the cylinder water must never exceed **125° Fahrenheit** or the rupture disc could open.

Refrigerant cylinders must always be approved by the Department of Transportation (**DOT).**

Relief valves are used to protect against excessively high systems pressures.

The following are common **types of pressure relief devices:** fusible plug, spring loaded relief valve, and rupture disc.

To prevent a possible explosion, never place a cap on a safety relief device.

Relief devices must never be in series with one another.

Relief devices can be installed on compressors, components, and on refrigerant piping.

When pressurizing a low-pressure chiller, **never exceed 10 PSIG** on the low side of the system. This will cause the ruptured disc to open and vent the refrigerant charge.

When charging a **low pressure chiller,** refrigerant is commonly added to the low side of the system in liquid form.

Fill limit devices are part of a recovery system. They can be sensor or float type devices used to shut down the recovery machine when a container reaches 80% full.

Empty **rechargeable refrigerant cylinders** must be evacuated before reusing.

Never use untested **refrigerant cylinders.**

To prevent personal injury, never use **dented or rusted** refrigerant cylinders.

Material safety data sheets (MSDS) describe the flammability, toxicity, reactance, and health problems of a particular refrigerant.

Any **refrigerant cylinder** should be stored in a vertical position.

Always secure the cylinder's **valve cap and hood cap** when a cylinder is being stored.

Always properly **secure a cylinder** to prevent it from falling or rolling around when being transported.

SOLDERING AND BRAZING SAFETY

Safety precautions should always be adhered to when using torches for soldering or brazing. Always be aware of the surroundings you are working in. The following is information that will help the technician to work in a safe manner.

Always have a **fire extinguisher** present when using a torch.

Always wear the proper **safety glasses** when soldering or brazing.

Use a **non-combustible heat shield** to protect combustible material when brazing or soldering.

When using a torch, always **wear gloves** to prevent burns.

To **prevent a burn,** never stand directly under material that is being brazed or soldered.

Always use **cadmium free** brazing materials.

To prevent asphyxiation, soldering and brazing must be done in a **well ventilated area.**

Oxyacetylene torches must be inspected frequently to make sure they are in proper working order.

Do not use short pieces of **brazing rods.** Burnt fingers or hands may result.

To prevent an injury, acetylene pressure must **never exceed 15 PSI.** Above this pressure, acetylene becomes unstable.

To prevent burns, **never use a lighter** when igniting a torch. Only use recommended strikers.

To prevent a refrigerant burn or possible explosion, never **solder or braze** a system that is under pressure.

Always **open the system** to atmospheric pressure when soldering or brazing.

ELECTRICAL SAFETY

Service work consists of testing and handling electrical wiring. The technician must be extremely careful when working with electrical circuits. If caution is not taken serious injury or death could occur. The following is information to help the technician to work in a safe manner. Always respect electricity.

After installing or servicing a system, **inspect all electrical wiring** for loose connections, improper grounding, and defective conductor insulation.

Always **observe precautions, literature, and labels** regarding electrical equipment.

All electrical installations and repairs made must meet the **National Electrical Code** (NEC).

Always be aware of the **electrical rating** of the equipment that is being serviced.

To prevent possible electrocution when repairing or replacing electrical components, always place the **electrical disconnect switch in the off position.**

Always use a voltmeter before servicing a system to determine if **voltage** is present.

A **power lock-out tag-out** should be placed on electrical disconnects to prevent the power from being turned on while the equipment is being serviced.

An **electrical shock** is when current flows through a body.

An electric shock across both arms of a technician can cause **heart damage.**

If skin is moist, it is more susceptible to an **electric burn.**

Electrocution from as low as + 25 **ma** (milliamps) can kill or cause a serious injury.

To prevent equipment damage or a personal injury, never leave an **electrical safety control** bypassed.

To prevent electrocution, always properly connect equipment **ground wires.**

A **ground fault circuit interrupter (GFCI)** is used to prevent electrocution due to an equipment ground fault.

When using power tools, use an extension cord protected by a GFCI outlet to prevent an electrocution.

Never remove the bottom prong off of a **three prong plug.** This will cause the equipment to be improperly grounded.

Always use a **ground adapter** for a two prong plug to properly ground the electrical equipment.

Never wear **jewelry** when servicing electrical equipment.

Never place both hands near electrical equipment when troubleshooting. This can cause a serious electrocution.

Never work in **damp areas** when servicing electrical equipment.

To prevent a burn or a possible electrocution, always **discharge capacitors** with a 20,000 ohm 2 watt resistor before touching their terminals.

To prevent a possible fire, **never exceed the rated amperage** of the equipment when replacing a fuse or circuit breaker. Always use a timed delay fuse on motor circuits.

To prevent a possible electrocution, use an insulated **fuse puller** to remove fuses from electrical disconnects.

The technician should know the functions of all electrical components that are in the system's electrical diagram.

To prevent electrocution, inspect hand tools for **adequate insulation** when performing an electrical repair.

Properly secure all hanging wires after an installation or during a service procedure to prevent electrocution or system damage.

Make sure electrical power tools are properly **grounded** before use to prevent electrocution.

The **electrical panels** must be tightly secured on all equipment after service has been performed.

All moving parts should be inspected by the technician to insure that safety guards are in their proper position.

Technicians should keep their **test instrumentation** properly calibrated.

The technician should carry a **first aid kit** in his work vehicle.

To prevent a personal injury, always use **adequate lighting** when servicing HVAC equipment.

When working on electrical panels, a technician should have thick **rubber soled shoes** to prevent grounding and possible electrocution.

To prevent a personal injury, **inspect and tighten** all fan motors and blades.

To prevent a personal injury, keep hands and fingers from **moving fan belts.**

To prevent a personal injury, never wear **loose clothing** when working on HVAC/R equipment.

To prevent a personal injury, always keep **long hair** pulled back or tucked in a cap when working on mechanical equipment.

EPA CERTIFICATION STUDY GUIDE

The Environmental Protection Agency (EPA) has mandated in Section 608 that all air conditioning and refrigeration technicians who will service and dispose of refrigeration and air conditioning equipment be certified. The certification also allows the technician to purchase refrigerants. The technician must be knowledgeable in general core type information, such as ozone depletion potential, general terms and methods of recovery, and recycling and reclaiming procedures. The technician must also pass a certification exam. There are three types of certification offered. The technician will test in the certification type that will be used in his work area. There is also a Universal Cerftificate offered for the technician who is able to pass Core, Type I, Type II, and Type III exams. The following information will assist technicians in passing EPA Recover, Recycle and Reclaim Examinations.

CORE SECTION

The **Montreal Protocol** is a treaty regulating the production of CFCs.

CFCs neither dissolve in water or break down into compounds that dissolve in water, so they do not rain out of the atmosphere.

It takes **one chlorine atom** to destroy 100,000 ozone molecules.

ODP stands for **Ozone Depletion Potential.** It is the measurement of the ability of CFC's and HCFC's to destroy ozone. CFC's have the highest ODP.

As of **July 1, 1992,** The Clean Air Act prohibits the venting of CFCs and HCFCs.

As of **November 15, 1995,** The Clean Air Act prohibits the venting of HFCs.

As of **November 14, 1994,** The Clean Air Act requires a service technician to be EPA certified.

When **disposing of a refrigerant cylinder,** recover the refrigerant and render the cylinder useless.

An appliance will need its **refrigerant recovered** before it can be discarded.

According to the EPA, it is legal to allow a small amount of refrigerant into the atmosphere. That may be caused by hose purging and equipment servicing. This is called *de minimis* releases.

Section 4 Indoor Air Quality (IAQ) and Safety

The **temperature glide** is the different refrigerant boiling and condensing points of a blended refrigerant.

Service technicians that violate the Clean Air Act may be **fined up to $25,000** per day per violation, may lose certification, and may be required to appear in Federal Court.

A **reward of $10,000** is given to a person that supplies information about the venting of refrigerants.

The pressure range of the **high-side gauge** is 0 PSIG to 500 PSIG.

The pressure range of the **low-side gauge** is 30" Hg. to 350 PSIG.

It is legal to add a **trace amount of R-22 and nitrogen** when leak testing a system.

CFCs were phased out of production on **December 31, 1995.**

Fractionation is the separation of blended refrigerants caused by different refrigerant vapor pressures.

A **Hygroscopic oil** has a high ability to absorb moisture.

The state of the refrigerant **leaving the compressor** is high-pressure superheated gas.

The state of the refrigerant **leaving the condenser** is high-pressure subcooled liquid.

The state of the refrigerant **leaving the metering device** is low-pressure saturated liquid.

The state of the refrigerant **leaving the evaporator** is low-pressure superheated gas.

The color code of **refrigerant gauges** is high-side red and low-side blue. The high-side red gauge reads pressure above atmospheric. The low-side blue gauge reads pressures above and below atmospheric (14.7 PSI).

Recovery is the removal of refrigerant in any condition from a system to store it in an external container.

Recycling is the cleaning of a refrigerant for the purpose of reusing. After recycling, the refrigerant is free of air, moisture, and acids.

Reclaiming is the process of cleaning refrigerant equal to new or virgin refrigerant. A chemical analysis must be done at a reclaiming station, and it must meet ARI 700 standards.

Mixing more than one refrigerant makes it **impossible to reclaim** refrigerant.

Refrigerant cylinders should be free of rust, undented, and undamaged.

The **electronic or ultrasonic leak detector** will help locate the general area of a system's leak. Soap bubbles are also used to pinpoint a system's leak.

Dehydration of a refrigeration system is done with a vacuum pump. It is done to remove air and moisture from a system.

If moisture is left in a refrigeration system, **hydrochloric and hydrofluoric acids** may form.

When **evacuating a system,** if the vacuum pump is turned off and the pressure rises to 0 PSIG, there is a leak present in the system.

When **evacuating a system,** if the vacuum pump is turned off when the pressure is at 29.92" Hg. and the pressure rises slightly and stops, there might be moisture present in the system. There might also be refrigerant mixed with the compressor oil.

The following will effect the **length of time** that it takes to evacuate a system:

The size of the equipment.

The ambient temperature.

The amount of moisture in the system.

The size of the vacuum pump.

The size and length of the refrigerant gauge lines.

To prevent an explosion, never fill a refrigerant cylinder to more than **80% full.**

Refrigerant can cause **suffocation** because it is heavier than air and it displaces oxygen.

R-11, R-12 and R-500 are **CFC** types of refrigerant that are commonly used with mineral oil.

R-123 and R-22 are **HCFC** types of refrigerant that are commonly used with alkylbenzene oil.

R-134A is an **HFC** type of refrigerant that is commonly used with polyolester or ester oil.

A **ternary blend** type of refrigerant is a three part blend alkylbenzene oil.

An **azeotropic blend** is a combination of refrigerants that, when blended, create one refrigerant that has its own characteristics.

A **System Dependent** (passive) recovery uses the appliance compressor to remove refrigerant from the system.

A **Self-Contained** (active) recovery uses an EPA approved recovery machine.

To **safely recover refrigerants,** mechanical float devices, electronic shut-off devices, and weight methods are used.

When **pressure testing** a system, never exceed the system's low-side test pressures.

The **Department of Transportation (DOT)** must approve refrigerant cylinders before they are transported.

To **prevent an explosion,** never pressure test a system using oxygen.

TYPE I CERTIFICATION

Type I Certification covers appliances that are charged at the factory with five pounds or less of refrigerant. These include refrigerators, freezers, room air conditioners, PTAC Units, dehumidifiers, ice makers, vending machines, and drinking water coolers.

A **small appliance,** according to the EPA, is a system containing 5 pounds of refrigerant or less including PTAC's (package terminal air conditioners).

Under conditions of ARI 740-1993, **recovery systems manufactured** *before* November 15, 1993 must be able to remove 80% of a refrigerant charge or achieve 4" Hg. vacuum.

Under conditions of ARI 740-1993, **recovery systems manufactured** *after* November 15, 1993 must be able to remove 90% of a refrigerant charge or achieve 4" Hg. vacuum if the systems compressor is in operation.

Under conditions of ARI 740-1993, **recovery systems manufactured** *after* November 15, 1993 must be able to remove 80% of a refrigerant charge or achieve 4" Hg. vacuum if the systems compressor is not operating.

Under conditions of ARI 740-1993, all **recovery systems** must be able to remove 80% of the refrigerant charge or achieve 4" Hg. vacuum, if the systems compressor not operating.

Technicians must **certify their recovery equipment** with the EPA after August 12, 1993.

It is necessary to install a **piercing valve** on both the high and low side of an appliance that has an inoperative compressor while recovering the refrigerant. This will decrease the time of the refrigerant recovery.

The EPA does not require a **leak in small appliances** to be repaired.

The **pressure of an R-12** cylinder at 70°F ambient temperature will be 70 PSIG.

When the pressure of an R-22 cylinder is 121 PSIG, its temperature is 70°F.

When the **temperature of an R-502** cylinder is 80°F, its pressure is 161 PSIG.

The following refrigerants **must not be recovered** with a current recovery system:

Ammonia

Methyl Formate

Methyl Chloride

Sulfur Dioxide

Hydrogen

Water (H_2O)

R-134A refrigerant is a replacement for R-12 refrigerant. However, at this time there is no drop in replacement for R-12 refrigerant where oil will not have to be changed or added, and expansion valves do not have to be adjusted.

When pressurizing a system with nitrogen, it is required to have a **pressure regulator and a relief valve** in line with the nitrogen cylinder.

Always leak test **piercing valves** after installing them onto a system.

If **refrigerants are exposed to open flames** and hot surface phosgene gas, hydrochloric and hydrofluoric acids are produced.

A burnt compressor may cause a **pungent odor** in a CFC/HCFC system.

A **Self-Contained Breathing Apparatus** is required if a large amount of R-12 has been released.

Relief valves must never be installed in series with each other.

The **color code** for refrigerant recovery tanks is yellow tops with gray bottoms.

TYPE II CERTIFICATION

Type II Certification covers all air conditioning equipment that uses more than five pounds of charge of any refrigerant, except R-11, R-123, and R-113. The most common refrigerant is R-22.

Type 2 technicians are certified to service high-pressure and very high-pressure equipment, excluding small appliances and motor vehicles.

After the installation of a refrigeration system, the technician should first **pressure test the system using nitrogen or nitrogen with trace amount of R-22.**

Section 4 Indoor Air Quality (IAQ) and Safety   87

The maintenance task that must always be done to recovery/recycle equipment is to change the system's **oil and filters.**

If an open compressor has sat for several months, it is most likely to **leak from the shaft seal.**

Two indications that a high-pressure system has a leak are **oil traces and high superheat.**

The EPA requires that all *comfort cooling* systems with **more than 50 pounds** of refrigerant be repaired when the annual leak rate exceeds **15%.**

The EPA requires that all *commercial and industrial process refrigeration systems* with **more than 50 pounds** of refrigerant be repaired when the annual leak rate exceeds **35%.**

The component that is directly after the condenser on a system that uses a TXV type metering device is the receiver.

A **recovery and recycle machine** is likely to overheat when drawing a deep vacuum because the refrigerant is needed to cool the compressor.

R-134 refrigerant requires a separate set of gauges, a vacuum pump, and a recovery machine when servicing.

The following methods help **reduce the recovery time** of a system:

Pack ice around the recovery tank.

Heat the appliance.

Recover liquid first.

During recovery, the refrigerant should be removed from **the condenser outlet** when the condenser is below the receiver.

When recovering refrigerant from a system that has a condenser on the roof and the evaporator on the first floor, the recovery should begin with the **liquid line entering the evaporator.**

After removing refrigerant from a system, the **refrigerant may be reused** in the same system or on another system owned by the same person.

Before transferring refrigerant into an empty refrigerant cylinder, the **cylinder must be evacuated.**

The use of **low loss fittings and hoses containing hand vales** commonly aids in minimizing the release of refrigerant to the atmosphere.

When using recovery and recycling equipment manufactured *after* **November 15, 1993,** technicians must evacuate an appliance containing **more than 200 pounds** of HCFC-22 to 10" Hg.

When using recovery and recycling equipment manufactured *before* **November 15, 1993,** technicians must evacuate an appliance containing **less than 200 pounds** of HCFC-22 to 0 PSIG.

If a leak in an appliance makes the prescribed evacuation level unattainable, the appliance can be evacuated to **0 PSIG.**

The repair or replacement of the following components is considered **major repair.**

Compressor

Condenser

Evaporator

Auxiliary Heat Exchanger

When servicing an open refrigeration system, always **replace the filter dryer.**

In a refrigeration system, **high-pressure liquid** exits in the condenser and receiver.

When evacuating a system, it is required that the vacuum pump pull a minimum of **500 microns.**

A technician must **never operate a hermetic compressor under a deep vacuum** or the compressor could overheat.

Non-condensables (air) cause **high head pressure** when they are trapped in the condenser.

A technician must **never front seat the discharge service valve** while the compressor is in operation.

ASHRAE stands for The American Society of Heating, Refrigeration, Air Conditioning Engineers.

Defined by ASHRAE Standard 15, a sensor and alarm are required for A1 refrigerants (R-12, R-22, R-11) to **sense oxygen depletion.**

A refrigeration system must always be protected by a **pressure relief device.**

To prevent freezing when charging R-12 into an evacuated system, it must be in the vapor form to raise the **pressure up to 33 PSIG.**

The sale of CFC and HCFC refrigerants has been restricted to certified technicians since **November 14, 1994.**

Refrigerators built before 1950 often contain **sulfur dioxide** (SO_2).

TYPE III CERTIFICATION

Type III Certfication covers low-pressure appliances such as centrifugal compressors. This includes chiller type systems.

If EPA **regulations change** after a technician has been certified, it is the technician's responsibility to comply with future changes.

It is easier to have **air enter gaskets and fittings** on a low-pressure system than a high-pressure system, because a low-pressure system operates in a vacuum.

A rupture disc opens at **15 PSIG** on a low-pressure system.

The easiest way to leak test a low-pressure system is to **add warm water over the coil** to raise the pressure to 10 PSIG.

It is important **not to exceed 10 PSIG** when leak testing a low-pressure system because the rupture disc, located on the evaporator, might fail.

When an electronic leak detector is used to test a water box for a leaks, **test at the drain valve** after the water has been removed.

A **hydrostatic test kit** is used when leak testing a water tube on a low-pressure chiller.

The EPA requires that all *comfort cooling systems* with more than 50 pounds of refrigerant must be repaired when the annual leak rate exceeds **15%**.

The EPA requires that all *commercial and industrial process refrigeration systems* with more than 50 pounds of refrigerant must be repaired when the annual leak rate exceeds **35%**.

A **high-pressure cut-out is set to 10 PSIG** on a recovery unit that is used on low-pressure systems.

When recovering R-11 or R-123 from a low-pressure system, **always recover liquid first** to decrease the recovery time.

To prevent a low-pressure system from freezing during evacuation, **water is commonly run through the chiller tubes.**

ASHRAE guideline 3-1990 states if a **1 mm Hg. vacuum level rises to 2.5 mm Hg.** during an evacuation, the system should be leak tested.

Refrigerant is added to a low-pressure chiller system after an evacuation by the use of the **evaporator charging valve.**

After a chiller system has had a deep system evacuation, pressurizing the system with vapor up to the **36 degree saturation temperature** is commonly done to decrease the possibility of system freeze-up prior to a liquid charge.

When using recovery or recycling equipment manufactured or imported before November 15, 1993, the required level of evacuation for low-pressure equipment is **25 inches Hg.**

When using recovery or recycling equipment manufactured or imported on or after November 15, 1993, the required level of evacuation for low-pressure equipment is **25 mm Hg. absolute.**

A **purge unit** is used on low-pressure chillers to remove non-condensables from the system.

The **purge unit** removes trapped air from a low-pressure system from the top of the condenser.

R-123 is considered a group **B-1** refrigerant. The equipment room where the low-pressure chiller is located requires a **refrigerant sensor.**

Since there are approximately **100 pounds of vapor in a common 350 ton R-11 chiller,** it is important to recover the vapor refrigerant in a chiller after the liquid has been removed.

Universal Type Certification is given to the technician who passes the Core questions as well as Types I, II, and III Certification Exams.

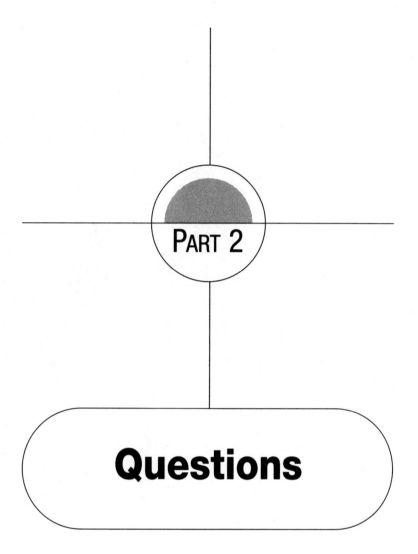

Section 1

System Components and Tools

HAND TOOLS

1. Which of the following tools has a six point star shaped head?
   A. a Phillips screwdriver
   B. a nut driver
   C. a spade bit
   D. a torx driver

2. Which of the following tools has a hexagon shaped head and is used to remove a screw with a six point hex head shape?
   A. a Phillips screwdriver
   B. a nut driver
   C. a spade bit
   D. a torx driver

3. Which tool has a variable open end that is used to remove a nut or a bolt?
   A. an allen wrench
   B. a flare nut wrench
   C. an adjustable wrench
   D. a box end wrench

4. Which tool is used to loosen or tighten set screws?
   A. an allen wrench
   B. a flare nut wrench
   C. an adjustable wrench
   D. a box end wrench

5. Which tool has a closed end and is used to loosen a nut or bolt?
   A. tongue and groove pliers
   B. a pipe wrench
   C. an adjustable wrench
   D. a box end wrench

6. Which tool is used to loosen or tighten a mechanical fitting on a refrigeration or air conditioning system?
   A. a flare nut wrench
   B. a pipe wrench
   C. an adjustable wrench
   D. a service wrench

7. Which tool is used to loosen or tighten a nut or a bolt from the side?
   A. a service wrench
   B. a box end wrench
   C. a socket wrench
   D. an open end wrench

8. Which is the proper tool to use when assembling or disassembling threaded pipe?
   A. a pipe adapting kit
   B. a pipe wrench
   C. a crescent wrench
   D. a torque wrench

9. Which is the correct tool to use when manipulating access valves on a refrigeration or air conditioning system?
   A. an adjustable wrench
   B. tongue and groove pliers
   C. a torque wrench
   D. a service valve wrench

10. What are the most common size openings for a service wrench used on light commercial refrigeration and air conditioning systems?
    A. $\frac{3"}{16}, \frac{1"}{4}, \frac{5"}{16}$, and $\frac{3"}{8}$
    B. $\frac{1"}{4}, \frac{1"}{2}, \frac{3"}{4}$, and $1"$
    C. $\frac{3"}{16}, \frac{5"}{16}, \frac{7"}{16}$, and $\frac{9"}{16}$
    D. $\frac{5"}{16}, \frac{3"}{8}, \frac{1"}{2}$, and $\frac{5"}{8}$

11. Which tool would most likely be used to fit on top of a $\frac{3"}{4}$ bolt?
    A. vise grip pliers
    B. socket wrench
    C. needle nose pliers
    D. service valve wrench

12. Which tool is used to apply a specific amount of force to a nut or a bolt?
    A. a pipe wrench
    B. an adjustable wrench
    C. a torque wrench
    D. a service wrench

13. The tool that is used to ensure that duct work and other work is level is
    A. a plumb bob.
    B. a carpenters square.
    C. a tape measure.
    D. a level.

14. Which of the following is an "L" shaped tool used to measure materials and check for uniformity?
    A. a plumb bob
    B. a carpenters square
    C. a tape measure
    D. a level

15. What type of saw is used to cut wood against the grain?
    A. a ripsaw
    B. a crosscut saw
    C. a hacksaw
    D. a keyhole saw

16. What type of saw is used to cut wood with the grain?
    A. a ripsaw
    B. a crosscut saw
    C. a hacksaw
    D. a keyhole saw

17. What type of saw is primarily used on metal?
    A. a ripsaw
    B. a crosscut saw
    C. a hacksaw
    D. a keyhole saw

18. What type of saw is most commonly used to cut drywall?
    A. a ripsaw
    B. a crosscut saw
    C. a hacksaw
    D. a keyhole saw

19. Which file has one row of teeth?
    A. a single cut file
    B. a double cut file
    C. a triple cut file
    D. a rasp

20. Which file has two rows of teeth that crisscross each other?
    A. a single cut file
    B. a double cut file
    C. a triple cut file
    D. a rasp

21. The three common drive sizes that a drill is available in are
    A. $\frac{3"}{8}, \frac{1"}{2}$, and $\frac{3"}{4}$.
    B. $\frac{1"}{4}, \frac{3"}{8}$, and $\frac{3"}{4}$.
    C. $\frac{5"}{16}, \frac{3"}{8}$, and $\frac{3"}{4}$.
    D. $\frac{1"}{4}, \frac{3"}{8}$, and $\frac{1"}{2}$.

22. The tool used with a drill to install or remove hex head sheet metal screws is called
    A. a hex head adapter.
    B. a screw adapter.
    C. a torx adapter.
    D. a hex head spade bit.

23. What tool is used to recess holes in wood and other material so that screws can be driven flush or slightly below the surface?
    A. a spade bit
    B. a countersink
    C. a punch
    D. a reamer bit

Section 1 System Components and Tools

24. Which bit is preferred for drilling holes in wood?
    A. a masonry bit
    B. a countersink
    C. a spade bit
    D. a keyhole bit

25. What tool has a corkscrew shape and is used to bore holes into wood?
    A. a masonry bit
    B. a spade bit
    C. a corkscrew bit
    D. an auger bit

26. What tool is used to bore holes in brick, block, concrete, stone, or mortar?
    A. a masonry bit
    B. a countersink
    C. a spade bit
    D. a keyhole bit

27. The proper tool to use when creating an indentation in sheet metal is
    A. a punch.
    B. a countersink.
    C. a torx driver.
    D. an auger bit.

28. What tool is used to cut a treaded hole in metal material?
    A. a tap
    B. a die
    C. a countersink
    D. a cutting tool

29. Which of the following would be the proper tool for threading gas pipe?
    A. a punch
    B. a tap
    C. a die
    D. a reamer

30. What tool is used to remove the burrs from the inside of tubing?
    A. a reamer
    B. a die
    C. a tube cutter
    D. a countersink

31. Which hammer would most likely be used to remove a nail from a 2 x 4 stud?
    A. a ball peen hammer
    B. a claw hammer
    C. a masonry hammer
    D. a sledge hammer

32. Which hammer, designed for metal work, has a flat back and a round front?
    A. a ball peen hammer
    B. a claw hammer
    C. a masonry hammer
    D. a sledge hammer

33. Which hammer has a large head that is designed for pounding block or concrete?
    A. a ball peen hammer
    B. a claw hammer
    C. a masonry hammer
    D. a sledge hammer

34. What care should be taken to prevent corrosion after using hand tools?
    A. Hand tools must be stored at 60° F to prevent corrosion and metal fatigue.
    B. Hand tools should be wiped down with machine oil.
    C. Hand tools should be cleaned with mineral spirits after each use.
    D. Hand tools should be wiped down with a salt water solution.

BASIC CONSTRUCTION

35. The two most common roof designs used on homes today are
    A. the flat roof and the truss built roof.
    B. the rafter built roof and the flat roof.
    C. the truss built roof and the rafter built roof.
    D. the gable roof design and the hip roof design.

36. The most common base used under shingles to strengthen the roof by adjoining the rafters or trusses together is
    A. $\frac{1}{4}"$ plywood.
    B. $\frac{1}{2}"$ OSB sheathing.
    C. $\frac{1}{2}"$ particle board.
    D. $\frac{1}{2}"$ pine board.

37. Trusses are commonly constructed of
    A. 2" x 8" or 2" x 10" lumber.
    B. 1" x 8" or 1" x 10" lumber.
    C. 2" x 2" or 2" x 4" lumber.
    D. 1" x 2" or 2" x 4" lumber.

38. What type of roof construction uses boards that connect the ridge of the roof and the exterior wall?
    A. a truss built roof
    B. a flat roof
    C. a hip roof
    D. a rafter built roof ✓

39. The material that is used for a standard ceiling is
    A. $\frac{1}{4}"$ drywall.
    B. $\frac{5}{8}"$ drywall. ✓
    C. $\frac{3}{4}"$ drywall.
    D. $\frac{7}{8}"$ drywall.

40. What would the effect of heat transfer be if the R-value on a roof were increased?
    A. It would increase the heat transfer.
    B. It would decrease the heat transfer. ✓
    C. It has no effect on heat transfer.
    D. It would reduce the thermal efficiency.

41. Common exterior residential walls are constructed of
    A. 2" x 2" or 2" x 4" studs.
    B. 1" x 2" or 1" x 4" studs.
    C. 2" x 4" or 2" x 6" studs. ✓
    D. 1" x 4" or 1" x 6" studs.

42. How far apart are the studs spaced in an interior and exterior wall?
    A. 12 inches on center
    B. 14 inches on center
    C. 16 inches on center ✓
    D. 18 inches on center

43. The bottom board located on a wall that the studs are nailed to is known as a
    A. lower rafter.
    B. horizontal stud.
    C. load barring joist.
    D. bottom plate. ✓

44. What is known as a load bearing wall?
    A. This is the wall that uses two studs to return air back to the furnace.
    B. This is the wall that always has a beam resting on top of it.
    C. This is the wall that has the weight of the roof on top of it. ✓
    D. This is the wall that should never have a beam resting on top of it.

45. What is the purpose of a vapor barrier?
    A. to keep moisture from entering and exiting a home ✓
    B. to protect the insulation of an exterior wall
    C. to retain steam created in bathrooms
    D. to remove moisture from a home

46. Name three common exterior wall insulation R-values.
    A. R-13, R-19, R-30
    B. R-5, R-11, R-13
    C. R-19, R-30, R-40
    D. R-11, R-13, R-19 ✓

47. What is the purpose of the 1" x 2" wood slats that are placed between the floor joists?
    A. They are used to decrease the squeak of a floor.
    B. They are bridging used to stiffen the floor. ✓
    C. They are used to support the floor joists.
    D. They are used to hold supply air ducts in place.

48. A sub-floor is commonly made with
    A. $\frac{1}{2}"$ tongue and groove plywood.
    B. $\frac{3}{4}"$ OSB tongue and groove sheathing. ✓
    C. $\frac{5}{8}"$ OSB sheathing.
    D. $\frac{5}{8}"$ plywood.

49. An underlayment floor is made of
    A. $\frac{1}{4}"$ plywood. ✓
    B. $\frac{1}{4}"$ OSB sheathing.
    C. $\frac{1}{2}"$ plywood.
    D. $\frac{1}{2}"$ OSB sheathing.

REFRIGERATION THEORY

50. Temperature is
    A. the intensity of molecular motion. ✓
    B. the measurement of heat in an liquid.
    C. the measurement of heat in a gas.
    D. the measurement of heat in an object.

51. What are the 4 common temperature scales?
    A. Rankine, Fahrenheit, Celsius, and Centigrade
    B. Centigrade, Fahrenheit, Celsius, and Kelvin
    C. Fahrenheit, Rankine, Celsius, and Kelvin ✓
    D. Absolute, Kelvin, Centigrade, and Fahrenheit

Section 1 System Components and Tools

52. Which of the following is the U.S. conventional standard scale?
    A. Celsius
    B. Fahrenheit
    C. Kelvin
    D. Rankine

53. Which of the following is the U.S. conventional absolute zero scale?
    A. Centigrade
    B. Fahrenheit
    C. Kelvin
    D. Rankine

54. Which temperature scale freezes water at 32 degrees and boils water at 212 degrees?
    A. Centigrade
    B. Fahrenheit
    C. Kelvin
    D. Absolute

55. To convert Fahrenheit to Rankine,
    A. subtract 273° F.
    B. add 273° F.
    C. subtract 460° F.
    D. add 460° F.

56. Which of the following is the SI unit standard scale?
    A. Celsius
    B. Fahrenheit
    C. Kelvin
    D. Rankine

57. Which of the following is the SI unit absolute zero scale?
    A. Centigrade
    B. Fahrenheit
    C. Kelvin
    D. Rankine

58. Which temperature scale freezes water at 0 degrees and boils water at 100 degrees?
    A. Centigrade
    B. Fahrenheit
    C. Kelvin
    D. Absolute

59. To convert Celsius to Kelvin
    A. subtract 273° Celsius.
    B. add 273° Celsius.
    C. subtract 460° Celsius.
    D. add 460° Celsius.

60. To convert Celsius to Fahrenheit, which of the following is the correct formula?
    A. $(1.8 \times C) + 32 = F$
    B. $(1.8 - C) + 32 = F$
    C. $(1.8 + C) - 32 = F$
    D. $(1.8 \times C) - 32 = F$

61. To convert Fahrenheit to Celsius, which of the following is correct?
    A. $.56 \times (F - 32) = C$
    B. $.56 \times (F + 32) = C$
    C. $.56 - (F \times 32) = C$
    D. $.56 + (F \times 32) = C$

62. What is the definition of specific heat?
    A. The heat that can be measured by a thermometer.
    B. The amount of heat required to raise one pound of water 1° F.
    C. The amount of heat required to raise one pound of a substance 1° F.
    D. The heat used to cause a change of state.

63. What is the definition of sensible heat?
    A. The heat that can be measured by a thermometer.
    B. The amount of heat required to raise one pound of water 1° F.
    C. The amount of heat required to raise one pound of a substance 1° F.
    D. The heat used to cause a change of state.

64. What is the definition of latent heat?
    A. The heat that can be measured by a thermometer.
    B. The amount of heat required to raise one pound of water 1° F.
    C. The amount of heat required to raise one pound of a substance 1° F.
    D. The heat used which causes a change of state.

65. What is the definition of a Btu?
    A. The heat that can be measured by a thermometer.
    B. The amount of heat required to raise one pound of water 1° F.
    C. The amount of heat required to raise one pound of a substance 1° F.
    D. The heat used to cause a change of state.

66. Heat energy is measured
    A. by a thermometer.
    B. by the temperature difference of a substance.
    C. in Btus.
    D. by the thermal mass of a substance.

67. To measure the amount of heat in a substance without the change of state, which formula is correct?
    A. temperature × weight × volume
    B. temperature × density × volume
    C. specific heat × temperature difference × weight
    D. specific heat × volume × temperature difference

68. If 2 pounds of water are raised from 40° F to 60° F how many Btus are added to the water?
    A. 20
    B. 30
    C. 40
    D. 60

69. The amount of heat that it takes to change one pound of 212° F water to 1 pound of 212° F steam is
    A. 144 Btus.
    B. 200 Btus.
    C. 746 Btus.
    D. 970 Btus.

70. The amount of heat required to change one pound of 32° F ice to 1 pound of 32° F water is
    A. 144 Btus.
    B. 200 Btus.
    C. 746 Btus.
    D. 970 Btus.

71. The three states of matter are
    A. solid, vapor, and gas.
    B. solid, liquid, and gas.
    C. solid, fluid, and liquid.
    D. fluid, vapor, and solid.

72. Boyles' Law states that
    A. at a constant pressure, the volume of a gas varies directly with an absolute temperature.
    B. the total pressure of a confined mixture of gasses is the sum of the pressures of each of the gasses in the mixture.
    C. the volume of a gas varies indirectly with an absolute temperature.
    D. a volume of gas varies inversely with the absolute pressure, if the temperature remains constant.

73. Charles' Law states that
    A. at a constant pressure, the volume of a gas varies directly with an absolute temperature.
    B. the total pressure of a confined mixture of gasses is the sum of the pressures of each of the gasses in the mixture.
    C. the volume of a gas varies indirectly with an absolute temperature.
    D. a volume of gas varies inversely with the absolute pressure, if the temperature remains constant.

74. Daltons' Law states that
    A. at a constant pressure, the volume of a gas varies directly with an absolute temperature.
    B. the total pressure of a confined mixture of gasses is the sum of the pressures of each of the gasses in the mixture.
    C. the volume of a gas varies indirectly with an absolute temperature.
    D. a volume of gas varies inversely with the absolute pressure, if the temperature remains constant.

75. The calculation for work is
    A. time × distance.
    B. mass × force.
    C. force × distance.
    D. force × rate.

76. One horsepower is equal to
    A. 33,000 ft-lb. of work in one minute.
    B. 12,000 ft-lb. of work in one minute.
    C. 746 ft-lb. of work in one minute.
    D. 200 ft-lb. of work in one minute.

77. One horsepower is equal to
    A. 33,000 watts.
    B. 746 watts.
    C. 200 watts.
    D. 3.41 watts.

78. One watt is equal to
    A. 33,000 Btus.
    B. 746 Btus.
    C. 200 Btus.
    D. 3.41 Btus.

79. Which method of heat transfer is known as conduction?
    A. The transfer of heat from liquids to vapors.
    B. The transfer of heat from one molecule to another.
    C. The transfer of heat from one place to another by a fluid, commonly air or water.
    D. The transfer of heat from the heat source to a solid object through space.

80. Which method of heat transfer is known as convection?
    A. The transfer of heat from liquids to vapors.
    B. The transfer of heat from one molecule to another.
    C. The transfer of heat from one place to another by a fluid, commonly air or water.
    D. The transfer of heat from the heat source to a solid object through space.

Section 1 System Components and Tools

81. Which method of heat transfer is known as radiation?
    A. The transfer of heat from liquids to vapors.
    B. The transfer of heat from one molecule to another.
    C. The transfer of heat from one place to another by a fluid, commonly air or water.
    D. The transfer of heat from the heat source to a solid object through space.

82. One ton of refrigeration is
    A. 33,000 Btus per hour.
    B. 12,000 Btus per hour.
    C. 746 Btus per hour
    D. 200 Btus per hour.

83. What type of heat is absorbed or extracted from a substance when the substance changes state?
    A. specific heat
    B. latent heat
    C. sensible heat
    D. super heat

84. When a substance changes from a liquid to a gas, this is known as
    A. latent heat of a condensation.
    B. sensible heat of a vapor.
    C. latent heat of sublimation.
    D. latent heat of vaporization.

85. When a substance changes from a gas to a liquid, this is known as
    A. latent heat of a condensation.
    B. sensible heat of a vapor.
    C. latent heat of sublimation.
    D. latent heat of vaporization.

86. When a substance changes from a solid to a gas, this is known as
    A. latent heat of a condensation.
    B. sensible heat of a vapor.
    C. latent heat of sublimation.
    D. latent heat of vaporization.

87. In a refrigeration system, heat is absorbed by
    A. the latent heat of condensation in the condenser.
    B. the latent heat of vaporization in the evaporator.
    C. the latent heat of vaporization in the condenser.
    D. the latent heat of condensation in the evaporator.

88. In a refrigeration system, heat is rejected by
    A. the latent heat of condensation in the condenser.
    B. the latent heat of vaporization in the evaporator.
    C. the latent heat of vaporization in the condenser.
    D. the latent heat of condensation in the evaporator.

AIR CONDITIONING COMPONENTS

89. The part of an air conditioning system that changes the refrigerant from a low-pressure gas to a high-pressure gas is the
    A. compressor.
    B. condenser.
    C. metering device.
    D. evaporator.

90. The part of an air conditioning system that changes the refrigerant from a high-pressure gas to a high-pressure liquid is the
    A. compressor.
    B. condenser.
    C. metering device.
    D. evaporator.

91. The part of an air conditioning system that changes the refrigerant from a low-pressure liquid to a low-pressure gas is the
    A. compressor.
    B. condenser.
    C. metering device.
    D. evaporator.

92. The part of an air conditioning system that changes the refrigerant from a high-pressure liquid to a low-pressure liquid is the
    A. compressor.
    B. condenser.
    C. metering device.
    D. evaporator.

93. Which part of the system stores refrigerant until it is needed in the evaporator?
    A. the accumulator
    B. the surge drum
    C. the receiver
    D. the heat exchanger

System Component Diagram

94. Which part of the system prevents liquid from entering the compressor?
    A. the moisture indicator
    B. the receiver
    C. the accumulator *(circled)*
    D. the surge drum

95. A split system air conditioning unit has
    A. the evaporator, receiver, and compressor connected by a refrigeration line set to the condenser and metering device.
    B. the metering device and compressor connected by a refrigeration line set to the condenser and evaporator.
    C. the receiver, compressor, and metering device connected by a refrigeration line set to the condenser and evaporator.
    D. the compressor and condenser connected by a refrigeration line set to the metering device and evaporator. *(circled)*

96. On the "System Component Diagram," match the following components with their proper names.
    1. B   6. J
    2. H   7. F
    3. C   8. G
    4. I   9. D
    5. E   10. A

    Components:
    A. accumulator   F. heat exchanger
    B. compressor   G. metering device
    C. condenser   H. oil separator
    D. evaporator   I. receiver
    E. filter dryer   J. sight glass

97. In a sealed system, as the pressure increases
    A. the temperature increases. *(circled)*
    B. the temperature decreases.
    C. the temperature stays the same.
    D. the refrigerant will always expand.

Section 1 System Components and Tools

Split System Diagram

98. On the "Split System Diagram," match the following components with their proper names.

1. A
2. D
3. B
4. C
5. H
6. F
7. I
8. E
9. J
10. G

Components:
A. compressor
B. condenser
C. condenser fan
D. discharge line
E. evaporator
F. filter dryer
G. indoor blower
H. liquid line
I. metering device (restrictor)
J. suction line

99. In the "Split System Diagram," where does the refrigerant change from a high-pressure gas to a high-pressure liquid?
A. 2
B. 3
C. 8
D. 9

100. In the "Split System Diagram," where does the refrigerant change from a low-pressure liquid to a low-pressure gas?
A. 2
B. 3
C. 8
D. 9

101. In the "Split System Diagram," where does the refrigerant change from a low-pressure gas to a high-pressure gas?
A. 1
B. 3
C. 7
D. 8

102. In the "Split System Diagram," where does the refrigerant change from a high-pressure liquid to a low-pressure liquid?
A. 1
B. 3
C. 7
D. 8

103. Why is it recommended to add a gravel fill under the pad of a condensing unit?
    A. This will help water to drain from inside the unit.
    B. This can prevent vibration of the condensing unit.
    C. This will prevent settling of the ground which can cause refrigerant line damage.
    D. This will allow the system to be free floating for installation flexibility.

104. Residential split system outdoor condensing units are commonly wired with a
    A. 208/240V single phase power supply.
    B. 120V single phase power supply.
    C. 208/240V three phase power supply.
    D. 208/240V polyphase power supply.

105. Before energizing a newly installed system, what test should always be done?
    A. check operating pressures
    B. check operating temperatures
    C. check for tight connections at electrical terminals
    D. check the resistance to the compressor terminals

106. A room thermostat should be mounted
    A. four feet from the floor.
    B. near a supply air register for accuracy.
    C. on an outside wall.
    D. near a return air grill.

107. What is the secondary voltage on a step down control transformer for a residential furnace?
    A. 24V
    B. 120V
    C. 220V
    D. 440V

108. What is the R terminal used for on the thermostat?
    A. the power from the transformer
    B. the blower relay coil
    C. the cooling contactor coil
    D. the heating control circuit

109. What is the W terminal used for on the thermostat?
    A. the power from the transformer
    B. the blower relay coil
    C. the cooling contactor coil
    D. the heating control circuit

110. What is the Y terminal used for on the thermostat?
    A. the power from the transformer
    B. the blower relay coil
    C. the cooling contactor coil
    D. the heating control circuit

111. What is the G terminal used for on the thermostat?
    A. the power from the transformer
    B. the blower relay coil
    C. the cooling contactor coil
    D. the heating control circuit

112. What is the purpose for insulating a suction line?
    A. to prevent excessive subcooling of the refrigerant
    B. to increase the refrigerant superheat entering the compressor
    C. to decrease the systems vibration
    D. to prevent condensation from forming

113. When installing a refrigeration line set through a masonry wall,
    A. the lines must be covered by PVC to prevent contact with the masonry material.
    B. the lines must use a flexible connector to protect against a foundation shift.
    C. the lines must never be installed through a masonry wall.
    D. the lines must be secured tightly against the top portion of the wall to prevent vibration.

114. To ensure proper oil return to the compressor, when installing a refrigeration line set
    A. the suction line must be pitched away from the condensing unit.
    B. the liquid line must be pitched away from the evaporator coil.
    C. the suction line must be pitched toward the condensing unit.
    D. the liquid line must be pitched toward the evaporator coil.

115. To insure proper oil return to the compressor, a coiled refrigeration lineset
    A. must be coiled horizontally and the vapor must flow from bottom to top.
    B. must be coiled horizontally and the vapor must flow from top to bottom.
    C. the lineset must never be coiled.
    D. must be coiled vertically.

116. What must be done to the system when a condensing unit is precharged for 25 feet of refrigerant line and uses 32 feet of refrigerant line?
    A. The operating high-side pressure must be reduced by 7 PSIG.
    B. The operating high-side pressure must be increased by 7 PSIG.
    C. Refrigerant must be added to the system.
    D. Refrigerant must be removed from the system.

Section 1   System Components and Tools

117. Quick connect fittings are commonly used on
    A. precharged refrigeration line sets.
    B. soft copper tubing with the use of a flare nut.
    C. hard drawn copper tubing with the use of an O-ring.
    D. None of the above.

118. What type of fitting uses an O-ring to help seal its joint?
    A. a compression fitting
    B. a quick connect fitting
    C. a flare fitting
    D. a sweat connection

119. When soldering a copper connection and solder is drawn into the joint, this is commonly called
    A. conduction.
    B. capillary action.
    C. siphon action.
    D. flow absorption.

120. When installing an evaporator coil, before soldering, the end of the evaporator lines must be opened to the atmosphere because
    A. the coil is pre-charged with CO^2.
    B. the coil is precharged with R-22.
    C. the coil is pre-charged with nitrogen.
    D. None of the above.

121. What type of metering device is used on a high efficiency split system air conditioning system?
    A. a capillary tube
    B. a piston type
    C. a TXV
    D. an AXV

122. If ductwork is located in an attic or crawl space, it must be
    A. supported every 18 inches.
    B. insulated.
    C. supported with joist hangers.
    D. a maximum of 15 feet in length.

123. An auxiliary drain pan must
    A. be a minimum of 3 inches deep.
    B. have a minimum perimeter of 5 inches larger than the unit being served.
    C. have a minimum perimeter of 3 inches larger than the unit being served.
    D. be made of aluminized steel.

124. A secondary drain line
    A. is always necessary in an auxiliary drain pan.
    B. must have no more than two 90 degree elbows.
    C. must not be less than 1 inch diameter.
    D. must be located in a conspicuous place to serve as an alarm.

125. A normally closed float switch that is connected to an auxiliary drain pan should be wired to control
    A. the blower motor.
    B. the contactor coil.
    C. the condensate pump.
    D. the inducer motor.

REFRIGERANTS

126. Which of the following refrigerant is considered an HCFC?
    A. R-12
    B. R-22
    C. R-502
    D. R-134A

127. What type of refrigerant is commonly used with alkylbenzene oil?
    A. R-12
    B. R-22
    C. R-410A
    D. R-134A

128. Refrigerant tanks must never be stored in an area where the temperature is above
    A. 100° F.
    B. 115° F.
    C. 120° F.
    D. 125° F.

129. A refrigerant that is being used to replace R-22 is
    A. R-408A.
    B. R-409A.
    C. R-410A.
    D. R-500.

130. The HFC refrigerant that is used as a replacement for R-12 is
    A. R-409A.
    B. R-134A.
    C. R-22.
    D. R-408A.

131. The blended refrigerant that is used as a replacement for R-12 is
    A. R-409A.
    B. R-134A.
    C. R-22.
    D. R-408A.

132. The blended refrigerant that is used as a replacement for R-502 is
    A. R-409A.
    B. R-134A.
    C. R-22.
    D. R-408A.

133. What type of refrigerant, when charged as a gas, will not cause fractionation?
    A. a ternary blend
    B. a zeotropic
    C. a single component
    D. a near-zeotropic

134. What type of refrigerant is considered a single component refrigerant?
    A. R-134A
    B. R-502
    C. R-409A
    D. R-408A

135. Which of the following is a mixture of two or more refrigerants that have the same boiling point and act as one refrigerant?
    A. an azeotropic
    B. a ternary blend
    C. a zeotropic
    D. a near-azeotropic

136. Which of the following is a mixture of two or more refrigerants that do not have the same boiling points and that act close to an azeotropic blend?
    A. an azeotropic
    B. a ternary blend
    C. a single component
    D. a near-azeotropic

137. Which of the following is a mixture of two or more refrigerants that do not have the same boiling points and that do not act as one refrigerant?
    A. an azeotropic
    B. a zeotropic
    C. a single component
    D. a ternary blend

138. Which of the following is three refrigerants blended together?
    A. an azeotropic
    B. a zeotropic
    C. a single component
    D. a ternary blend

139. Zeotropic and near-azeotropic refrigerants must be charged into a system
    A. as a gas.
    B. as a liquid.
    C. as a liquid or gas.
    D. using the superheat method only.

140. What type of refrigerant is R-410A?
    A. a single component
    B. an azeotropic
    C. a near-azeotropic
    D. a zeotropic

141. What type of refrigerant is considered an azeotropic refrigerant?
    A. R-12
    B. R-22
    C. R-502
    D. R-409A

142. What type of refrigerant is considered a single component refrigerant?
    A. R-12
    B. R-410A
    C. R-502
    D. R-409A

143. What type of refrigerant is considered a near-azeotropic refrigerant?
    A. R-410A
    B. R-408A
    C. R-134A
    D. R-503

144. What type of refrigerant is a zeotropic refrigerant?
    A. R-22
    B. R-502
    C. R-409A
    D. R-407A

145. Which of the following is *not* considered a ternary refrigerant?
    A. R-407C
    B. R-502
    C. R-409A
    D. None of the above.

146. When charging a system using the superheat method, which value located on a pressure temperature chart would be used with R-409A?
    A. the dewpoint line value
    B. the bubble line value
    C. the temperature glide value
    D. the superheat value

147. When charging a system using the subcool method, which value located on a pressure temperature chart would be used with R-408A?
    A. the dewpoint line value
    B. the bubble line value
    C. the temperature glide value
    D. the superheat value

148. What is known as the difference between the vapor and liquid state of a near-azeotropic, a zeotropic, and a ternary blend refrigerant that occurs during evaporation and condensation at a constant pressure?
    A. bubble line value
    B. dewpoint value
    C. temperature glide
    D. ternary glide

Section 1 System Components and Tools

TABLE 1–1: Pressure Temperature Chart

PSIG	Temperature in Degrees F.		
	R-12	R-22	R-502
25	25	1	-7
26	27	2	-6
27	28	4	-5
28	29	5	-3
29	31	6	-2
30	32	7	-1
31	33	8	0
32	34	9	1
33	35	10	2
34	37	11	3
35	38	12	4
36	39	13	5
37	40	14	6
38	41	15	7
39	42	16	8
40	43	17	9
42	45	19	11
44	47	21	13
46	49	23	15
48	51	24	16
50	53	26	18
52	55	28	20
54	57	29	21
56	58	31	23
58	60	32	24
60	62	34	26
62	64	35	27
64	65	37	29
66	67	38	30
68	68	40	32
70	70	41	33
72	72	42	34
74	73	44	36

149. An azeotropic refrigerant has a temperature glide of
   A. 0° F.
   B. above 0° up to 10° F.
   C. above 10° up to 20° F.
   D. over 20° F.

150. A near-azeotropic refrigerant has a temperature glide of
   A. 0° F.
   B. above 0° up to 10° F.
   C. above 10° up to 20° F.
   D. over 20° F.

151. A zeotropic refrigerant has a temperature glide of
   A. 0° F.
   B. above 0° up to 10° F.
   C. above 10° F.
   D. None of the above.

152. What does the term fractionation mean?
   A. to separate a refrigerant mixture improperly by charging as a liquid
   B. to separate a refrigerant mixture improperly by charging as a gas
   C. to blend two types of refrigerants into a system for an increase of efficiency
   D. to add small parts of refrigerant oil into the system as it is in operation

153. Using the pressure temperature chart, what pressure is R-502 at 26° F?
   A. -6 PSIG
   B. 26 PSIG
   C. 50 PSIG
   D. 60 PSIG

154. Using the pressure temperature chart, what is the temperature of R-22 at 60 PSIG?
   A. 26° F
   B. 34° F
   C. 58° F
   D. 62° F

LUBRICANTS

155. The properties of mineral oil are very common to the properties of
   A. petroleum oil.
   B. polyol ester oil.
   C. alkylbenzene oil.
   D. PAG oil.

156. Which refrigerant is not miscible with mineral oil?
   A. R-12
   B. R-11
   C. R-409A
   D. R-134A

157. Which lubricant is most commonly used with CFC refrigerants?
   A. polyol ester
   B. mineral
   C. polyalkylene glycol
   D. synthetic

158. Which lubricant is most commonly used with HFC-134A refrigerant?
    A. polyol ester
    B. mineral
    C. alkylbenzene
    D. petroleum based oils

159. What are two compounds that are used to compose an alkylbenzene lubricant?
    A. alcohol and benzene
    B. glycol and alcohol
    C. propylene and glycol
    D. propylene and benzene

160. What types of refrigerants commonly use alkylbenzene lubricants?
    A. CFCs and HFCs
    B. HCFCs and HFCs
    C. CFCs and HCFCs
    D. HFCs only

161. What systems commonly use PAG lubricants?
    A. automotive air conditioning
    B. residential air conditioning systems
    C. cascade refrigeration systems
    D. chillers

162. What does PAG stand for?
    A. polyurethane alkylene glycol
    B. polymorphous alkylene glycol
    C. polyalkylene glycol
    D. None of the above.

163. Hygroscopic oil
    A. has a high attraction to moisture.
    B. should be stored at ambient temperatures.
    C. can be left open to the atmosphere.
    D. has a low attraction to moisture.

164. Corrosion will start to form inside of an air conditioning system if the moisture content exceeds
    A. 10 PPM.
    B. 100 PPM.
    C. 200 PPM.
    D. 250 PPM.

165. What are two good characteristics of oil?
    A. high floc point and low hygroscopic ability
    B. low carbon forming tendency and high hygroscopic ability
    C. high floc point and high carbon forming tendencies
    D. low floc point and low hygroscopic ability

166. Which of the following refers to the temperature at which oil starts to produce a waxy substance?
    A. floc point
    B. pour point
    C. sludge point
    D. flow point

167. Which of the following refers to the ability of oil to flow at very low temperatures?
    A. viscous point
    B. flow point
    C. pour point
    D. floc point

168. A miscible fluid has
    A. the ability to mix well with another fluid.
    B. the ability to flow at low temperatures.
    C. extremely thin qualities.
    D. extremely dense qualities.

169. Why is it necessary to increase the dehydration time of a system using a synthetic oil compared to a system using a mineral oil?
    A. Because a synthetic oil has a higher evaporation point than a mineral oil.
    B. Because a mineral oil has a higher evaporation point than a synthetic oil.
    C. Because a synthetic oil has a higher hygroscopic capability.
    D. Because a mineral oil has a higher hygroscopic capability.

170. Which oil has the ability to absorb the most moisture?
    A. a PAG oil
    B. a polyolester oil
    C. a mineral oil
    D. Both A and B.

171. Polyester oil will tolerate a mixture of
    A. 8% of mineral oil.
    B. 3 to 5% of mineral oil.
    C. 10 to 20% of mineral oil.
    D. 40 to 50% of mineral oil.

172. High viscosity oil
    A. is used on low temperature systems only.
    B. is a thin oil.
    C. is a thick oil.
    D. is used on medium temperature systems only.

173. What is the point at which the vapors of an oil will ignite but the flame will not persist?
    A. burn point
    B. flash point
    C. floc point
    D. pour point

Section 1 System Components and Tools

COMPRESSORS

174. What is the purpose of the compressor?
    A. It takes a low-pressure liquid and changes it into a high-pressure liquid.
    B. It takes a high-pressure gas and changes it into a high-pressure liquid.
    C. It takes a low-pressure gas and changes it into a high-pressure gas.
    D. It takes a low-pressure liquid and changes it into a high-pressure gas.

175. A crankcase heater is used on a compressor to
    A. thin the oil for increased lubrication.
    B. evaporate liquid refrigerant out of the compressor oil.
    C. warm the oil for increased lubrication of moving compressor components.
    D. keep snow from building up on the compressor during winter months.

176. What are the two most common types of compressors used in residential and light commercial air conditioning systems?
    A. reciprocating and rotary
    B. rotary and scroll
    C. centrifugal and reciprocating
    D. reciprocating and scroll

177. Which of the following is considered a non-positive displacement compressor?
    A. reciprocating
    B. rotary
    C. screw
    D. centrifugal

178. Which type of compressor is completely sealed and must be replaced if it has internal damage?
    A. a semi-hermetic
    B. a hermetic
    C. an open type
    D. a centrifugal type

179. Name two types of compressor lubrication systems.
    A. forced and gravity feed lubrication
    B. forced and splash lubrication
    C. gravity feed and splashed lubrication
    D. plate and gravity feed lubrication

180. How are compressors cooled in residential and light commercial air conditioning systems?
    A. air motion and water
    B. refrigerant and water
    C. refrigerant and air
    D. refrigerant and oil

181. What are the two types of compressor drives?
    A. belt and pulley
    B. direct and belt
    C. direct and indirect
    D. indirect and pulley

182. Which compressor is not internally driven by a motor?
    A. the hermetic
    B. the semi-hermetic
    C. the open type
    D. the closed type

183. What compressor can handle some liquid refrigerant floodback?
    A. the reciprocating hermetic type
    B. the reciprocating semi-hermetic type
    C. the scroll type
    D. the rotary type

184. How is the compressor ratio determined?
    A. discharge PSIA / suction PSIA
    B. suction PSIA / discharge PSIA
    C. discharge PSIG / suction PSIG
    D. suction PSIG / discharge PSIG

185. What is commonly used with a two speed compressor?
    A. a tachometer
    B. a double riser
    C. a two stage thermostat
    D. a dual pressure relief valve

AIR COOLED CONDENSERS

186. What is the purpose of the condenser?
    A. It takes a low-pressure liquid and changes it into a high-pressure liquid.
    B. It takes a high-pressure gas and changes it into a high-pressure liquid.
    C. It takes a low-pressure gas and changes it into a high-pressure gas.
    D. It takes a high-pressure liquid and changes it into a high-pressure gas.

187. Out of what materials are the tubes on a residential or light commercial air conditioning condenser fabricated?
    A. aluminum and steel
    B. iron and steel
    C. steel and copper
    D. aluminum or copper

188. Where in the system is the excess superheat added that must be removed before condensation in the condenser can take place?
    A. at the beginning of the evaporator and at the end of the receiver
    B. at the end of the receiver and at the beginning of the TXV
    C. at the beginning of the evaporator and in the compressor
    D. at the end of the evaporator and in the compressor

189. What section of the condenser coil is most likely to become dirty on a unit that has a side air discharge with the fan pushing the air over the coil?
    A. the outside of the coil
    B. the top section of the coil
    C. the inside of the coil
    D. the bottom section of the coil

190. What is a common temperature difference from the air entering a condenser and the refrigerant condensing temperature?
    A. 5°–10° F
    B. 10°–15° F
    C. 20°–30° F
    D. 35°–45° F

191. A split system condensing unit contains
    A. the compressor, condenser, and metering device.
    B. the condenser, compressor, and service valves.
    C. the condenser, service valves, and metering device.
    D. the condenser, evaporator, and service valves.

192. If air is trapped inside a condenser coil, what symptoms would be present?
    A. high discharge pressure, low amperage, and a decrease of efficiency
    B. low discharge pressure, high amperage, and excessive heat build up
    C. a decrease in efficiency, a decrease of heat build up, and high amperage
    D. an increase of heat build up, high amperage, and high discharge pressure

WATER COOLED CONDENSERS

193. What two factors effect the efficiency of a water cooled condenser?
    A. the high-side pressure of the condenser and the inlet water temperature
    B. the high-side pressure of the condenser and the outlet water temperature
    C. the quantity of water and the temperature difference between the refrigerant and the water
    D. the quantity of water and the refrigerant condensing point compared to the water boiling point

194. The capacity of a water cooled condenser increases
    A. as the temperature difference between the refrigerant and the water decreases.
    B. as the temperature difference between the refrigerant and the water increases.
    C. as the condenser outlet temperature increases.
    D. as the condenser outlet temperature decreases.

195. If the water quantity on a water cooled condenser was increased, what effect would this have on the system?
    A. It decreases the power consumption of the compressor.
    B. It increases the power consumption of the compressor.
    C. It has no effect on the power consumption of the compressor.
    D. It increases the high-side pressure.

196. When using a water cooled condenser, a decrease of water
    A. decreases the power consumption of a compressor.
    B. increases the fouling factor of the water.
    C. decreases operating pressures.
    D. increases the power consumption of the compressor.

197. The fouling factor of the water used in a water cooled condenser
    A. increases as water quantity increases.
    B. is the measurement of heat transfer characteristics of the water.
    C. decreases as water quantity decreases.
    D. can be controlled by the water regulating valve.

Section 1 System Components and Tools

198. What must be done when the water of a water cooled condenser has a high fouling factor?
   A. The water regulating valve must be turned counterclockwise.
   B. The flow rate of the water must be decreased.
   C. The flow rate of the water must be increased.
   D. The systems flow rate must not be adjusted.

199. The water of a water cooled condenser
   A. should never be discharged to storm drains or sewer drains.
   B. should never be reused.
   C. should always be reused.
   D. should be discharged to storm drains and storm sewers if local codes permit.

200. What is used to control the flow of water on a water cooled condenser?
   A. a check valve
   B. a pressure reducing valve
   C. a water regulating valve
   D. a float valve

201. Where on a water cooled condenser should a water regulating valve be located?
   A. at the inlet of the condenser
   B. at the outlet of the condenser
   C. at the inlet or the outlet of the condenser
   D. Water regulating valves are not used on water cooled condensers.

202. What two pressures control a water regulating valve?
   A. the outlet water pressure and the high-side pressure
   B. the inlet water pressure and the high-side pressure
   C. the high-side pressure and the crankcase pressure
   D. the high-side pressure and the spring pressure

203. Which of the following is the opening force on a water regulating valve?
   A. the water outlet pressure
   B. the water inlet pressure
   C. the high-side pressure
   D. the spring pressure

204. Which of the following is the closing force on a water regulating valve?
   A. the crankcase pressure
   B. the water inlet pressure
   C. the high side pressure
   D. the spring pressure

205. What is a common temperature difference between the high-side refrigerant temperature and the temperature of the leaving water in a water cooled system?
   A. 10° F
   B. 20° F
   C. 30° F
   D. 35° F

206. What are the effects of mineral and scale deposits on the inside of a water cooled condenser?
   A. It causes the high-side pressure to decrease and a decrease of water flow.
   B. It causes the high-side pressure to increase and a decrease of water flow.
   C. It causes the leaving water temperature to decrease and an increase of water flow.
   D. It causes an increase of water flow and high-side pressure.

207. Name three common water cooled condensers.
   A. a coiled tube type, a vertical tube type, and a tube within a tube type
   B. a shell and tube type, a vertical tube type, and a horizontal tube type
   C. a tube with in a tube type, a shell and coil type, and a shell and tube type
   D. a shell and coil type, a horizontal tube type, and a vertical tube type

208. The tube within a tube condenser uses
   A. the outer tube for refrigerant and the inner tube for water.
   B. the inner tube for refrigerant and the outer tube for water.
   C. the inner tube for refrigerant on coil type condensers and the inner tube for water on horizontal tube types.
   D. the inner tube for refrigerant on horizontal tube type condensers and the inner tube for water on coil tube types.

209. The counterflow principle is
   A. where the cooler water is in contact with the cooler refrigerant liquid and the warmer water is in contact with the hot gas.
   B. where the warmer water is in contact with the cooler refrigerant liquid and the cooler water is in contact with the hot gas.
   C. where the hot gas refrigerant meets the cool city water at its inlet.
   D. where the subcooled refrigerant meets the leaving water at its outlet.

210. What are two common types of tube within a tube condensers?
    A. A sealed coil type and a flanged vertical tube type.
    B. A sealed tube type and an open coil type.
    C. A flanged horizontal tube type and a sealed coil type.
    D. A sealed horizontal tube type and flanged vertical tube type.

211. What type of tube within a tube condenser must always be chemically cleaned?
    A. the sealed coil type
    B. the sealed horizontal tube type
    C. the sealed vertical tube type
    D. the open coil type

212. What type of tube within a tube condenser can be manually or chemically cleaned?
    A. the flanged coil type
    B. the flanged horizontal tube type
    C. the flanged vertical tube type
    D. the sealed coil type

213. The shell and coil type water cooled condenser
    A. has the refrigerant in the coil, and water flows over the coil inside of the shell.
    B. has subcooled refrigerant leaving the coil and superheated gas entering the coil.
    C. has the refrigerant in the shell, and water flows through the coil located inside the shell.
    D. can have refrigerant or water flowing inside of the coil.

214. The shell and coil condenser
    A. must have the coil removed in order to be properly cleaned.
    B. can be manually or chemically cleaned.
    C. cannot be chemically cleaned. It must be manually cleaned.
    D. cannot be manually cleaned. It must be chemically cleaned.

215. The shell and tube type water cooled condenser
    A. has the refrigerant in the tubes, and water flows over the tubes inside of the shell.
    B. has subcooled refrigerant leaving the tubes and superheated gas entering the tubes.
    C. has the refrigerant in the shell, and water flows through the tubes located inside the shell.
    D. can have refrigerant or water flowing inside of the tubes.

216. The shell and tube condenser
    A. must be chemically cleaned.
    B. can be manually or chemically cleaned.
    C. cannot be manually cleaned.
    D. is completely sealed with no flanges and must be chemically cleaned.

217. What is another name for the end cap of a shell and tube water cooled condenser?
    A. puddle plates
    B. water plates
    C. water box
    D. water trap

218. The shell of a shell and tube or shell and coil condenser acts as a
    A. liquid receiver.
    B. water storage tank.
    C. purge unit for non-condensables.
    D. subcool tank.

INDOOR EVAPORATOR COILS

219. What is the purpose of the evaporator coil in a split system air conditioner?
    A. It removes heat from the system.
    B. It absorbs heat into the system.
    C. It decreases air velocity.
    D. It is used to help increase the relative humidity.

220. An indoor evaporator coil must be installed
    A. at least 1 inch above the heat exchanger.
    B. at least 2 to 3 inches above the heat exchanger.
    C. at least 4 to 5 inches above the heat exchanger.
    D. at least 6 inches above the heat exchanger.

221. A direct expansion evaporator coil
    A. has the controlled medium in contact with the coil.
    B. never uses a TXV type metering device.
    C. has the controlled medium in contact with a fluid that is cooled by the coil.
    D. always uses a TXV type metering device.

222. Air conditioning evaporator coils are commonly made of
    A. stainless steel and aluminum.
    B. copper and stainless steel.
    C. galvanized steel and copper.
    D. copper and aluminum.

Section 1 System Components and Tools

223. What is the purpose of a distributor?
    A. It feeds refrigerant to the compressor from the evaporator coil.
    B. It feeds refrigerant to the evaporator coil.
    C. It feeds refrigerant to the condenser coil.
    D. It feeds refrigerant to a solenoid valve.

224. When an uncased evaporator coil is installed, metal plates are commonly installed around the coil. What is the purpose for these metal plates?
    A. to increase the quantity of air in the system
    B. to direct air flow around the coil
    C. to decrease the flow of air through the coil
    D. to force the air to flow through the coil

225. Why is it important to ensure that the evaporator coil on a split system air conditioner is installed level?
    A. to allow the proper flow of liquid refrigerant in the coil
    B. to help the system properly return the oil back to the compressor
    C. to allow for proper drainage
    D. to direct the refrigerant gas back to the compressor at the proper rate

226. Split system evaporator coils are commonly installed
    A. on the positive pressure side of the blower before the heat exchanger.
    B. on the positive pressure side of the blower after the heat exchanger.
    C. on the negative pressure side of the blower before the heat exchanger.
    D. on the negative pressure side of the blower after the heat exchanger.

227. What type of evaporator coil is commonly used in a horizontal air handling systems?
    A. an "A" coil
    B. an "N" coil
    C. an "H" coil
    D. a slant coil

228. What type of evaporator coil is commonly used in residential upflow systems?
    A. an "A" coil
    B. an "N" coil
    C. an "H" coil
    D. a slant coil

229. What type of evaporator coil is commonly used in residential upflow systems with limited plenum space?
    A. an "A" coil
    B. an "N" coil
    C. an "H" coil
    D. a slant coil

230. Rooftop package unit evaporator coils are commonly installed
    A. on the positive pressure side of the blower before the heat exchanger.
    B. on the positive pressure side of the blower after the heat exchanger.
    C. on the negative pressure side of the blower before the heat exchanger.
    D. on the negative pressure side of the blower after the heat exchanger.

231. When is a "P" trap installed?
    A. when an evaporator coil is located after a heat exchanger
    B. when the evaporator is an "A" type coil
    C. when the evaporator is an "H" type coil
    D. when an evaporator coil is located on the negative side of the blower

232. What is the purpose of a "P" trap?
    A. to prevent blockage of a condensate drain
    B. to prevent condensation from entering the duct system
    C. to aid in cleaning a condensate drain
    D. to trap debris that enters a condensate drain

METERING DEVICES

233. What condition is the refrigerant in before it enters the metering device?
    A. a high-pressure high-temperature saturated liquid
    B. a high-pressure low-temperature saturated liquid
    C. a high-pressure high-temperature subcooled liquid
    D. a low-pressure high-temperature subcooled liquid

234. What is the definition of flash gas?
    A. The rapid condensation of a refrigerant gas.
    B. The rapid evaporation of a refrigerant liquid.
    C. The rapid sublimation of a refrigerant liquid.
    D. The rapid evaporation of a refrigerant gas.

235. What component will help decrease flash gas?
    A. a heat exchanger
    B. an accumulator
    C. a filter dryer
    D. a liquid injecting valve

236. What are three common types of metering devices that are used in residential and light commercial air conditioning?
    A. TXV, AXV, Low Side Float
    B. High Side Float, Low Side Float, and Capillary Tube
    C. Low Side Float, Piston Type and AXV
    D. Capillary Tube, Piston Type and TXV

237. What controls the pressure drop of a capillary tube?
    A. the type and length of the tubing
    B. the outer diameter and the type of the tubing
    C. the length and the inner diameter of the tubing
    D. the inner diameter and the outer diameter of the tubing

238. What are the operating pressures of a TXV?
    A. spring pressure, TXV inlet pressure, and sensing bulb pressure
    B. TXV inlet pressure, TXV outlet pressure, and sensing bulb pressure
    C. spring pressure, evaporator pressure, and sensing bulb pressure
    D. spring pressure, TXV inlet pressure, and TXV outlet pressure

239. What is the purpose of the external equalizer on a TXV?
    A. to decrease the pressure drop across the TXV
    B. to increase the pressure drop across a TXV
    C. to drop the pressure across the evaporator
    D. to offset for the pressure drop across the evaporator

240. The external equalizer line should be located
    A. next to the TXV sensing bulb.
    B. 6 inches to 8 inches upstream from the TXV sensing bulb.
    C. 6 inches to 8 inches downstream from the TXV sensing bulb.
    D. None of the above.

241. If a system using a TXV has a suction line of $\frac{3}{4}"$ diameter, where should the sensing bulb be placed?
    A. at the 4 o'clock and 8 o'clock position
    B. at the 12 o'clock position
    C. at the 6 o'clock position
    D. either the 6 o'clock or 12 o'clock position

242. If a system using a TXV has a suction line of $\frac{7}{8}"$ diameter, where should the sensing bulb be placed?
    A. at the 4 o'clock and 8 o'clock position
    B. at the 12 o'clock position
    C. at the 6 o'clock position
    D. either the 6 o'clock or 12 o'clock position

243. What is the formula for calculating superheat?
    A. suction line temperature + evaporator boiling point temperature
    B. liquid line temperature − evaporator boiling point temperature
    C. suction line temperature − evaporator boiling point temperature
    D. liquid line temperature + evaporator boiling point temperature

244. How is superheat increased on a TXV system?
    A. Superheat cannot be increased.
    B. by turning the TXV valve stem counter clockwise
    C. by turning the TXV valve stem clockwise
    D. by placing the sensing bulb farther down stream from the evaporator

SERVICE VALVES

245. Where are service valves commonly located on a split air conditioning system?
    A. the evaporator coil
    B. the condensing unit
    C. the indoor air handling system
    D. the discharge and suction line

246. How will back seating a two position service valve effect a system?
    A. It will allow the system to operate normally and obtain gauge readings.
    B. It will pump a system down.
    C. It will block the gauge port.
    D. None of the above.

247. How will front seating a two position liquid line service valve effect a system?
    A. It will allow the system to operate normally and obtain gauge readings.
    B. It will pump a system down.
    C. It will block the gauge port.
    D. None of the above.

248. How can the refrigerant in a split air conditioning system be stored in the condenser?
    A. by back seating the suction line service valve
    B. by front seating the discharge line service valve
    C. by front seating the liquid line service valve
    D. by front seating the suction line service valve

249. What is the purpose of a schreader valve?
    A. It blocks the refrigerant flow through a system.
    B. It seals refrigerant in a system and connects a manifold gauge set.
    C. Being located on a refrigerant tank, it aids in charging a refrigeration system.
    D. It allows refrigerant to flow in one direction.

250. When using a three position liquid line service valve, when is the back seated position used?
    A. when the manifold gauges are disconnected
    B. when the manifold gauges are connected
    C. when the system is to be pumped down
    D. when checking the compressor valves or rings

Section 1 System Components and Tools

251. When using a three position suction line service valve, when is the mid-seated position used?
    A. when the manifold gauges are disconnected
    B. when the manifold gauges are connected
    C. when the system is to be pumped down
    D. when checking the compressor valves or rings

252. When using a three position suction line service valve, when is the front seated position used?
    A. when the manifold gauges are disconnected
    B. under normal operation
    C. when the system is to be pumped down
    D. when checking the compressor valves or rings

253. When using a three position liquid line service valve, when is the front seated position used?
    A. when the manifold gauges are disconnected
    B. under normal operation
    C. when the system is to be pumped down
    D. when checking the compressor valves or rings

254. Where is a king valve located?
    A. at the outlet of the compressor
    B. at the inlet of the compressor
    C. at the outlet of the receiver
    D. at the inlet of the receiver

255. Where is a queen valve located?
    A. at the outlet of the compressor
    B. at the inlet of the compressor
    C. at the outlet of the receiver
    D. at the inlet of the receiver

256. Where is a discharge service valve located?
    A. at the outlet of the compressor
    B. at the inlet of the compressor
    C. at the outlet of the receiver
    D. at the inlet of the receiver

257. Where is a suction service valve located?
    A. at the outlet of the compressor
    B. at the inlet of the compressor
    C. at the outlet of the receiver
    D. at the inlet of the receiver

LEAK DETECTION

258. What is an indication of a refrigerant leak in an air conditioning system?
    A. an oil trace around copper fittings
    B. excessively high discharge pressures
    C. high discharge line temperatures
    D. low suction line temperatures

259. What is commonly used to pressurize an air conditioning system for the purpose of leak detection?
    A. air
    B. R-22 with a trace amount of nitrogen
    C. R-12 with a trace amount of nitrogen
    D. oxygen with a trace amount of nitrogen

260. When testing for a leak in an air conditioning system,
    A. the system must be pressurized and electrically de-energized.
    B. the system must be in full operation except for the condenser fan.
    C. the system must be in full operation except for the indoor blower.
    D. only the compressor must be energized.

261. If air is trapped in an air conditioning system it has tendency to stay in the
    A. evaporator.
    B. compressor.
    C. condenser.
    D. metering device.

262. When pressurizing an air conditioning system for the purpose of leak detection, the pressure in the system must never exceed
    A. the high-side test pressure rating.
    B. the low-side test pressure rating.
    C. 500 PSIG.
    D. 500 PSIA.

263. When pressurizing a system during a leak test, what must never be used?
    A. carbon dioxide
    B. nitrogen
    C. R-22
    D. oxygen

264. What must be used when pressure testing a system with nitrogen?
    A. a vacuum pump and charging cylinder
    B. a recovery system
    C. a pressure regulator and an in-line pressure relief valve
    D. No accessories are needed to use nitrogen on a system.

265. How long must an air conditioning system operate after adding a refrigerant dye for leak detection?
    A. 1 hr.
    B. 10 hrs.
    C. 15 hrs.
    D. 24 hrs.

266. Which leak detector is the most accurate and is used on systems with extremely small R-22 refrigerant leaks?
    A. halide leak detector
    B. soap bubbles
    C. smoke test method
    D. electronic leak detector

267. When using an electronic leak detector
    A. the probe must be held below the tubing being tested.
    B. the probe must be held above the tubing being tested.
    C. the probe must be held pointing in the down position for an accurate reading.
    D. the probe must be held pointing in the *up* position for an accurate reading.

268. When is it inadvisable to use a halide leak detector?
    A. on a system that uses halogenated fluorocarbons
    B. on an inoperative system
    C. near flammable materials
    D. on a system that contains 10 pounds of refrigerant or more

269. When using a halide leak detector, what color will the flame be when it is in the presence of a small refrigerant leak?
    A. blue
    B. green
    C. red
    D. purple

270. An ultrasonic leak detector senses a leak by
    A. radar.
    B. excessive refrigerant concentration.
    C. sound waves.
    D. a chemical reaction.

271. An ultrasonic leak detector senses a leak on a system that
    A. is under a positive pressure only.
    B. is under a negative pressure only.
    C. is under a positive or a negative pressure.
    D. uses CFC refrigerants only.

272. What is commonly added to a system when using an ultraviolet light leak detector?
    A. a UV additive
    B. nitrogen
    C. a red dye additive
    D. a fluorescent additive

RECOVERY EQUIPMENT

273. The term recovery means
    A. to remove and store refrigerant from an appliance.
    B. to remove and clean refrigerant from an appliance.
    C. to remove, clean, and store refrigerant from an appliance.
    D. to catch refrigerant in the air as it is being released.

274. What are the three types of recovery methods that are being used today?
    A. charge migration, use of the appliance compressor, and use of a refrigerant recovery unit
    B. use of appliance compressor, use of refrigerant recovery unit, and vacuum pump procedure
    C. use of refrigerant recovery unit, charge migration, and liquid recovery
    D. vapor recovery, liquid recovery, and use of appliance compressor

275. What are the three types of recovery machines that are being used today?
    A. vapor/liquid recovery units, 10 CFM systems, and 15 CFM systems
    B. 5 CFM systems, 10 CFM systems, and 15 CFM systems
    C. vapor recovery units, liquid recovery units, and vapor/liquid recovery units
    D. liquid recovery units, vapor recovery units, and 10 CFM systems

276. What are the three classifications of recovery equipment?
    A. very high-pressure, high-pressure, and medium-pressure equipment
    B. high-pressure, medium-pressure, and low-pressure equipment
    C. very low-pressure, low-pressure, and high-pressure equipment
    D. very high-pressure, high-pressure, and low-pressure equipment

277. What type of equipment uses refrigerant that boils below -58° F at atmospheric pressure?
    A. very high-pressure
    B. high-pressure
    C. medium-pressure
    D. low-pressure

Section 1   System Components and Tools   113

278. What type of equipment uses refrigerant that boils between -58° and 50° F at atmospheric pressure?
   A. very high-pressure
   B. high-pressure
   C. medium-pressure
   D. low-pressure

279. What type of equipment uses refrigerant that boils above 50° F at atmospheric pressure?
   A. very high-pressure
   B. high-pressure
   C. medium-pressure
   D. low-pressure

280. The term recycle means
   A. to remove and store refrigerant from an appliance.
   B. to remove and clean refrigerant from an appliance.
   C. to remove and reuse refrigerant in another appliance.
   D. to remove and take to a reclaim center.

281. After November 11, 1993, when recovering R-22 that is more than 200 pounds,
   A. the system must be taken down to 0 PSIG.
   B. the system must be taken down to 4 PSIG.
   C. the system must be taken down to 4" Hg.
   D. the system must be taken down to 10" Hg.

282. After November 15, 1993, when recovering R-22 that is less than 200 pounds,
   A. the system must be taken down to 0 PSIG.
   B. the system must be taken down to 4 PSIG.
   C. the system must be taken down to 4" Hg.
   D. the system must be taken down to 10" Hg.

283. Recovery cylinders must be approved by
   A. the Environmental Protection Agency.
   B. the American Society of Heating, Refrigeration, and Air Conditioning Engineers.
   C. the Cylinder Inspectors of America.
   D. the Department of Transportation.

284. A recovery cylinder must not be filled to more than
   A. 60% of its capacity.
   B. 70% of its capacity.
   C. 80% of its capacity.
   D. 90% of its capacity.

285. What component that is commonly used in a recycling machine is not found in a recovery machine?
   A. a filter
   B. an oil separator
   C. an expansion tank
   D. an automatic shut off system

SOLDERING AND BRAZING

286. A copper joint must never be soldered or brazed
   A. when nitrogen is present in the tubing.
   B. when the system is opened to the atmosphere.
   C. when the system is under pressure.
   D. when air is present in the system.

287. At what temperature does Stay Brite #8 solder melt?
   A. 360° F
   B. 535° F
   C. 635° F
   D. 660° F

288. At what temperature does 95-5 solder melt?
   A. 150° F
   B. 250° F
   C. 450° F
   D. 650° F

289. Using flux while soldering will help
   A. keep the joint that is being soldered clean.
   B. retain heat on the joint that is being soldered.
   C. stop oxides from forming.
   D. absorb the solder into the joint.

290. What are the common melting points of brazing material and silver solder?
   A. 500° to 800° F
   B. 1100° to 1480° F
   C. 2000° to 2500° F
   D. 2500° to 3000° F

291. Leaks in coils made from aluminum are commonly repaired with
   A. silver solder.
   B. silver brazing material.
   C. brass brazing material.
   D. epoxy.

292. The term capillary action means to
   A. spray water on a joint after it has been soldered.
   B. expand a joint as it is being heated by a torch.
   C. draw solder into a joint as it is being heated.
   D. add flux to the male part of a joint that is to be soldered.

293. The gauges that are located on the oxyacetylene torch measure
   A. the tank pressure (in PSIG) and the regulated pressure (in PSIG) to the torch tip.
   B. the tank pressure (in PSIA) and the regulated pressure (in PSIA) to the torch tip.
   C. the tank weight (in lbs.) and the regulated pressure (in PSIG) to the torch tip.
   D. the volumetric pressure (in PSIG) and the regulated pressure (in PSIA) to the torch tip.

294. The operating pressure of an acetylene regulator ranges from
    A. 0 to 15 PSIG.
    B. 0 to 20 PSIG.
    C. 0 to 40 PSIG.
    D. 0 to 50 PSIG.

295. The operating pressure of an oxygen regulator ranges from
    A. 0 to 20 PSIG.
    B. 0 to 40 PSIG.
    C. 0 to 50 PSIG.
    D. 0 to 100 PSIG.

296. When adjusting pressure on a regulator,
    A. turn the regulator clockwise to increase pressure for oxygen and counter clockwise to increase pressure for acetylene.
    B. turn the regulator counter clockwise to increase pressure for oxygen and clockwise to increase pressure for acetylene.
    C. turn the regulator clockwise to increase pressure for oxygen and clockwise to increase pressure for acetylene.
    D. turn the regulator counter clockwise to increase pressure for oxygen and counter clockwise to increase pressure for acetylene.

297. What should be used to purge an air conditioning system's tubing while brazing?
    A. air with a regulator
    B. air
    C. nitrogen with a regulator
    D. nitrogen

298. What is the purpose for purging an air conditioning system while brazing?
    A. to keep moisture out of the system
    B. to prevent oxidation
    C. to draw the heat from the brazed joint
    D. to aid the chemical reaction of the brazing compound

299. What could cause an explosion if it is used to purge an air conditioning system while brazing?
    A. oxygen
    B. air
    C. nitrogen
    D. carbon dioxide

300. The purpose for a reaming tool is to
    A. prepare the copper tubing before cutting.
    B. expand the copper tubing after cutting.
    C. smooth the outer edge of the tubing for a tighter fitting joint.
    D. remove the burrs from the tubing after cutting.

301. What is used to expand the diameter of soft copper tubing?
    A. a swaging tool
    B. a flaring block
    C. a flaring tool
    D. an expansion joint

302. What would be the minimum bending radius on $\frac{1}{2}"$ soft copper tubing?
    A. 1 inch
    B. $1\frac{1}{2}$ inches
    C. 2 inches
    D. $2\frac{1}{2}$ inches

303. When using a spring type tube bender,
    A. it must only be placed on the inside of the tubing.
    B. it must only be placed on the outside of the tubing.
    C. it can be placed on the inside or on the outside of the tubing.
    D. it must be oiled before each use.

304. What is the maximum bend that a lever bender can make?
    A. 45 degrees
    B. 90 degrees
    C. 160 degrees
    D. 180 degrees

305. Which tool holds the copper tubing in place when making a mechanical joint?
    A. the flaring tool
    B. the flaring block
    C. the swaging tool
    D. the reaming tool

306. When making a flare, what can be done to prevent the copper from splitting?
    A. cool the copper to 40° F
    B. heat the copper to 105° F
    C. tighten and loosen the flaring tool during the procedure
    D. quickly turn the flaring tool as the copper expands

307. Which material will create a stronger joint?
    A. brazing material
    B. $\frac{50}{50}$ solder
    C. silver solder
    D. 45% silver

Section 1 System Components and Tools

MANIFOLD GAUGES

308. How is a temperature measurement obtained on a manifold gauge set?
    A. by adding 14.7 to the pressure and comparing it to gauge temperature
    B. by subtracting 14.7 from the pressure and comparing it to gauge temperature
    C. by taking the temperature directly from the inner colored numbers
    D. by taking the temperature directly from the outer black numbers

309. What would be the proper procedure for testing pressures on an air conditioning system?
    A. connect the blue hose to the smaller line, the red hose to the larger line, and close the gauge valves
    B. connect the blue hose to the larger line, the red hose to the smaller line, and open the gauge valves
    C. connect the blue hose to the smaller line, the red hose to the larger line, and open the gauge valves
    D. connect the blue hose to the larger line, the red hose to the smaller line, and close the gauge valves

310. A manifold gauge set is used to check operating pressures in
    A. PSIA.
    B. PSIG.
    C. Inches WC.
    D. PSIG and Inches WC.

311. To convert Pounds per Square Inch Gauge (PSIG) to Pounds per Square Inch Absolute (PSIA),
    A. add 14.7 inches of WC.
    B. add 14.7 PSIG.
    C. subtract 14.7 PSIA.
    D. subtract 14.7 inches WC.

312. What is the purpose of the center hose on a manifold gauge set?
    A. It is used to measure the pressure difference in a system.
    B. It is used to connect to a heat pump system for reversing pressures.
    C. It is used to charge a system.
    D. It is used to measure the pressure on a suction line filter dryer.

313. On a manifold gauge set, the low-side pressure gauge measures
    A. 0 to 300 PSIG.
    B. 0 to 250 PSIG.
    C. 30" Hg. to 350 PSIG.
    D. 30" Hg. to 500 PSIG.

314. On a manifold gauge set, the high-side pressures gauge measures
    A. 0 to 500 PSIG.
    B. 0 to 550 PSIG.
    C. 30" Hg. to 700 PSIG.
    D. 30" Hg. to 700 PSIG.

315. What is the procedure for adjusting a manifold gauge set?
    A. Connect the gauges to a R-12 refrigerant tank and adjust it to 70 PSIG.
    B. After connecting the gauges to a system, adjust each gauge to 32° F.
    C. Close gauges from the atmosphere and adjust to zero.
    D. Open the gauges to the atmosphere and adjust to zero.

316. What is commonly used with refrigerant gauges to prevent excessive refrigerant loss?
    A. low loss fittings
    B. service port caps
    C. refrigerant line purge device
    D. pressure reducing valve

317. What can cause excessively high discharge pressures on an air conditioning system?
    A. insufficient air flow over the A-coil
    B. non-condensables in the system
    C. restricted metering device
    D. a kink in the suction line

318. What can cause a decrease in suction pressure on an air conditioning system?
    A. a dirty condenser coil
    B. too large of a TXV
    C. a dirty evaporator coil
    D. re-circulated air over the condenser

SYSTEM EVACUATION

319. What is the purpose of evacuating an air conditioning system?
    A. to remove air and moisture from the system
    B. to de-pressurize the system for the expansion of refrigerant
    C. to draw oil from the compressor during the dehydration process
    D. to increase the rate of refrigerant flow through the expansion valve

320. When must a system be evacuated?
    A. when the system has dirty evaporator and condenser coils
    B. when the system has excessively high discharge pressures
    C. when the system has been opened to the atmosphere
    D. as a routine maintenance to ensure no moisture is present in the system

321. How often should the oil in a vacuum pump be replaced?
    A. after every other evacuation
    B. after every evacuation
    C. after it has operated for 20 hours
    D. after it has operated for 10 hours

322. Why must the oil of a vacuum pump be replaced?
    A. The vacuum in the pump causes the oil to become thinner which will cause excessive wear.
    B. The refrigerant's pressure causes a break down of the vacuum pump oil.
    C. The oil becomes full of moisture and acids which will decrease the capability of the vacuum pump.
    D. The high speed of the vacuum pump will cause a break down of vacuum pump oil.

323. What is commonly used to measure a deep vacuum?
    A. a mercury column located on the low pressure manifold gauge
    B. a micrometer gauge
    C. a micron gauge
    D. a manometer

324. What will decrease the evacuation time of an air conditioning system?
    A. the removal of a schrader valve core, smaller diameter hoses, and higher ambient temperatures
    B. the removal of a schrader valve core, larger diameter hoses, and lower ambient temperatures
    C. the removal of a schrader valve core, smaller diameter hoses, and lower ambient temperatures
    D. the removal of a schrader valve core, larger diameter hoses, and higher ambient temperatures

325. When a system incorporates a three way service valve,
    A. it must be in the mid position during the evacuation.
    B. it must be in the back seated position during evacuation.
    C. it must be in the front seated position during evacuation.
    D. All of the above.

326. During a system evacuation, always
    A. open the high-side gauge and close the low-side gauge.
    B. open the low-side gauge and close the high-side gauge.
    C. open both the high-side and the low-side gauge.
    D. close both the high-side and the low-side gauge.

327. A minimum evacuation level for an air conditioning system is
    A. 5000 microns.
    B. 1000 microns.
    C. 500 microns.
    D. 100 microns.

328. A triple evacuation is done by evacuating a system to
    A. 25" Hg. and charging the system to 50 PSIG three times.
    B. 28" Hg. and charging the system to 0 PSIG three times.
    C. 25" Hg. and charging the system to 0 PSIG three times.
    D. 28" Hg. and charging the system to 50 PSIG three times.

Section 1 System Components and Tools

329. After evacuating a system, the manifold gauges are closed, then the vacuum pump is turned off. Using a micron gauge, the system is drawn down to 500 microns, the system is left off, and the pressure increases slightly then stops. What is the problem?
   A. the gauges are leaking
   B. moisture is in the system
   C. expansion of oil
   D. a systems leak is present

330. When using a micron gauge, what is the proper level of vacuum that must be drawn on a system?
   A. 5000 microns
   B. 500 microns
   C. 1000 microns
   D. 3000 microns

REFRIGERANT CHARGING

331. When an electronic scale is being used to charge a system, a button must be pushed after turning the scale on. What is the button used for?
   A. to set the scale to zero
   B. to change the scale from pounds to ounces
   C. to adjust for refrigerant pressure
   D. to adjust for refrigerant type

332. What is a charging cylinder used for?
   A. to charge a system by weight
   B. to charge a system by quantity
   C. to charge a system by sight
   D. to charge a system by temperature

333. What component can aid a technician while charging an air conditioning system?
   A. a filter dryer
   B. a sight glass
   C. a TXV
   D. a receiver

334. When liquid charging a system,
   A. the system must be de-energized and have refrigerant added into the low side of the system through the low side manifold gauge.
   B. the system must be in operation and have refrigerant added into the high side of the system through the high side manifold gauge.
   C. the system must be de-energized and have refrigerant added into the high side of the system through the high side manifold gauge.
   D. the system must be in operation and have refrigerant added into the low side of the system through the low side manifold gauge.

335. When gas charging a system,
   A. the system must be de-energized and have refrigerant added into the low side of the system through the low side manifold gauge.
   B. the system must be in operation and have refrigerant added into the high side of the system through the high side manifold gauge.
   C. the system must be de-energized and have refrigerant added into the high side of the system through the high side manifold gauge.
   D. the system must be in operation and have refrigerant added into the low side of the system through the low side manifold gauge.

336. How will an overcharge effect a capillary tube system?
   A. It will have a decrease of superheat.
   B. It will have an increase of superheat.
   C. It will not effect the superheat.
   D. It will cause a starved evaporator coil.

337. What type of air conditioning system must be charged using the superheat method?
   A. a TXV system
   B. an AXV system
   C. a piston type metering device
   D. a pitot-tube system

338. When charging an air conditioning system using the superheat method, what factors need to be known?
   A. The indoor wet-bulb temperature and the outdoor dry-bulb temperature.
   B. The indoor dry-bulb temperature and the outdoor wet-bulb temperature.
   C. The indoor dry-bulb temperature and the outdoor relative humidity.
   D. The indoor wet-bulb temperature and the outdoor relative humidity.

339. To measure superheat,
   A. subtract the high side condensing temperature from the suction line temperature.
   B. subtract the suction line temperature from the ambient temperature.
   C. subtract the low side boiling point temperature from the suction line temperature.
   D. subtract the ambient temperature from the suction line temperature.

340. When charging a system using the subcool method, what type of metering device must be used on the system?
   A. a capillary tube
   B. a piston
   C. an AXV
   D. a TXV

341. To measure subcool,
    A. subtract the liquid line temperature from the high side condensing temperature.
    B. subtract the high side condensing temperature from the liquid line temperature.
    C. subtract the ambient temperature from the high side condensing temperature.
    D. subtract the ambient temperature from the liquid line temperature.

DEFROST SYSTEMS

342. What type of evaporator coil is commonly used with a manual shut down type of defrost system?
    A. spiral type
    B. finned type
    C. plate type
    D. finned and plate type

343. When using a manual shutdown defrost system, the evaporator coil usually operates
    A. at 34° to 35° F.
    B. below 32° F.
    C. at 40° to 45° F.
    D. at 50° F.

344. What control is commonly used to initiate a defrost on a pressure operated defrost system?
    A. a low pressure control
    B. a high pressure control
    C. a reverse acting pressure control
    D. a temperature pressure control

345. When adjusting a low pressure switch that is used to control the frost build up on an evaporator coil,
    A. adjust the control to cut-in at a minimum of 33° F saturation temperature.
    B. adjust the control to cut-in at a maximum of 33° F saturation temperature.
    C. adjust the control to cut-in at 20° F saturation temperature.
    D. adjust the control to cut-out at 32° F saturation temperature.

346. The typical storage temperature of a system that uses a pressure operated defrost system is
    A. 0° F.
    B. 10° F.
    C. 25° F.
    D. above 35° F.

347. What is the method of defrost when using a pressure operated defrost system?
    A. an electric heater
    B. a hot gas solenoid
    C. warm water
    D. evaporator fans

348. What is the length of the defrost period when using a pressure operated defrost system?
    A. 20 minutes
    B. 30 minutes
    C. 40 minutes
    D. The length is determined by the cut-in and cut-out settings of the control.

349. When using a time shutdown defrost, what is commonly used to start and stop the defrost cycle?
    A. a temperature control
    B. trip pins
    C. a pressure control
    D. an electric heater

350. What type of switch contacts are located in the defrost timer and are used to control the condensing unit?
    A. a "normally closed" pressure type
    B. a "normally opened" pressure type
    C. a "normally closed" cam operated type
    D. a "normally opened" cam operated type

351. What should the length of time be when adjusting the stop and start times of a time shutdown defrost system?
    A. always use the manufacturer's recommendation
    B. always set it for 20 minutes
    C. always set it for 30 minutes
    D. always set it for 40 minutes

352. What are three common supplementary heat methods used to aid in the removal of frost on an evaporator coil?
    A. electric, radiant, and warm air convection
    B. hot gas, electric, and radiant
    C. warm water, electric, and warm air convection
    D. warm water, electric, and hot gas

353. When supplementary heat is used to aid frost removal during a defrost cycle, the
    A. products are commonly kept below 40° F.
    B. products are commonly kept above 32° F.
    C. products are commonly kept below 29° F.
    D. None of the above.

Section 1 System Components and Tools

354. On an electric defrost system using a 24-hour defrost timer, what do the normally open contacts control?
    A. the defrost heater
    B. the evaporator fan
    C. the timer motor
    D. the condensing unit

355. On a hot gas defrost system using a 24-hour defrost timer, what do the normally closed contacts control?
    A. the hot gas solenoid
    B. the evaporator fan
    C. the timer motor
    D. the condensing unit

356. When using a 24-hour defrost timer that is used on systems with supplementary heat, the
    A. trip pins are used to start the defrost and an adjustable setting is used to end the defrost.
    B. trip pins are used to end the defrost and an adjustable setting is used to start the defrost.
    C. trip pins are used to start and end the defrost.
    D. adjustable settings are used to start and end the defrost without the use of trip pins.

357. What is the maximum amount of trip pins that can be installed in a 24-hour defrost timer at one time?
    A. 12
    B. 24
    C. 36
    D. 48

358. On a 24-hour timer, the length of the defrost cycle ranges from
    A. 20 to 40 minutes.
    B. 30 to 40 minutes.
    C. 10 to 90 minutes.
    D. 2 to 110 minutes.

359. During an electric, hot gas, or water defrost, what load must always be de-energized?
    A. the compressor
    B. the defrost timer
    C. the evaporator fan
    D. None of the above.

360. On a hot gas defrost system, where does the defrost line connect?
    A. from the outlet of the condenser to the outlet of the evaporator
    B. from the inlet of the condenser to the outlet of the evaporator
    C. from the outlet of the condenser to the inlet of the evaporator
    D. from the inlet of the condenser to the inlet of the evaporator

361. What component is commonly added to a hot gas defrost system that is not on an electric or warm water defrost system?
    A. a reversing valve
    B. a drain valve
    C. a crankcase pressure regulator
    D. an accumulator

362. When using a time initiated temperature terminated defrost system what control is commonly used to de-energize the defrost circuit?
    A. a solenoid valve
    B. a switching relay
    C. a reverse acting high-pressure control
    D. a termination thermostat

363. When a defrost system incorporates a termination thermostat, what temperature commonly controls its operation?
    A. the evaporator coil temperature
    B. the supply air temperature of the evaporator
    C. the suction line temperature
    D. the return air temperature of the evaporator

364. What are two common types of termination thermostats that are used with a time initiated temperature terminated defrost system?
    A. single pole double throw and double pole single throw types of controls
    B. single pole single throw and single pole double throw types of controls
    C. single pole single throw and double pole single throw types of controls
    D. double pole single throw and single pole double throw types of controls

365. When a termination thermostat is used with a time initiated temperature terminated defrost system, what is commonly used to override the timer motor?
    A. a pressure control
    B. a holding relay
    C. a timer release solenoid
    D. an override relay

366. When using a single pole double throw termination thermostat, what loads are controlled by the switch contacts?
    A. The "close on rise" contacts control the evaporator fan and the "open on rise" contacts control the compressor.
    B. The "open on rise" contacts control the compressor and the "close on rise" contacts control the timer release solenoid.
    C. The "open on rise" contacts control the defrost heater and the "close on rise" contacts control the defrost timer.
    D. The "close on rise" contacts control the timer release solenoid and the "open on rise" contacts control the evaporator fan.

367. On a time initiated temperature terminated defrost system, what would be the action of the system that is set for a 35 minute defrost but the coil is defrosted after 20 minutes?
    A. The system is switched into the cooling mode after 20 minutes and the evaporator fan is delayed for a short period of time.
    B. The system is switched into the cooling mode after 35 minutes energizing the compressor, condenser fan, and evaporator fan.
    C. The system is switched into the cooling mode after 20 minutes, energizing the compressor, condenser fan, and evaporator fan.
    D. The system is switched into the cooling mode after 35 minutes and the evaporator fan is delayed for a short period of time.

FUNDAMENTALS OF GAS COMBUSTION

368. What fuel is heavier then air?
    A. natural gas
    B. L.P. gas
    C. ethane
    D. manufactured gas

369. What is the heating value of natural gas?
    A. 800 Btu/cu. ft.
    B. 1000 Btu/cu. ft.
    C. 1500 Btu/cu. ft.
    D. 2500 Btu/cu. ft.

370. What is the heating value of L.P. gas?
    A. 800 Btu/cu. ft.
    B. 1000 Btu/cu. ft.
    C. 1500 Btu/cu. ft.
    D. 2500 Btu/cu. ft.

371. What are the three requirements needed for combustion?
    A. carbon, oxygen, and ignition
    B. air, carbon, and fuel
    C. fuel, carbon, and ignition
    D. oxygen, fuel, and ignition

372. Complete combustion is accomplished when
    A. one part methane is combined with two parts oxygen.
    B. two parts methane are combined with one part oxygen.
    C. one part carbon is combined with two parts methane.
    D. two parts carbon is combined with one part methane.

373. What are two products of complete combustion?
    A. CO and H_2O
    B. CO_2 and H_2O
    C. CH_4 and O_2
    D. CH_4 and H_2O

374. Air is approximately
    A. 21% oxygen and 79% nitrogen.
    B. 51% oxygen and 49% nitrogen.
    C. 70% oxygen and 21% nitrogen.
    D. 49% oxygen and 51% nitrogen.

375. The proper air and fuel mixture for complete combustion is
    A. 1 part fuel and 1 part air.
    B. 1 part fuel and 4 parts air.
    C. 1 part fuel and 8 parts air.
    D. 1 part fuel and 10 parts air.

376. To ensure complete combustion
    A. an additional 2 parts of excess air is added to the fuel air mixture.
    B. an additional 5 parts of excess air is added to the fuel air mixture.
    C. an additional 10 parts of excess air is added to the fuel air mixture.
    D. an additional 15 parts of excess air is added to the fuel air mixture.

Section 1 System Components and Tools 121

377. What can cause carbon monoxide during the combustion process?
    A. incomplete combustion and flame impingement on the cold surface of a heat exchanger
    B. not enough fuel mixture and leaks in the vent pipe
    C. too much primary air and incomplete combustion
    D. too much primary and secondary air

378. Natural gas is also known as
    A. butane.
    B. propane.
    C. methane.
    D. None of the above.

379. What are three methods of heat transfer?
    A. evaporation, radiation, and convection
    B. convection, air motion, and conduction
    C. conduction, convection, and radiation
    D. condensation, convection, and conduction

GAS FURNACE CONTROLS AND COMPONENTS

380. What is the purpose of a pilot?
    A. It ignites the main burners.
    B. It proves the existence of a flame.
    C. It sends a signal to the gas valve.
    D. It opens a pilot valve.

381. What type of pilot sensor is used on a standing pilot furnace?
    A. a mercury flame sensor
    B. a flame rod
    C. a hot surface ignitor
    D. a thermocouple

382. What is the purpose of a thermocouple?
    A. It energizes the system using 750 mv (millivolts).
    B. It ignites the main burner.
    C. It proves the existence of a pilot flame.
    D. It ignites the pilot.

383. How far must the tip of a thermocouple be placed in the pilot flame?
    A. $\frac{1}{8}"$ to $\frac{1}{4}"$
    B. $\frac{3}{8}"$ to $\frac{1}{2}"$
    C. $\frac{3}{4}"$ to $1"$
    D. $1\frac{1}{2}"$

384. How many millivolts does a thermocouple produce?
    A. 750
    B. 550
    C. 30
    D. 15

385. How many millivolts does a power-pile produce?
    A. 750
    B. 350
    C. 150
    D. 30

386. What three components make up a spark ignition system?
    A. a cable, an electrode, and a sensor
    B. an ignition transformer, a cable, and an electrode
    C. an electrode, an ignition transformer, and a sensor
    D. a cable, a sensor, and an ignition transformer

387. In a furnace that uses an intermittent pilot,
    A. the pilot is continually lit.
    B. the pilot is lit with each cycle of the furnace.
    C. is considered a standing pilot system.
    D. often uses a thermocouple.

388. An intermittent pilot system uses
    A. a spark ignitor to light the pilot.
    B. a hot surface ignitor to light the pilot.
    C. a match to light the pilot.
    D. a spark or hot surface ignitor to light the pilot.

389. What is another name for a hot surface ignitor?
    A. a cadmium ignitor
    B. a silicon carbide ignitor
    C. a tungsten ignitor
    D. Both A and B.

390. Which of the following would be the quietest ignition system?
    A. an indirect spark ignition
    B. a direct spark ignition
    C. a hot surface ignitor
    D. an intermittent spark ignitor

391. The voltage of a hot surface ignitor can be
    A. 120 volts.
    B. 50 volts.
    C. 24 volts.
    D. Both A and C.

392. What type of pilot is commonly used on a direct ignition system?
    A. a non-aerated pilot only
    B. an aerated pilot only
    C. an aerated or non-aerated pilot
    D. Pilots are not used on direct ignition systems.

393. What is the setting of the spark gap on a spark ignition system?
   A. $\frac{1}{8}"$
   B. $\frac{1}{4}"$
   C. $\frac{3}{8}"$
   D. $\frac{1}{2}"$

394. How far must the tip of a flame ignitor/sensor be placed from the main burner or pilot flame?
   A. $\frac{1}{8}"$ to $\frac{1}{4}"$
   B. $\frac{2}{8}"$ to $\frac{1}{2}"$
   C. $\frac{3}{4}"$ to $1"$
   D. $1\frac{1}{2}"$

395. The voltage created by an ignition control module is commonly in the range of
   A. 2,000 to 3,000 volts.
   B. 7,000 to 15,000 volts.
   C. 20,000 to 50,000 volts.
   D. None of the above.

396. What are four common furnace designs?
   A. downflow, lowboy, highboy, and multi-positional
   B. upflow, lowboy, downflow, and multi-positional
   C. horizontal, lowboy, highboy, and multi-positional
   D. upflow, downflow, horizontal, and multi-positional

397. Which of the following are not located on a combination gas valve?
   A. a pilot solenoid and a main valve solenoid
   B. a pressure regulator and a pilot solenoid
   C. a main shutoff and a pilot adjustment
   D. a spark ignitor and a sensor wire

398. What is the purpose of the pressure regulator that is located on a combination gas valve?
   A. It takes an incoming gas pressure of 3.5" WC and increases it to 7" WC.
   B. It takes an incoming gas pressure of 7" WC and reduces it to 3.5" WC.
   C. It increases the pressure when the screw is turned clockwise.
   D. Both B and C.

399. What component is located in the gas manifold that is used to distribute gas to each burner?
   A. a distributor
   B. an equalizer
   C. an orifice
   D. None of the above.

400. What is the purpose of primary air?
   A. It is air needed for proper combustion. It mixes with the main burner fuel at the point of ignition.
   B. It is air needed for proper combustion. It mixes with the main burner fuel before the point of ignition.
   C. It is air needed for proper combustion. It mixes with the main burner fuel after the point of ignition.
   D. Both A and B.

401. What is secondary air?
   A. It is air that mixes with the main burner fuel at the point of ignition. This ensures proper combustion.
   B. It is air that mixes with the main burner fuel before the point of ignition. This ensures proper combustion.
   C. It is air that mixes with the main burner fuel after the point of ignition. This ensures proper combustion.
   D. Both A and B.

402. In what part of the burner does air and fuel mix before combustion?
   A. the cross over section
   B. the venturi
   C. the orifice
   D. the outlet

403. If gas burners are not installed properly, what problem could this cause?
   A. erratic limit cycles
   B. improper heat exchanger contraction
   C. noisy ignition
   D. decreased blower operation

404. Which of the following are not considered a type of gas burner design?
   A. blast tube
   B. slotted
   C. ribbon
   D. inshot

405. Which part of a gas furnace allows the air within the furnace location to blend with the products of combustion?
   A. the draft diverter
   B. the vent connector
   C. the barometric damper
   D. the flue pipe

Section 1 System Components and Tools

406. The purpose of a fan control is to
  A. start the blower.
  B. stop the blower.
  C. start and stop the blower.
  D. start the inducer motor.

407. What type of fan control operates the blower by temperature only?
  A. a bimetal fan control
  B. a timed delay fan control
  C. a solid state fan control
  D. Both A and B.

408. What type of fan control uses an electric heater as a means of operating the blower motor?
  A. a bimetal fan control
  B. a timed delay fan control
  C. a solid state fan control
  D. Both A and B.

409. A pressure switch on a non-condensing furnace is used to
  A. prove the operation of the indoor blower.
  B. prove the operation of the induced draft blower.
  C. sense the gas pressure to the gas valve.
  D. sense the gas pressure to the pilot valve.

410. A limit control is designed to
  A. cycle the system on and off depending on space temperature.
  B. operate the blower when the thermostat calls for heat.
  C. de-energize the gas valve if the furnace overheats.
  D. limit the operation of the induced draft motor.

411. Which control is commonly found on downflow and multi-positional furnaces, but not normally used on upflow furnaces?
  A. an air pressure control
  B. a primary limit control
  C. a pilot safety control
  D. an auxiliary limit control

412. What control is commonly used on high efficiency, non-condensing furnaces to protect against high stack temperatures?
  A. a limit control
  B. a flame roll out control
  C. a vent safety control
  D. an auxiliary limit control

413. What control is commonly used on high efficiency, non-condensing furnaces to protect against a plugged heat exchanger?
  A. a limit control
  B. a flame roll out control
  C. a vent safety control
  D. an auxiliary limit control

414. The part of the system that contains the products of combustion primarily for a thermal transfer is the
  A. combustion blower.
  B. heat exchanger.
  C. burner.
  D. accumulator.

415. Two common methods used to seal a sectional heat exchanger are
  A. mechanical crimp and iron banding.
  B. welded seam and Pittsburgh seam.
  C. iron banding and Pittsburgh seam.
  D. welded seam and mechanical crimp.

416. A bimetal flame sensor
  A. proves the existence of a pilot flame.
  B. proves the existence of the main gas flame.
  C. uses a normally open set of contacts to control the pilot valve.
  D. uses a normally closed set of contacts to control the main valve.

417. A flame rod sensor
  A. proves the presence of a pilot flame only.
  B. proves the presence of a main burner flame only.
  C. proves the presence of a pilot or main burner flame.
  D. proves the presence of the spark from an ignition system.

418. A normally open float switch should be wired to control
  A. the blower motor.
  B. the contactor coil.
  C. the condensate pump.
  D. the inducer motor.

419. An induced draft blower used in a non-condensing gas furnace
  A. produces a positive pressure in the vent system when properly installed.
  B. produces a positive pressure in the heat exchanger.
  C. pulls the products of combustion from the heat exchanger and pushes them into the venting system.
  D. pushes the products of combustion into the heat exchanger then into the venting system.

420. In what type of furnace is a secondary heat exchanger used?
    A. a high efficiency non-condensing type
    B. a high efficiency condensing type
    C. a standing pilot system
    D. a mid efficiency non-condensing type

421. What is the purpose of the secondary heat exchanger?
    A. to increase the temperature of the products of combustion for added efficiency
    B. to increase the quantity of the products of combustion for added efficiency
    C. to increase the velocity of the products of combustion for added efficiency
    D. to condense the products of combustion for added efficiency

422. Which control is used to prevent the products of combustion from mixing with supply air?
    A. the limit control
    B. the auxiliary limit control
    C. the pressure switch
    D. the interlock switch

423. What control is used to start the operation of the induced draft blower of a high efficiency non-condensing furnace?
    A. the pressure switch
    B. the auxiliary switch
    C. the thermostat
    D. the end switch

424. What control is used to energize the ignition control on the start up of a high efficiency non-condensing furnace?
    A. the end switch
    B. the auxiliary limit control
    C. the pressure switch
    D. the pilot safety switch

COMBUSTION AIR

425. Dilution air is
    A. air that mixes with the fuel for combustion.
    B. air that is brought into a return air duct system.
    C. air that mixes with combustion gases in a natural draft furnace.
    D. air that mixes with combustion gases in an induced draft non-condensing furnace.

426. Infiltration is
    A. air that is never allowed to mix with the combustion air.
    B. air that is exhausted out of a gas fired appliance area.
    C. air that is brought into tight constructed homes through a combustion air system.
    D. air that is brought into loose constructed homes around windows and doors.

427. How is infiltration measured?
    A. by cubic foot per minute
    B. by foot per minute
    C. by air changes per hour
    D. by calculating the square footage of the home by a factor of .5

428. How much combustion air is required for every one cubic foot of natural gas used on a gas fired appliance?
    A. 5 cu. ft.
    B. 10 cu. ft.
    C. 15 cu. ft.
    D. 20 cu. ft.

429. How much dilution air is required for every one cubic foot of gas used on a natural draft appliance?
    A. 6 cu. ft.
    B. 8 cu. ft.
    C. 10 cu. ft.
    D. 14 cu. ft.

430. Which statement is true?
    A. A fan assisted draft furnace only requires 5 cu. ft. of combustion air for every 1 cu. ft. of natural gas.
    B. A fan assisted draft furnace does not use dilution air.
    C. A fan assisted draft furnace does not require combustion air.
    D. A fan assisted draft furnace only requires 5 cu. ft. of combustion air for every 1 cu. ft. of natural gas.

431. What is a common cause of heat exchanger and vent system corrosion on an induced draft non-condensing furnace?
    A. not enough dilution air for the vent system
    B. too much outdoor air brought in for combustion
    C. household chemicals
    D. not enough outdoor air brought in for combustion

Section 1   System Components and Tools   **125**

432. An unconfined space is
    A. a space whose volume is greater than 50 cu. ft. / 1000 Btu/h of total input of all gas fired appliances.
    B. a space whose volume is less than 50 cu. ft. / 1000 Btu/h of total input of all gas fired appliances.
    C. a space whose volume is greater than 50 cu. ft. / 4000 Btu/h of total input of all gas fired appliances.
    D. a space whose volume is less than 50 cu. ft. / 4000 Btu/h of total input of all gas fired appliances.

433. Combustion air
    A. can be taken from an unconfined space if the construction of the home is tight.
    B. can be taken from an unconfined space if the construction of the home is loose.
    C. can be taken from an unconfined space only if a mechanical ventilation system installed.
    D. can never be taken from an unconfined space.

434. When introducing outdoor combustion air into an unconfined space,
    A. one 100 sq. in. opening is generally required.
    B. two equal sized openings are required.
    C. never place the opening near the ceiling.
    D. never place the opening near the floor.

435. When introducing outdoor combustion air into an unconfined space,
    A. each opening must have the total free area of at least 1 sq. in. per 1000 Btu/h input and not less than 100 sq. in.
    B. each opening must have the total free area of at least 1 sq. in. per 2000 Btu/h input and not less than 100 sq. in.
    C. each opening must have the total free area of at least 1 sq. in. per 4000 Btu/h input and not less than 100 sq. in.
    D. each opening must have the total free area of at least 1 sq. in. per 5000 Btu/h input and not less than 100 sq. in.

436. When horizontal ducts are used to introduce outdoor air into an unconfined space,
    A. the openings must be 1 sq. in. for every 1000 Btu/h input and not less than 100 sq. in.
    B. the openings must be 1 sq. in. for every 2000 Btu/h input and not less than 100 sq. in.
    C. the openings must be 1 sq. in. for every 4000 Btu/h input and not less than 100 sq. in.
    D. the openings must be 1 sq. in. for every 5000 Btu/h input and not less than 100 sq. in.

437. What is the total free area of a 10" x 15" combustion air opening used with a grill that has a 70% free area?
    A. 70 sq. in.
    B. 105 sq. in.
    C. 120 sq. in.
    D. 150 sq. in.

438. A confined space is
    A. a space whose volume is greater than 50 cu. ft. / 1000 Btu/h of total input of all gas fired appliances.
    B. a space whose volume is less than 50 cu. ft. / 1000 Btu/h of total input of all gas fired appliances.
    C. a space whose volume is greater than 50 cu. ft. / 4000 Btu/h of total input of all gas fired appliances.
    D. a space whose volume is less than 50 cu. ft. / 4000 Btu/h of total input of all gas fired appliances.

439. When introducing combustion air into a confined space,
    A. two equal openings must be provided, within 12 inches of the floor.
    B. two equal openings must be provided, within 12 inches of the ceiling.
    C. at least one opening must be provided near the ceiling or the floor.
    D. two equal openings must be provided, one at or within 12 inches of the ceiling and the other at or within 12 inches from the floor.

440. When introducing outdoor combustion air into a confined space,
    A. each opening must have the total free area of at least 1 sq. in. per 1000 Btu/h input and not less than 100 sq. in..
    B. each opening must have the total free area of at least 1 sq. in. per 2000 Btu/h input and not less than 100 sq. in.
    C. each opening must have the total free area of at least 1 sq. in. per 4000 Btu/h input and not less than 100 sq. in.
    D. each opening must have the total free area of at least 1 sq. in. per 5000 Btu/h input and not less than 100 sq. in.

441. When horizontal ducts are used to introduce outdoor air into a confined space,
    A. the openings must be 1 sq. in. for every 1000 Btu/h input and not less than 100 sq. in.
    B. the openings must be 1 sq. in. for every 2000 Btu/h input and not less than 100 sq. in.
    C. the openings must be 1 sq. in. for every 4000 Btu/h input and not less than 100 sq. in.
    D. the openings must be 1 sq. in. for every 5000 Btu/h input and not less than 100 sq. in.

442. The minimum dimension of a horizontal duct that is used to introduce combustion air into a confined space must not be less than
    A. 1 inch.
    B. 2 inches.
    C. 3 inches.
    D. 5 inches.

443. When a vertical duct is used to introduce outdoor combustion air into a confined space, the opening must be
    A. 1 sq. in. for every 1000 Btu/h total input and not less than 100 sq. in.
    B. 1 sq. in. for every 2000 Btu/h total input and not less than 100 sq. in.
    C. 1 sq. in. for every 4000 Btu/h total input and not less than 100 sq. in.
    D. 1 sq. in. for every 5000 Btu/h total input and not less than 100 sq. in.

444. When a vertical duct is used to introduce combustion air into a confined space, it must terminate
    A. 1 foot above the floor.
    B. 4 feet above the floor.
    C. 2 feet below the ceiling.
    D. 1 foot below the ceiling.

445. What should never be allowed to mix in a confined space when installing a furnace?
    A. the dilution with the combustion air
    B. the supply with the return air
    C. the combustion with the supply and return air
    D. the dilution and the exhaust air

446. Direct vent high efficiency condensing furnaces
    A. must be vented according to manufacturer's recommendation only.
    B. must be vented by using AGA vent tables only.
    C. must use AGA vent tables and manufacturer's recommendation.
    D. must never use polyvinyl chloride pipe for combustion air.

447. Most direct vent high efficiency condensing furnaces
    A. only allow for vertical venting.
    B. only allow for horizontal venting.
    C. allow for vertical and horizontal venting.
    D. must be vented into a chimney whenever possible.

448. When installing combustion air for a direct vent high efficient condensing furnace, the combustion air and vent pipes must be supported at least every
    A. 5 feet.
    B. 10 feet.
    C. 15 feet.
    D. 20 feet.

449. The combustion air and vent pipes used on a direct vent high efficiency furnace must be located
    A. at two different areas to prevent the system from operating under the same air pressure zones.
    B. at two different areas to prevent the system from operating under different air pressure zones.
    C. near the same area to prevent the system from operating under the same air pressure zones.
    D. near the same area to prevent the system from operating under different air pressure zones.

VENTS AND CHIMNEYS

450. What gases are commonly found in products of combustion?
    A. CO_2, H, N, O_2
    B. CL, O_2, H, CO
    C. H_2O, CO, AL
    D. None of the above.

451. What two problems can cause carbon monoxide?
    A. too long of an on cycle and soot
    B. too much primary air and flame impingement
    C. flame impingement and incomplete combustion
    D. incomplete combustion and water vapor

452. What is meant by the term buoyancy?
    A. It is the increase of flue gas temperatures during a furnace start up.
    B. It is the sooting of a heat exchanger.
    C. It is the lifting of the flame off the burners.
    D. It is the moving of gases through a vent system due to high temperatures.

Section 1 System Components and Tools

453. Buoyancy
    A. increases as combustion gas temperature increases.
    B. increases as combustion gas temperature decreases.
    C. is not affected by combustion gas temperatures.
    D. is controlled by the primary air adjustment.

454. For a natural draft venting system to operate, there must be a difference in _____ between the ambient air and the flue gas.
    A. volume
    B. density
    C. velocity
    D. viscosity

455. What can cause a vent to have a minimized static pressure, reducing its effectiveness?
    A. increase of flue gas velocity
    B. heat loss through the vent
    C. the use of a type "B" vent
    D. having too short of a lateral run

456. A category 1 venting system
    A. operates in a positive pressure, less than 140° F above flue gas dewpoint temperature, and requires a fan powered air tight vent to remove the flue gasses.
    B. operates in a negative pressure, at least 140° F above flue gas dewpoint temperature, and depends on a natural draft to remove the flue gasses.
    C. operates in a positive pressure, below 140° F and requires a fan powered air tight vent to remove the flue gasses.
    D. operates in a negative pressure, less than 140° F above flue gas dewpoint temperature, and depends on a natural draft to remove the flue gasses.

457. A category 2 venting system
    A. operates in a positive pressure, less than 140° F above flue gas dewpoint temperature, and requires a fan powered air tight vent to remove the flue gasses.
    B. operates in a negative pressure, at least 140° F above flue gas dewpoint temperature, and depends on a natural draft to remove the flue gasses.
    C. operates in a positive pressure, below 140° F and requires a fan powered air tight vent to remove the flue gasses.
    D. operates in a negative pressure, less than 140° F above flue gas dewpoint temperature, and depends on a natural draft to remove the flue gasses.

458. A category 4 venting system
    A. operates in a positive pressure, less than 140° F above flue gas dewpoint temperature, and requires a fan powered air tight vent to remove the flue gasses.
    B. operates in a negative pressure, at least 140° F above flue gas dewpoint temperature, and depends on a natural draft to remove the flue gasses.
    C. operates in a positive pressure, below 140° F and requires a fan powered air tight vent to remove the flue gasses.
    D. operates in a negative pressure, less than 140° F above flue gas dewpoint temperature, and depends on a natural draft to remove the flue gasses.

459. What is the most common venting system used with a direct vent high efficiency condensing furnace?
    A. a category 1
    B. a category 2
    C. a category 3
    D. a category 4

460. What is the most common venting system used with an induced draft non-condensing furnace?
    A. a category 1
    B. a category 2
    C. a category 3
    D. a category 4

461. What type of material is most commonly used for venting a category 4 type system?
    A. type "B" vent only
    B. standard single wall vent pipe only
    C. standard single wall or type "B" vent
    D. schedule 40 PVC

462. What does the term "wet-time" refer to?
    A. It is the middle of a furnace cycle on an induced draft non-condensing furnace.
    B. It is the end of a furnace cycle where a cool down process could cause vent condensation.
    C. It is the beginning cycle of a furnace where condensate could form in a vent system.
    D. Both B and C.

463. "Wet-time" usually lasts no more than
    A. 30 seconds.
    B. 2 minutes.
    C. 5 minutes.
    D. 10 minutes.

464. When dilution air is brought into the draft diverter on a natural draft furnace, what effect does this have on the products of combustion?
    A. It decreases the dewpoint temperature.
    B. It increases the dewpoint temperature.
    C. It increases the velocity of the gasses leaving the vent.
    D. There is no effect on flue gasses.

465. What is the main purpose of a type "B" vent pipe?
    A. It aids in helping to keep flue gasses from reaching their dewpoint temperature.
    B. It decreases heat transfer from the flue gasses.
    C. It is used when a vent system passes through floors and ceilings.
    D. All of the above.

466. The outer wall of a type "B" vent is constructed of
    A. galvanized steel.
    B. aluminum.
    C. galvanized metal.
    D. aluminized steel.

467. The inner wall of a type "B" vent is constructed of
    A. galvanized steel.
    B. aluminum.
    C. galvanized metal.
    D. aluminized steel.

468. The added air space located in the middle of the inner and outer wall of type "B" vent
    A. decreases heat loss by 25% compared to standard single wall vent pipe.
    B. decreases heat loss by 50% compared to standard single wall vent pipe.
    C. allows for the formation of condensation to occur.
    D. decreases heat loss by 75% compared to standard single wall vent pipe.

469. Common clearances between type "B" vent and combustible materials are
    A. 5 inches minimum.
    B. $\frac{1}{2}$ to 1 inch.
    C. 4 to 6 inches.
    D. 1 to 3 inches.

470. What extra precautions are taken when a type "B" vent passes through a ceiling or floor?
    A. A smoke detector must be placed near the vent system.
    B. The vent system must always be placed outside of the building whenever possible.
    C. A fire stop must be added around the vent.
    D. A type "B" vent must never pass through a ceiling or floor.

471. Common clearances between standard single wall vent and combustible material are
    A. 3 inches.
    B. 4 inches.
    C. 5 inches.
    D. 6 inches.

472. Vent connectors must have a minimum pitch of
    A. $\frac{1}{2}$ inch per 1 foot of horizontal run.
    B. 1 inch per 1 foot of horizontal run.
    C. $2\frac{1}{2}$ inches per 10 feet of horizontal run.
    D. 1 inch per 10 feet of horizontal run.

473. A 6 inch single wall vent connector must be at least
    A. 30 gauge galvanized metal.
    B. 24 gauge galvanized metal.
    C. 22 gauge galvanized metal.
    D. 30 gauge aluminized steel.

474. A 10 inch single wall vent connector must be at least
    A. 30 gauge galvanized metal.
    B. 26 gauge galvanized metal.
    C. 22 gauge galvanized metal.
    D. 30 gauge aluminized steel.

475. The maximum length of a 6 inch single wall vent connector that is used on a category 1 appliance is
    A. 6 feet.
    B. $7\frac{1}{2}$ feet.
    C. 9 feet.
    D. 12 feet.

476. All vent connectors entering a chimney must
    A. never enter more than $\frac{3}{4}"$ into the chimney.
    B. be cemented at the chimney with an approved cement.
    C. be constructed of a 30 gauge galvanized metal (minimum thickness).
    D. always be made of type "B" venting.

477. One 3 inch 90 degree elbow has the same resistance as
    A. 2 feet of 3 inch pipe.
    B. 3 feet of 3 inch pipe.
    C. 4 feet of 3 inch pipe.
    D. 5 feet of 3 inch pipe.

478. What will happen to a system that has too small of a vertical vent when using an induced draft motor?
    A. An excessive draft will be in the vent pipe.
    B. Condensation will form inside of the vent pipe.
    C. Pressurization will form inside the vent pipe.
    D. This will prevent down drafts.

Section 1   System Components and Tools   129

479. What can cause condensation to form in a venting system?
   A. too low of a temperature rise across the heat exchanger
   B. too high of a temperature rise across the heat exchanger
   C. too large of a vent system
   D. Both A and C.
   E. Both B and C.

480. What can cause pressurization in a venting system?
   A. too long of a lateral run
   B. too short of a lateral run
   C. blower speed set too high
   D. Both A and C.
   E. Both B and C.

481. Which is not acceptable when installing a main vent system?
   A. a straight vertical pipe with no elbows
   B. a vertical pipe with two 45 degree elbows
   C. a vertical pipe with two 90 degree elbows
   D. All of the above.

482. What is the minimum height of a main vent from the nearest connected appliance?
   A. 3 feet
   B. 5 feet
   C. 8 feet
   D. 10 feet

483. The main vent must be
   A. 2 feet higher than any vertical wall within 15 feet away.
   B. 3 feet higher than any vertical wall within 10 feet away.
   C. 2 feet higher than any vertical wall within 10 feet away.
   D. 3 feet higher than any vertical wall within 5 feet away.

484. What is the minimum height of a main vent that is used on a $\frac{7}{12}$ pitched roof? Use the chart on page 129 to help determine your answer.
   A. 1' 3"
   B. 1' 6"
   C. 2'
   D. 2' 6"

485. When venting a direct vent high efficiency condensing furnace horizontally, the vent pipe should
   A. be pitched toward the furnace.
   B. be pitched away from the furnace.
   C. be insulated.
   D. never be schedule 40 PVC.

Minimum Height Above Roof is Dependant on Pitch

Roof Pitch	Minimum Height From High Point of Roof Penetration To Lowest Discharge Opening of Cap Feet - Inches
Flat to 6/12	1' 0"
Over 6/12 to 7/12	1' 3"
Over 7/12 to 8/12	1' 6"
Over 8/12 to 9/12	2' 0"
Over 9/12 to 10/12	2' 6"
Over 10/12 to 11/12	3' 3"
Over 11/12 to 12/12	4' 0"
Over 12/12 to 14/12	5' 0"
Over 14/12 to 16/12	6' 0"
Over 16/12 to 18/12	7' 0"
Over 18/12 to 20/12	7' 6"
Over 20/12 to 21/12	8' 0"

486. When venting a direct vent high efficiency condensing furnace horizontally, the vent pipe should
   A. be pitched a minimum of $\frac{1}{2}$" per foot of length.
   B. be pitched a minimum of $\frac{1}{4}$" per foot of length.
   C. be pitched a minimum of 1" per foot of length.
   D. be completely level.

487. When venting a direct vent high efficiency condensing furnace horizontally, the vent pipe should
   A. have an air tight seal and be supported every 5 feet.
   B. have an air tight seal and be supported every 10 feet.
   C. have an air tight seal and be supported every 15 feet.
   D. have an air tight seal and be supported every 20 feet.

488. What is the minimum height of a masonry chimney?
   A. 1 foot above the peak
   B. 2 feet above the peak.
   C. 3 feet above the peak.
   D. 4 feet above the peak.

489. If a masonry chimney is to be used for a gas appliance,
   A. it must have a clay tile liner.
   B. it must always have an aluminum liner.
   C. it must be located in center of the home away from an outside wall.
   D. only one appliance is permitted per chimney.

TABLE 1–3: CAPACITY OF TYPE B DOUBLE WALL VENTS WITH SINGLE WALL METAL CONNECTORS SERVING A SINGLE CATEGORY 1 APPLIANCE

VENT AND CONNECTOR DIAMETER IN INCHES

Height H (ft)	Lateral L (ft)	3" Fan Min	3" Fan Max	3" Nat Max	4" Fan Min	4" Fan Max	4" Nat Max	5" Fan Min	5" Fan Max	5" Nat Max	6" Fan Min	6" Fan Max	6" Nat Max
6	0	38	77	45	59	151	85	85	249	140	126	373	204
	2	39	51	36	60	96	66	85	156	104	123	231	156
	4	NR	NR	33	74	92	63	102	152	102	146	225	152
	6	NR	NR	31	83	89	60	114	147	99	163	220	148
8	0	37	83	50	58	164	93	83	273	154	123	412	234
	2	39	56	39	59	108	75	83	176	119	121	261	179
	5	NR	NR	37	77	102	69	107	168	114	151	252	171
	8	NR	NR	33	90	95	64	122	161	107	175	243	163
10	0	37	87	53	57	174	99	82	293	165	120	444	254
	2	39	61	41	59	117	80	82	193	128	119	287	194
	5	52	56	39	76	111	76	105	185	122	148	277	186
	10	NR	NR	34	97	100	68	132	171	112	188	261	171
15	0	36	93	57	56	190	111	80	325	186	116	499	283
	2	38	69	47	57	136	93	80	225	149	115	337	224
	5	51	63	44	75	128	86	102	216	140	144	326	217
	10	NR	NR	39	95	116	79	128	201	131	182	308	203
	15	NR	NR	NR	NR	NR	72	158	186	124	220	290	192

Appliance Input Rating in Thousands of Btu/h

490. Which statement is false?
    A. The maximum lateral run of a 5 inch vent connector is $7\frac{1}{2}$ feet.
    B. An induced draft non-condensing furnace must have a chimney liner when used in a masonry chimney.
    C. Chimneys with clay tile liners can be used to vent gas appliances.
    D. The space surrounding a chimney liner can be used to vent an existing appliance.

491. When common venting a 100,000 Btu/h furnace with a 35,000 Btu/h water heater,
    A. the furnace vent must be the shortest of the two and located above the water heater vent.
    B. the furnace vent must be the longest of the two and located above the water heater vent.
    C. the water heater vent must be the shortest of the two and located above the furnace vent.
    D. the water heater vent must be the longest of the two and located above the furnace vent.

492. Using Table 1–3, determine what the first column under height and lateral represent?
    A. the height of the main vent and the height of the vent connector
    B. the length of the chimney and the height of the vent connector
    C. the height of the chimney and the length of the vent connector
    D. the height of the main vent and the length of the vent connector

493. Using Table 1–3, determine what the numbers under FAN MIN, FAN MAX, and NAT MAX represent?
    A. appliance input rating of thousands of Btu/h.
    B. appliance output rating of hundreds of Btu/h.
    C. appliance input rating of hundreds of Btu/h.
    D. appliance output rating of thousands of Btu/h.

Section 1 System Components and Tools

494. When venting a 150,000 Btu/h induced draft non-condensing furnace that has a 10 ft. single wall lateral connector that is connected to a 15 ft. vent, what size should the main vent be? Use Table 1–3 to find your answer.
   A. 4"
   B. 5"
   C. 6"
   D. This cannot be accomplished according to the vent tables.

495. Using Table 1–4 on page 132, determine what the first column under height and rise represents.
   A. the height of the main vent and the height of the vent connector
   B. the length of the chimney and the height of the vent connector
   C. the height of the chimney and the length of the vent connector
   D. the height of the main vent and the length of the vent connector

496. When common venting two or more appliances, the vent connector capacity tables that are used assume
   A. that the length of the vent connector is no more than 3 feet.
   B. that the maximum length of the vent connector is no more than 1½ feet per inch of connector diameter.
   C. that the maximum length of the vent connector is no more than 2 feet per inch of connector diameter.
   D. that the minimum length of the vent connector is no more than 1 foot.

497. According to Table 1–4 on page 132, what is the minimum vent diameter that can be used on a 150,000 Btu/h induced draft non-condensing furnace using a 3' connector rise and a main vent that is 20' high?
   A. 4"
   B. 5"
   C. 6"
   D. This cannot be accomplished according to the vent tables.

498. Using Table 1–4 on page 132, determine what the minimum vent diameter that can be used on a 40,000 Btu/h natural draft water heater using a 2' connector rise if the main vent is 20' high?
   A. 4"
   B. 5"
   C. 6"
   D. This cannot be accomplished according to the vent tables.

499. According to the tables on page 132, what is the minimum diameter of a 20' main vent when common venting a 150,000 Btu/h induced draft non-condensing furnace using a 3' connector rise with a 40,000 Btu/h natural draft water heater using a 2' connector rise?
   A. 4"
   B. 5"
   C. 6"
   D. This cannot be accomplished according to the vent tables.

GAS PIPING

500. A drip leg should be installed
   A. on the vertical gas line.
   B. on a horizontal gas line.
   C. before the main gas shut-off.
   D. at the gas meter.

501. A manual gas cock valve shut-off should be installed at which part of the furnace?
   A. internal
   B. external
   C. at the main gas line
   D. at the gas meter

502. A common pressure of _____ WC is needed from the meter to the furnace when using natural gas.
   A. 3.5 inches
   B. 9 inches
   C. 7 inches
   D. 11 inches

503. A common pressure of _____ WC is needed from the gas regulator to the burner orifices for natural gas.
   A. 3.5 inches
   B. 9 inches
   C. 7 inches
   D. 11 inches

TABLE 1–4: CAPACITY OF TYPE "B" DOUBLE WALL VENTS WITH TYPE "B" DOUBLE WALL CONNECTORS SERVING TWO OR MORE CATEGORY 1 APPLIANCES

* Vent Connector Capacity Chart   VENT CONNECTOR DIAMETER IN INCHES

Connector Height H (ft)	Rise L (ft)	3" Fan Min	3" Fan Max	3" Nat Max	4" Fan Min	4" Fan Max	4" Nat Max	5" Fan Min	5" Fan Max	5" Nat Max	6" Fan Min	6" Fan Max	6" Nat Max
8	1	22	40	27	35z	72	48	49	114	76	64	176	109
	2	23	44	32	36	80	57	51	128	90	66	195	129
	3	24	47	36	37	87	64	53	139	101	67	210	145
10	1	22	43	28	34	78	50	49	123	78	65	189	113
	2	23	47	33	36	86	59	51	136	93	67	206	134
	3	24	50	37	37	92	67	52	146	104	69	220	150
15	1	21	50	30	33	89	53	47	142	83	64	220	120
	2	22	53	35	35	96	63	49	153	99	66	235	142
	3	24	55	40	36	102	71	51	163	111	68	248	160
20	1	21	54	31	33	99	56	46	157	87	62	246	125
	2	22	57	37	34	105	66	48	167	104	64	259	149
	3	23	60	42	35	110	74	50	176	116	66	271	168

Appliance Input Rating in Thousands of Btu/h

TABLE 1–5: COMMON VENT DIAMETER IN INCHES

* Common Vent Capacity Chart

Vent Height (ft)	4" FAN +FAN	4" FAN +NAT	4" NAT +NAT	5" FAN +FAN	5" FAN +NAT	5" NAT +NAT	6" FAN +FAN	6" FAN +NAT	6" NAT +NAT
8	101	90	73	155	129	114	224	178	163
10	110	97	79	169	141	124	243	194	178
15	125	112	91	195	164	144	283	228	206
20	136	123	102	215	183	160	314	255	229

Section 1 System Components and Tools

504. Liquefied petroleum gas (LPG) must maintain a pressure of _____ WC from the regulator at the tank when it is being used.
 A. 3.5 inches
 B. 9 inches
 C. 7 inches
 D. 11 inches

505. _____ refers to the relative weight of a gas as compared to the weight of an equal amount of air.
 A. Specific gravity
 B. Relative humidity
 C. Specific heat
 D. Density

506. The specific gravity of natural gas is approximately
 A. 0.25. sp. gr.-× 4
 B. 0.65.
 C. 0.75.
 D. 2.01.

507. To install or alter fuel gas piping,
 A. a license is always required to obtain a permit.
 B. a permit is required by most municipalities.
 C. a permit is not required by most municipalities.
 D. none of the above are true.

508. The gas supply line to a mobile home is commonly
 A. $\frac{3}{4}"$.
 B. $\frac{1}{2}"$.
 C. 1".
 D. $1\frac{1}{4}"$.

509. The gas piping to supply mobile homes should be installed
 A. at least 10 inches below grade.
 B. at least 12 inches below grade.
 C. at least 18 inches below grade.
 D. at least 16 inches below grade.

510. When installing gas pipe near structures such as porches, steps, breezeways, etc., gas pipe installations should be kept at least
 A. 4 inches above grade or structure.
 B. 6 inches above grade or structure.
 C. 8 inches above grade or structure.
 D. 16 inches above grade or structure.

511. All gas piping must be
 A. vacuum tested.
 B. tested with CO_2.
 C. tested with air pressure.
 D. tested with oxygen.

512. Gas piping is commonly constructed of
 A. wrought iron.
 B. steel.
 C. black pipe.
 D. All of the above.

513. Gas piping that is not protected can be
 A. at least 10 inches below grade.
 B. embedded in 4 inches of cement.
 C. embedded in 6 inches of cement.
 D. installed above grade in recesses or channels that are approved.

514. Gas shut-off valves should be installed
 A. 6 feet above floor level.
 B. 8 feet above floor level.
 C. on a horizontal pipe.
 D. within 3 feet of the appliance.

515. Leaks in gas piping should be located with
 A. nitrogen.
 B. oxygen.
 C. a halide torch.
 D. a soapy solution.

516. All manual gas cock valves need to be
 A. constructed of brass.
 B. approved for gas fuel only.
 C. constructed of aluminum casting.
 D. approved by "Gama."

517. Never use an open flame
 A. while using a soap solution.
 B. to locate gas leaks.
 C. around a electronic leak detector.
 D. around a combustible leak detector.

518. Plastic gas piping should be
 A. at least 18 inches below grade.
 B. at least 10 inched below grade.
 C. at least 12 inches below grade.
 D. at least 14 inches below grade.

519. Supports for hanging gas piping should be constructed of
 A. plastic straps.
 B. thin wire.
 C. copper hooks.
 D. metal straps or hooks at approved intervals.

520. Underground gas piping that is ferrous and metallic should be at least
 A. 6 inches below grade.
 B. 8 inches below grade.
 C. 10 inches below grade.
 D. 12 inches below grade.

TABLE 1–6: PIPE EQUIVALENT CHART

Length of Pipe in Feet

Nominal Iron Pipe Size in Inches	10'	20'	30'	40'	50'	60'
$\frac{1}{4}"$	32	22	18	15	14	12
$\frac{3}{8}"$	72	49	40	34	30	27
$\frac{1}{2}"$	132	92	73	63	56	50
$\frac{3}{4}"$	278	190	152	130	115	105
1"	520	350	285	245	215	195
$1\frac{1}{4}"$	1,050	730	590	500	440	400
$1\frac{1}{2}"$	1,600	1,100	890	760	670	610

Maximum Capacity of Pipe in Cubic Feet Per Hour

521. Underground piping should be supported by
 A. compacted soil or sand.
 B. concrete.
 C. top soil.
 D. gravel and soil.

522. On the Gas Piping diagram provided, what is the total amount of gas volume used in the system?
 A. 250 sq. ft.
 B. 250,000 Btu/h
 C. 275 sq. ft.
 D. 275,000 Btu/h

Section 1 System Components and Tools

523. On the Gas Piping diagram provided, what length of gas pipe will be used to size the system?
    A. The sections vary depending where the appliance is placed in the system.
    B. 40'
    C. 50'
    D. 60'

524. When sizing pipe section "C" to "F" in the system, what length will be used on the chart provided?
    A. 10'
    B. 20'
    C. 50'
    D. 60'

525. Using the Gas Piping Chart and diagram, determine the correct size for the main gas pipe from points "A" to "B"?
    A. $\frac{3}{4}"$
    B. 1"
    C. $1\frac{1}{4}"$
    D. $1\frac{1}{2}"$

526. Using the Gas Piping Chart and diagram, determine the correct size for the main gas pipe from points "B" to "C"?
    A. $\frac{1}{2}"$
    B. $\frac{3}{4}"$
    C. 1"
    D. $1\frac{1}{4}"$

527. Using Table 1–6, determine the correct size for the main gas pipe from points "C" to "D"?
    A. $\frac{1}{2}"$
    B. $\frac{3}{4}"$
    C. 1"
    D. $1\frac{1}{4}"$

528. Using Table 1–6, determine the correct size for the gas pipe from points "D" to "E"?
    A. $\frac{1}{2}"$
    B. $\frac{3}{4}"$
    C. 1"
    D. $1\frac{1}{4}"$

529. Using Table 1–6, determine the correct size for the gas pipe from points "B" to "G"?
    A. $\frac{1}{2}"$
    B. $\frac{3}{4}"$
    C. 1"
    D. $1\frac{1}{4}"$

530. Using Table 1–6, determine the correct size for the gas pipe from points "C" to "F"?
    A. $\frac{1}{2}"$
    B. $\frac{3}{4}"$
    C. 1"
    D. $1\frac{1}{4}"$

OIL FURNACES

531. What is the heating value of a number 2 heating oil?
    A. 80,000 Btus / gallon
    B. 100,000 Btus / gallon
    C. 140,000 Btus / gallon
    D. 80,000 Btus / gallon

532. What is the common operating pressure of a high-pressure oil burner?
    A. 25 PSIG
    B. 50 PSIG
    C. 75 PSIG
    D. 100 PSIG

533. What is the secondary voltage of the ignition transformer used on a high-pressure oil burner?
    A. 5000 volts
    B. 10,000 volts
    C. 15,000 volts
    D. 20,000 volts

534. What is the purpose of an electrode in a high-pressure gun burner?
    A. It is used in conjunction with the blower to create a static electrical charge for ignition.
    B. It is used to generate the voltage to the ignition transformer.
    C. It is used to carry the spark to the point of ignition.
    D. Electrodes are only used in low-pressure gun burners.

535. What is the measurement of the spark gap on an electrode assembly?
    A. $\frac{1}{8}"$
    B. $\frac{5}{16}"$
    C. $\frac{1}{4}"$
    D. $\frac{3}{8}"$

536. The capacity of a high-pressure oil burner nozzle is rated in
    A. GPH (gallons per hour).
    B. GPM (gallons per minute).
    C. Btu/h (British thermal units per hour).
    D. PPG (pounds per gallon).

537. What are the three different spray patterns that are used on a nozzle?
    A. solid, semi-solid, hollow
    B. solid, semi-hollow, hollow
    C. uniform, semi-solid, hollow
    D. non-uniform, semi-hollow, hollow

538. The spray angle of an oil nozzle ranges from
   A. 60 degrees to 80 degrees.
   B. 45 degrees to 80 degrees.
   C. 60 degrees to 90 degrees.
   D. 30 degrees to 90 degrees.

539. What type of motor is used on a residential high-pressure oil burner?
   A. shaded pole
   B. permanent split capacitor
   C. split phase
   D. capacitor start capacitor run

540. What is the speed of the burner motor on a high-pressure gun burner?
   A. 1725 RPM
   B. 3450 RPM
   C. 1725 or 3450 RPM
   D. 1050 RPM

541. What is a common capacity of an indoor oil tank?
   A. 200 gallons
   B. 275 gallons
   C. 325 gallons
   D. 375 gallons

542. What is the size of a fill pipe on an oil storage tank?
   A. $1\frac{3}{4}$" diameter
   B. 2" diameter
   C. $2\frac{1}{2}$" diameter
   D. 3" diameter

543. What is the size of an air vent on an oil storage tank?
   A. $\frac{3}{4}$" diameter
   B. 1" diameter
   C. $1\frac{1}{4}$" diameter
   D. $1\frac{1}{2}$" diameter

544. What safety feature is used on an inside oil storage tank that will protect against a house fire?
   A. a fuse type globe valve
   B. a rupture disc
   C. a pressure relief valve
   D. a fusible plug

545. What effect, if any, would a decrease of primary air have on combustion?
   A. It would decrease efficiency.
   B. It would decrease CO_2 percentage.
   C. It would increase CO_2 percentage.
   D. There would be no effect on CO_2 percentage.

546. What effect, if any, would an increase of primary air have on combustion?
   A. It would increase efficiency.
   B. It would decrease CO_2 percentage.
   C. It would increase CO_2 percentage.
   D. There would be no effect on CO_2 percentage.

547. If an oil has a high viscosity,
   A. it mixes easily with water.
   B. it does not mix easily with water.
   C. it is a thicker oil under low temperatures.
   D. it is a thinner oil under low temperatures.

548. What type of a system uses a single stage fuel oil pump?
   A. an aboveground one pipe system
   B. an underground one pipe system
   C. an aboveground two pipe system
   D. an underground two pipe system

549. What type of a system uses a two stage fuel oil pump?
   A. an aboveground one pipe system
   B. an underground one pipe system
   C. an aboveground two pipe system
   D. an underground two pipe system

550. When a cad-cell is located in the light, its resistance is
   A. above 100,000 ohms of resistance.
   B. below 1,500 ohms of resistance.
   C. above 1,600 ohms of resistance.
   D. zero resistance.

551. When a cad-cell is located in the dark, its resistance is
   A. above 100,000 ohms of resistance.
   B. below 1,600 ohms of resistance.
   C. below 100,000 ohms of resistance.
   D. zero resistance.

552. The orange wire on a cad-cell relay is used to
   A. feed line voltage to the relay.
   B. feed the common line to the relay.
   C. connect the cad cell to the circuit.
   D. feed line voltage to the burner motor.

553. The number 3 terminal on the stack mount relay is used to
   A. feed line voltage to the relay.
   B. feed the common line to the relay.
   C. connect the cad-cell to the circuit.
   D. connect the burner motor circuit to the relay.

Section 1   System Components and Tools

554. What is the purpose of the barometric damper used on an oil fired furnace?
    A. It measures the pressure created during operation.
    B. It controls the draft through the system.
    C. It controls the quantity of primary air during the combustion process.
    D. It restricts the flow of flue gasses through the vent.

555. What is the most common size oil line used on a residential oil fired system?
    A. $\frac{1}{4}$" OD
    B. $\frac{3}{8}$" OD
    C. $\frac{1}{2}$" OD
    D. $\frac{5}{8}$" OD

556. What type of lubrication should be used in the burner motor of a high-pressure gun burner?
    A. 10 weight detergent type oil
    B. 30 weight non-detergent type oil
    C. 10 weight non-detergent type oil
    D. 30 weight detergent type oil

557. What is the purpose of a delayed oil valve?
    A. It delays the ignition of a gun burner.
    B. It delays the oil pressure to the nozzle.
    C. It shuts down the oil flow if there is a house fire.
    D. It delays the oil pressure to the oil pump.

558. How is a delayed oil valve wired in a heating circuit?
    A. It is connected across the black and white wire on the cad-cell relay.
    B. It is connected across terminals 2 and 3 on a stack mount relay.
    C. It is connected across terminals 1 and 2 on a stack mount relay.
    D. It is connected across the limit control.

559. What are three products produced in the combustion process?
    A. NH_3, H_2O, and CO
    B. O_3, H_2O, and CO_2
    C. H_2O, CO_2, and N
    D. NH_3, CO_2, and N

560. A high CO_2 reading in an oil fired furnace indicates
    A. a hot fire.
    B. a cool fire.
    C. too much excess air.
    D. a high O_2 reading also.

561. A common CO_2 measurement for an oil fired furnace is from
    A. 8.5% to 9.5%.
    B. 10% to 12%.
    C. 13% to 15%.
    D. 15% to 20%.

562. A common CO_2 measurement for a gas fired furnace is from
    A. 8.5% to 9.5%.
    B. 10% to 12%.
    C. 13% to 15%.
    D. 15% to 20%.

563. A common fire draft reading for an oil furnace would be
    A. -1" WC to -2" WC.
    B. -.05" WC to -.1" WC.
    C. -.03" WC to -.06" WC.
    D. -.01" WC to -.02" WC.

564. A common breech draft for an oil furnace is
    A. -1" WC to -2" WC.
    B. -.05" WC to -.1" WC.
    C. -.03" WC to -.06" WC.
    D. -.01" WC to -.02" WC.

565. If the draft through a furnace is too high, what adjustment should be made?
    A. increase the primary air adjustment
    B. adjust the barometric damper to more of an open position
    C. adjust the barometric damper to more of a closed position
    D. decrease the burner motor adjustment

566. A common smoke reading for an oil fired furnace is
    A. #0.
    B. #1 to #2.
    C. #3 to #4.
    D. #10.

567. What does a high smoke reading indicate on an oil fired furnace?
    A. too much of an over the fire draft
    B. not enough of an over the fire draft
    C. too much primary air
    D. not enough primary air

568. What does a low smoke reading indicate on an oil fired furnace?
    A. too much of an over the fire draft
    B. not enough of an over the fire draft
    C. too much primary air
    D. not enough primary air

Heat Pump
Diagram #1

569. A stack thermometer commonly has a range from
   A. 0° to 500° F.
   B. 300° to 500° F.
   C. 200° to 1000° F.
   D. 500° to 1000° F.

570. Common stack temperatures for an oil fired furnace are
   A. 600° to 800° F.
   B. 800° to 1000° F.
   C. 200° to 450° F.
   D. 500° to 550° F.

571. If combustion air must be supplied to an oil furnace, there must
   A. not be less than 1 sq. in. per 5000 Btu/h.
   B. not be less than 2 sq. in. per 5000 Btu/h.
   C. not be less than 1 sq. in. per 10,000 Btu/h.
   D. not be less than 2 sq. in. per 10,000 Btu/h.

HEAT PUMPS

572. On the Heat Pump Diagram #1, match the following components to their proper names.
   1. _____   5. _____
   2. _____   6. _____
   3. _____   7. _____
   4. _____   8. _____

   Components:
   A. accumulator   E. indoor coil
   B. check valve   F. metering device
   C. compressor   G. outdoor coil
   D. de-ice coil   H. reversing valve

573. What is the purpose of the reversing valve on an air to air heat pump?
   A. It reverses the direction of the compressor.
   B. It changes the flow of the refrigerant making the outdoor coil the condenser during the winter months.
   C. It changes the flow of the refrigerant making the indoor coil the condenser during the winter months.
   D. It changes the flow of the refrigerant making the indoor coil the condenser during the summer months.

574. On a reversing valve solenoid, what is connected to the top tube?
   A. the compressor discharge line
   B. the compressor suction line
   C. the indoor coil
   D. the outdoor coil

575. On a reversing valve solenoid, what is connected to the bottom center tube?
   A. the compressor discharge line
   B. the compressor suction line
   C. the indoor coil
   D. the outdoor coil

576. What pressure difference must be between the low side and the high side of the a heat pump system for the reversing valve to operate properly?
   A. 25 PSIG
   B. 50 PSIG
   C. 75 PSIG
   D. 100 PSIG

Section 1 System Components and Tools

577. Where are the check valves located on an air to air heat pump?
    A. on the outdoor coil
    B. on the indoor coil
    C. on the outdoor coil and indoor coil
    D. on the suction line

578. How are the check valves connected in a heat pump system?
    A. on the inlet of the accumulator
    B. in parallel with both metering devices
    C. in series with one metering device and parallel with the other
    D. Both B and C.

579. A ball type check valves must be installed
    A. horizontally, with the seat on the outlet of the valve.
    B. vertically, with the seat on the bottom of the valve.
    C. horizontally, with the seat on the inlet of the valve.
    D. vertically, with the seat on the top of the valve.

580. When installing a ball type check valve, what must be done to prevent warping?
    A. A heat sink must be used when brazing it into the line.
    B. The ball must be removed when brazing it into the line.
    C. A brass coupling must be used if connecting it to the liquid line.
    D. The valve must be cooled quickly after brazing it into the liquid line.

581. Where is the accumulator located on a heat pump system?
    A. between the bottom center tube of the reversing valve and the compressor
    B. between the top tube of the reversing valve and the compressor
    C. between the bottom right tube of the reversing valve and the compressor
    D. between the bottom left tube of the reversing valve and the compressor

582. When would you most likely find liquid in the vapor line on a heat pump?
    A. during the cooling mode
    B. during the heating mode
    C. during the beginning of a defrost
    D. during the heating mode after a defrost

583. What component must always be de-energized during the defrost cycle on a heat pump?
    A. the reversing valve solenoid
    B. the outdoor fan
    C. the indoor fan
    D. the defrost relay coil

584. Why are the auxiliary strip heaters energized during the defrost cycle?
    A. to add an extra heat load to the indoor coil
    B. to increase the pressure of the system for maximum defrost efficiency
    C. to increase supply air temperature for comfort
    D. The auxiliary strip heaters are not energized during the defrost cycle.

585. What position must the reversing valve be in during the defrost cycle?
    A. the cooling mode
    B. the heating mode
    C. in the open position
    D. in the closed position

586. When a heat pump switches into the defrost cycle, it must
    A. remain in defrost until the temperature rise across the indoor coil reaches 50° F.
    B. remain in defrost until the indoor coil has reached a pre-determined temperature.
    C. remain in defrost until the outdoor coil has reached a pre-determined temperature.
    D. remain in defrost until the indoor thermostat is satisfied.

587. A defrost termination thermostat is commonly located
    A. on the vapor line, and opens at 55° F.
    B. on the liquid line, and closes at 55° F.
    C. on the indoor coil and closes at 55° F.
    D. on the outdoor coil and opens at 55° F.

588. What type of system uses a temperature and a pressure sensor to control the defrost cycle?
    A. a time-initiation/temperature termination defrost system
    B. a temperature initiation/temperature-termination defrost system
    C. a pressure initiation/temperature-termination defrost system
    D. a pressure-time initiation/temperature-termination defrost system

589. A pressure switch, used on the defrost cycle of a heat pump, commonly closes at a preset pressure difference of
   A. 1" to 2" WC.
   B. .5" to .65" WC.
   C. .02" to .05" WC.
   D. 2" to 2.45" WC.

590. Temperature split refers to
   A. the temperature difference of the air across the outdoor coil.
   B. the temperature difference between the indoor and the outdoor of the home.
   C. the temperature difference of the air across the indoor coil.
   D. the temperature difference between the air entering a coil and the temperature of the coil.

591. The outdoor ambient thermostat is commonly used to
   A. energize the first stage of heating in an air to air heat pump.
   B. de-energize the compressor during the second stage of heating in an air to air heat pump.
   C. energize the auxiliary strip heaters during the first stage of heating.
   D. energize the auxiliary strip heaters during the second stage of heating.

592. A hold back thermostat is used to
   A. hold back compressor operation during the defrost cycle.
   B. keep the system energized during a defrost cycle, when the room thermostat is satisfied.
   C. prevent the operation of the auxiliary strip heaters during the defrost cycle.
   D. energize the auxiliary strip heaters during the heating cycle.

593. A discharge temperature thermostat is used
   A. on air to air heat pump systems that use electric auxiliary strip heaters.
   B. on air to air heat pump systems that use fossil fuel.
   C. on air to air heat pump systems that use electric auxiliary strip heaters or fossil fuel.
   D. on air to air heat pump systems to control the defrost sequence.

594. When sizing a heat pump, it must initially be sized for
   A. the heating load.
   B. the cooling load.
   C. the heating and cooling load.
   D. the total capacity.

595. When sizing for the auxiliary heat for a heat pump system, the auxiliary heat must
   A. be capable of handling the entire heat load of the structure minus the heat pump capacity.
   B. be capable of handling the entire heat load of the structure.
   C. be electric strip heaters only.
   D. be fossil fuel only.

596. Auxiliary strip heaters are
   A. used to defrost the outdoor coil during low ambient conditions.
   B. located at the indoor unit upstream from the indoor coil.
   C. located at the indoor unit down stream from the indoor coil.
   D. are wired in parallel with the compressor.

597. What controls the auxiliary strip heaters in a heat pump system?
   A. the first stage of a two stage thermostat
   B. the second stage of a two stage thermostat
   C. the air pressure switch
   D. the auxiliary switch on the contactor

598. What is the purpose of the de-ice coil on an air to air heat pump?
   A. It prevents the formation of ice on the indoor coil during the cooling cycle.
   B. It prevents the formation of ice on the indoor coil during the heating cycle.
   C. It prevents the formation of ice on the upper section of the outdoor coil.
   D. It prevents the formation of ice on the lower section of the outdoor coil.

599. The coefficient of performance (COP) is
   A. a heat pump's rating in the cooling cycle, comparing the heat input by the electrical input.
   B. a heat pump's rating in the heating cycle, comparing the heat input by the electrical output.
   C. a heat pump's rating in the cooling cycle, comparing the heat output by the electrical output.
   D. a heat pump's rating in the heating cycle, comparing the heat output by the electrical input.

600. When installing a heat pump system, the unit should always be located as close to the indoor unit as possible because
   A. during the cooling cycle a longer liquid line will cause the system to operate less efficiently.
   B. during the cooling cycle a longer vapor line will cause the system to operate less efficiently.
   C. during the heating cycle a longer liquid line will cause the system to operate less efficiently.
   D. during the heating cycle a longer vapor line will cause the system to operate less efficiently.

Section 1 System Components and Tools

601. When installing an outdoor coil, it must
   A. always be placed under the overhang of the building.
   B. never be placed under the overhang of the building.
   C. have at least a one inch of clearance around all sides.
   D. have a maximum distance from the exterior wall of six inches.

602. What are common supply air static pressures for a heat pump duct system?
   A. 2" WC
   B. 1" WC
   C. .5" WC
   D. .15" WC

603. What is a common temperature rise over the indoor coil of a heat pump during the heating cycle?
   A. 70° F
   B. 50° F
   C. 35° F
   D. 20° F

604. When installing an indoor unit, a clearance that is commonly required for service accessibility is
   A. 20 inches.
   B. 16 inches.
   C. 10 inches.
   D. 6 inches.

605. How much of an air space is required between the supply air duct of a heat pump and combustible material?
   A. 6 inches
   B. 4 inches
   C. 2 inches
   D. 1 inch

606. What is required on a down flow type indoor unit used with an air to air heat pump?
   A. a smaller return air duct system
   B. a non-combustible floor base
   C. an air pressure switch located in the return air duct
   D. a 70° F temperature rise

607. What is required on a horizontal flow type indoor unit used with an air to air heat pump?
   A. a filter rack down stream from the auxiliary heaters
   B. a backward curved blower
   C. an auxiliary drain pan
   D. a de-ice coil

608. How many strand and what size wire is commonly used for the 24 volt control circuit on an air to air heat pump?
   A. 4 strand 18 gauge
   B. 6 strand 18 gauge
   C. 6 strand 22 gauge
   D. 8 strand 18 gauge

609. On a common heat pump thermostat, what connects to the "Y1" terminal?
   A. the fan relay coil
   B. the auxiliary heaters
   C. the reversing valve
   D. the compressor contactor

610. On a common heat pump thermostat, what connects to the "W1" terminal?
   A. the fan relay coil
   B. the auxiliary heaters
   C. the reversing valve
   D. the compressor contactor

611. On a common heat pump thermostat, what connects to the "W2" terminal?
   A. the thermostat trouble light
   B. the auxiliary heaters
   C. the reversing valve
   D. the compressor contactor

612. On a common heat pump thermostat, what connects to the "X" terminal?
   A. the thermostat trouble light
   B. the auxiliary heaters
   C. the reversing valve
   D. the emergency heat light

613. On a common heat pump thermostat, what connects to the "E" terminal?
   A. the thermostat trouble light
   B. the auxiliary heaters
   C. the reversing valve
   D. the emergency heat light

Section 2

Electrical Theory and Application

ELECTRICAL THEORY

614. The negatively charged particle of an atom is the
    A. proton.
    B. electron.
    C. neutron.
    D. nucleus.

615. The positively charged particle of an atom is the
    A. proton.
    B. electron.
    C. neutron.
    D. nucleus.

616. Like charged particles
    A. attract each other.
    B. repel each other.
    C. have a negative charge.
    D. have a positive charge.

617. Which material would be the best conductor of electrons?
    A. glass
    B. wood
    C. air
    D. copper

618. Which material would be the best insulator?
    A. aluminum
    B. copper
    C. rubber
    D. gold

619. When a conductor is moved through a magnetic field,
    A. magnetism is increased.
    B. reactance is decreased.
    C. a power supply is induced.
    D. frequency is reduced.

620. Electrical pressure is measured in
    A. amps.
    B. ohms.
    C. volts.
    D. watts.

621. The symbol used for electrical pressure is
    A. "E".
    B. "I".
    C. "R".
    D. "P".

622. Electron flow is measured in
    A. amps.
    B. ohms.
    C. volts.
    D. watts.

623. The symbol used for electron flow is
   A. "E".
   B. "I".
   C. "R".
   D. "P".

624. Electrical power is measured in
   A. amps.
   B. ohms.
   C. volts.
   D. watts.

625. The symbol used for electrical power is
   A. "E".
   B. "I".
   C. "R".
   D. "P".

626. Electrical resistance is measured in
   A. amps.
   B. ohms.
   C. volts.
   D. watts.

627. The symbol used for electrical resistance is
   A. "E".
   B. "I".
   C. "R".
   D. "P".

628. A millivolt is best expressed as
   A. 1000 volts.
   B. 100 volts.
   C. 1/1000 of a volt.
   D. 1/100 of a volt.

629. A microamp is best expressed as
   A. 100,000 amps.
   B. 1,000,000 amps.
   C. 1/100,000 of an amp.
   D. 1/1,000,000 of an amp.

630. A megaohm is best expressed as
   A. 100,000 ohms.
   B. 1,000,000 ohms.
   C. 1/100,000 of an ohm.
   D. 1/1,000,000 of an ohm.

631. One amp is defined as
   A. one coulomb per second.
   B. one watt per second.
   C. one volt multiplied by one ohm.
   D. one watt divided by one ohm.

632. Power is best described as
   A. a large rate of inrush current.
   B. the rate that electrons do work.
   C. amperage in an electrical circuit.
   D. resistance of the load in a circuit.

633. One horsepower is the same as
   A. 3.412 watts.
   B. 3142 watts.
   C. 746 watts.
   D. 3.142 watts.

634. One watt is equivalent to
   A. 3.412 Btus.
   B. 3412 Btus.
   C. 746 Btus.
   D. 3.142 Btus.

635. One kilowatt is equivalent to
   A. 3412 Btus.
   B. 3142 Btus.
   C. 1000 Btus.
   D. 3.142 Btus.

MAGNETISM

636. All magnets have
   A. stator poles.
   B. north and south poles.
   C. stator, north, and south poles.
   D. reversing poles.

637. If a north pole of one magnet was placed near the north pole of another magnet, the poles would
   A. repel each other.
   B. attract each other.
   C. there would be no effect.
   D. a north pole does not exist on a magnet.

638. What flows between the two ends of a magnet?
   A. current
   B. voltage
   C. electrical charges
   D. lines of force

639. What is known as magnetic lines of force?
   A. flux
   B. permeability
   C. capacitance
   D. eddy current

640. A permanent magnet is
   A. a coil of wire that has current flowing through it.
   B. magnetism created by induction.
   C. a piece of material that has been magnetized.
   D. created by striking two pieces of soft iron material together.

Section 2   Electrical Theory and Application

641. An electromagnet is produced by
   A. running current through a coiled wire.
   B. running current through an iron core.
   C. moving an iron core through a coiled wire.
   D. moving a coiled wire over an iron core.

642. How is the magnetic field intensified in an electrical circuit?
   A. by increasing the size of the conductor
   B. by decreasing the size of the conductor
   C. by adding more turns to the conductor
   D. by decreasing the turns in the conductor

643. When current flows through the conductor of a solenoid coil, what happens to the iron core?
   A. It tries to move away from the magnetic field.
   B. It tries to center itself in the magnetic field.
   C. No action occurs in the iron core.
   D. The magnetic field tries to center itself around the iron core.

644. What component uses magnetism for operation?
   A. a pressure control
   B. an electric heating element
   C. a motor
   D. a thermal overload protector

DIRECT CURRENT

645. In a DC circuit, current flows
   A. from negative to positive.
   B. from positive to negative.
   C. slower if the voltage is reduced in a circuit.
   D. faster if the resistance is increased in the circuit.

646. The three components needed in an electrical circuit are the
   A. path, volts, and power.
   B. current, path, and load.
   C. power source, path, and load.
   D. switch, power source, and load.

647. On a direct current power source
   A. black represents negative and green represents positive.
   B. black represents positive and the green represents negative.
   C. red represents positive and the black represents negative.
   D. red represents negative and the black represents positive.

648. Increasing the length of the path of an electrical circuit will
   A. decrease the total circuit resistance.
   B. increase the current.
   C. increase resistance.
   D. increase wattage.

649. The load in an electrical circuit is used to
   A. create electrical energy.
   B. convert electrical energy to heat, light, or magnetism.
   C. convert mechanical energy to electrical energy.
   D. create heat, light, or magnetism.

650. A switch in an electrical circuit is used to
   A. control current to a load.
   B. control the resistance of an electrical load.
   C. control the flow of voltage.
   D. control the wattage of a load.

651. When placed in an electrical circuit, an open switch will
   A. make the path to a load.
   B. break the path to a load.
   C. has no effect on the path to the load.
   D. will temporarily interrupt voltage flow.

652. When placed in an electrical circuit, a closed switch will
   A. make the path to a load.
   B. break the path to a load.
   C. have no effect on the path to the load.
   D. temporarily interrupt voltage flow.

653. An example of a manually operated switch would be a
   A. temperature control.
   B. pressure control.
   C. defrost timer.
   D. toggle switch.

654. An example of an automatically operated switch would be a
   A. lever type.
   B. push button.
   C. high-pressure control.
   D. single pole double throw toggle switch.

655. On what type of an electrical circuit can Ohm's Law be used?
   A. an inductive circuit
   B. a capacitive circuit
   C. a resistive circuit
   D. an inductive, capacitive, and a resistive circuit

656. Which is used to find amperage using Ohm's Law?
A. E = IR
B. I = E/R
C. R = E/I
D. I = RE

657. Which is used to find resistance using Ohm's Law?
A. E = IR
B. R = EI
C. R = E/I
D. I = RE

658. Which is used to find voltage using Ohm's Law?
A. E = IR
B. E = I/R
C. R = E/I
D. I = RE

659. Which is used to find power using Watt's law?
A. P = E/R
B. P = EI
C. P = E^2/I
D. P = E/R

660. Which is used to find power using Watt's law?
A. P = E/I
B. P = E^2/I
C. P = I^2 R
D. P = E/R

661. When three 15 ohm resistors are connected in series using a 120 volt power source, what is the total amperage of the circuit?
A. 8 amps
B. 3.2 amps
C. 2.7 amps
D. 375 milliamps

662. When two 20 ohm resistors are connected in parallel and the source voltage is 240 volts, what is the total amperage of the circuit?
A. 12 amps
B. 24 amps
C. 3 amps
D. 10 amps

663. When a 24 volt power supply is used in a circuit with 12 ohms of resistance, what is the total power rating of the circuit?
A. 2 watts
B. 288 watts
C. 24 watts
D. 48 watts

664. What is the total resistance of a circuit when a 5 ohm resistor, a 10 ohm resistor, and a 15 ohm resistor are wired in parallel with each other?
A. 30
B. 15
C. 4.55
D. 2.73

665. In a series circuit, which would be true?
A. Current is the same throughout the circuit.
B. Current increases as loads are added.
C. Resistance decreases as loads are added.
D. Voltage increases as loads are added.

666. In a series circuit, which would not be true?
A. Resistance decreases as loads are added.
B. Resistance increases as loads are added.
C. Voltage drops across loads wired in series.
D. Current is the same throughout the circuit.

667. In a series circuit that is using an electric heater as the load, what would cause an increase in circuit current?
A. a decrease in resistance
B. an increase in resistance
C. a decrease in voltage
D. a burnt contact

668. In a parallel circuit, which would be true?
A. Voltage is the same across each load.
B. Current is always the same across each load.
C. Total resistance increases as loads are added.
D. Voltage increases as loads are added.

669. In a parallel circuit, which would not be true?
A. Total resistance decreases as loads are added.
B. Voltage is the same across each load.
C. Current decreases as loads are added.
D. Current increases as loads are added.

ALTERNATING CURRENT

670. How does alternating current get its name?
A. Alternating current uses different paths for electron flow.
B. Alternating current constantly changes direction in a circuit.
C. An alternating current circuit has a constant voltage fluctuation.
D. An alternating current circuit has a magnetic field when it flows through a circuit.

Section 2   Electrical Theory and Application   147

671. How is alternating current generated?
   A. by coiling a wire around an iron core
   B. by moving current though a conductor
   C. by moving a conductor through a magnetic field
   D. by moving a conductor near an iron core

672. What determines the frequency in an AC circuit?
   A. The amount of cycles that occur in a circuit in one second.
   B. The amount of current that flows from a given source.
   C. The force of the generator that is being used.
   D. The speed of a generator divided by four.

673. Frequency is measured in
   A. henrys.
   B. reactance.
   C. farads.
   D. hertz.

674. What is the resistance of the passing of magnetic flux lines through a substance called?
   A. inductance
   B. impedance
   C. reluctance
   D. inductive reactance

675. What is the ability to store electrons in an electrical circuit called?
   A. inductance
   B. capacitance
   C. reluctance
   D. impedance

676. What is the resistance in an AC circuit due to the change of current flow called?
   A. reactance
   B. impedance
   C. reluctance
   D. inductance

677. What is the opposition to change in voltage in an AC capacitive circuit called?
   A. capacitive reactance
   B. capaance
   C. impedance
   D. reactance

678. What is the generation of a voltage by the variance of the current flow rate in an AC circuit called?
   A. reactance
   B. capacitance
   C. impedance
   D. inductance

679. What is the opposition to current flow in an AC inductive circuit caused by magnetic lines of flux called?
   A. inductance
   B. impedance
   C. inductive reactance
   D. reactance

680. What is the total resistance of an AC circuit, caused by inductive reactance, resistance, and capacitive reactance called?
   A. inductance
   B. impedance
   C. reactance
   D. reluctance

681. What is the relationship of current and voltage in an inductive circuit?
   A. voltage leads current
   B. current leads voltage
   C. voltage is in phase with current
   D. there is no relationship

682. What is the relationship of current and voltage in a capacitive circuit?
   A. voltage leads current
   B. current leads voltage
   C. voltage is in phase with current
   D. there is no relationship

683. What is the relationship of current and voltage in an resistive circuit?
   A. voltage leads current
   B. current leads voltage
   C. voltage is in phase with current
   D. there is no relationship

684. One complete sine wave represents
   A. the force of an AC circuit.
   B. the action of the magnetic flux lines in an AC circuit.
   C. a complete cycle of a generator.
   D. the direction of the current through a conductor.

685. If the current of an AC circuit changes direction 100 times in one second, the frequency of the power source would be
   A. 100 Hz.
   B. 200 Hz.
   C. 60 Hz.
   D. 50 Hz.

686. On a 120V, 60 Hz power supply, what leg must be fused?
   A. the hot leg
   B. the neutral leg
   C. the hot and neutral leg
   D. A fuse is not needed.

687. On a 240V, 60 Hz single phase power supply, where is a fuse necessary?
A. in one hot leg
B. in both hot legs
C. in both hot legs and a neutral leg
D. in one hot leg and one neutral leg

688. When placing a voltmeter across two hot legs of a 240V three phase power supply, what voltage should be present?
A. 120V
B. 480V
C. 240V
D. 277V

689. How many degrees are any two hot legs of a three-phase electrical system out of phase with each other?
A. 90 degrees
B. 100 degrees
C. 120 degrees
D. 360 degrees

690. On a three-phase power supply, what legs must be fused?
A. L1, L2, and neutral
B. L1, L2, and L3
C. L1, L3, and neutral
D. T1, T2, and T3

691. The condition indicating the direction of electron flow is known as
A. current direction.
B. voltage direction.
C. flux condition.
D. polarity.

692. On a 120 volt power supply, what color of wire is the hot leg?
A. white
B. black
C. red
D. green

693. On a 120 volt power supply, what color of wire is the neutral leg?
A. white
B. black
C. red
D. green

694. On a 120 volt power supply, what color of wire is the grounded leg?
A. white
B. black
C. red
D. green

695. The purpose of connecting a ground wire to a load is to
A. allow a secondary path for electrons during normal conditions.
B. protect the equipment against short circuits.
C. protect the equipment against an electrical short to ground.
D. allow a secondary path for electrons during a grounded condition.

POWER DISTRIBUTION

696. For residential areas, substations reduce the voltage from 120,000 volts down to
A. 8400 volts.
B. 4800 volts.
C. 5200 volts.
D. 2500 volts.

697. For commercial areas, substations reduce the voltage from 120,000 volts down to
A. 34,000 volts.
B. 43,000 volts.
C. 52,000 volts.
D. 25,000 volts.

698. A step down transformer will reduce the voltage to a common low voltage residential application of
A. 240 volts, single-phase, 60 Hz.
B. 240 volts, three-phase, 60 Hz.
C. 480 volts, single-phase, 60 Hz.
D. 460 volts, three-phase, 60 Hz.

699. Using a delta transformer, commercial applications will step the voltage down to
A. 240 volts, single-phase, 60 Hz.
B. 240 volts, three-phase, 60 Hz.
C. 480 volts, single-phase, 60 Hz.
D. 208 volts, three-phase, 60 Hz.

700. Using a wye transformer, commercial applications will step the voltage down to
A. 240 volts, single-phase, 60 Hz.
B. 240 volts, three-phase, 60 Hz.
C. 480 volts, single-phase, 60 Hz.
D. 208 volts, three-phase, 60 Hz.

701. When measuring voltage using a voltmeter, across any two hot legs of a low voltage Wye transformer should read
A. 120 volts.
B. 240 volts.
C. 208 volts.
D. 460 volts.

Section 2   Electrical Theory and Application

Low Voltage
Wye Transformer

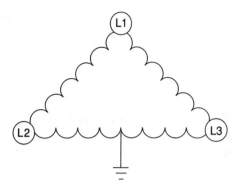

Low Voltage Delta
Transformer

702. When measuring the voltage on a low voltage delta system, what reading would be obtained across any two hot legs?
   A. 120 volts
   B. 240 volts
   C. 208 volts
   D. 480 volts

703. Which three-phase system, when measuring from L1, L2, and L3 to ground, will read 120 volts?
   A. A 120 volt, three-phase delta system
   B. A 208 volt, three-phase delta system
   C. A 480 volt, three-phase wye system
   D. A 208 volt, three-phase wye system

704. Which three-phase electrical system will measure 120 volts when testing two of the hot leads to ground and 208 volts on the third hot leg to ground?
   A. A 460 volt, three-phase delta system
   B. A 240 volt, three-phase delta system
   C. A 480 volt, three-phase wye system
   D. A 208 volt, three-phase wye system

705. On the Wye transformer diagram, what voltage reading should be from L1 to L2?
   A. 240 volts
   B. 120 volts
   C. 208 volts
   D. 277 volts

706. On the Wye transformer diagram, what voltage reading should be from L1 to ground?
   A. 240 volts
   B. 120 volts
   C. 208 volts
   D. 480 volts

707. On the delta transformer diagram, what voltage reading should be from L1 to L2?
   A. 240 volts
   B. 120 volts
   C. 208 volts
   D. 277 volts

708. On the delta transformer diagram, what voltage reading should be from L1 to ground?
   A. 240 volts
   B. 120 volts
   C. 208 volts
   D. 480 volts

709. When placing a voltmeter across any two legs on a 480V three-phase system, what voltage reading should be obtained?
   A. 240 volts
   B. 208 volts
   C. 480 volts
   D. 277 volts

710. What is the maximum voltage output of a 240 volt single phase power supply?
   A. 250 volts
   B. 240 volts
   C. 277 volts
   D. 264 volts

AC MOTORS

711. An AC motor is used to
   A. create electrical energy.
   B. convert electrical energy to mechanical energy.
   C. convert mechanical energy to electrical energy.
   D. create mechanical energy.

712. What is the maximum voltage that can be used on a 120V motor?
    A. 108V
    B. 120V
    C. 132V
    D. 135V

713. What is the minimum voltage that can be used on a 120V motor?
    A. 100V
    B. 108V
    C. 123V
    D. 135V

714. The shaft of a motor is connected to the
    A. stator.
    B. shell.
    C. drum.
    D. rotor.

715. The poles of a motor are also known as the
    A. stator.
    B. rotor.
    C. iron core.
    D. armature.

716. Increasing the poles of a motor
    A. increases the speed of the motor.
    B. decreases the speed of the motor.
    C. has no effect on the speed of the motor.
    D. changes the frequency of the motor.

717. A 240V, 60 Hz motor with four poles will operate at
    A. 3600 RPM.
    B. 1800 RPM.
    C. 1200 RPM.
    D. 1050 RPM.

718. A 240V, 60 Hz motor with six poles will operate at
    A. 3600 RPM.
    B. 1800 RPM.
    C. 1200 RPM.
    D. 1050 RPM.

719. The definition for horsepower is
    A. The amount of power to raise 550 lbs. in one second.
    B. The amount of power to raise 550 lbs. in one minute.
    C. The amount of power to raise 746 lbs. in one second.
    D. The amount of power to raise 746 lbs. in one minute.

720. One horsepower is equivalent to
    A. 550 watts.
    B. 550 kilowatts.
    C. 746 watts.
    D. 746 kilowatts.

721. On a single phase motor, where would the highest resistance be found?
    A. run to common
    B. start to common
    C. run to start
    D. T1 to T2

722. On a single phase motor, where would the lowest resistance be found?
    A. run to common
    B. start to common
    C. run to start
    D. T1 to T2

723. When is the start winding removed from the electrical circuit?
    A. When the motor reaches 100% of its running speed.
    B. When the motor reaches 90% of its running speed.
    C. When the motor reaches 75% of its running speed.
    D. When the motor reaches 50% of its running speed.

724. In a single phase motor the start winding is
    A. 45 degrees out of phase with the run winding.
    B. 90 degrees out of phase with the run winding.
    C. 120 degrees out of phase with the run winding.
    D. in phase with the run winding.

725. In a single phase motor the start and run winding are wired
    A. in parallel with each other.
    B. in series with each other.
    C. in series parallel with each other.
    D. in a delta configuration.

726. When testing the windings of a single phase motor with an ohmeter, the meter should always be set on
    A. R x 1.
    B. R x 100.
    C. R x 1,000.
    D. R x 10,000.

Section 2 Electrical Theory and Application   151

727. When checking the resistance of a single phase compressor motor, a reading is obtained with an ohmmeter from T1 to T2 is 3 ohms, from T2 to T3 is 5 ohms and from T1 to T3 is 8 ohms. Which terminals are RUN, START, and COMMON? Use the example below to answer this question.
   A. T1 is Common
      T2 is Run
      T3 is Start
   B. T1 is Start
      T2 is Common
      T3 is Run
   C. T1 is Run
      T2 is Start
      T3 is Common
   D. T1 is Run
      T2 is Common
      T3 is Start

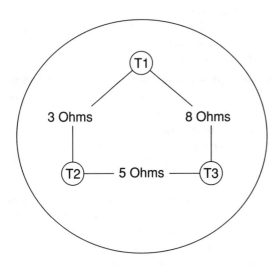

728. When measuring an open run winding, the ohmmeter will read
   A. infinite ohms from Start to Common.
   B. zero ohms from Run to Common.
   C. infinite ohms from Run to Common.
   D. zero ohms from Start to Common.

729. When measuring resistance on a compressor motor that has a grounded condition, the ohmmeter will read
   A. infinite resistance from ground to T1 and T2 on the contactor.
   B. a measurable resistance from ground to the bare copper wire at the power supply.
   C. infinite resistance from ground to the compressor terminals.
   D. a measurable resistance from ground to the compressor terminals.

730. When measuring a shorted start winding, the ohmmeter will read
   A. infinite ohms from Start to Common.
   B. lower resistance from Run to Common than Start to Common.
   C. infinite ohms from Run to Common.
   D. lower resistance from Start to Common than Run to Common.

731. When troubleshooting a single phase motor that has a shorted run winding, it is most likely that
   A. the motor will hum and not start.
   B. there will be voltage to the motor and the motor will not run.
   C. the motor will cycle on the external overload.
   D. there will be no voltage to the motor.

732. The contacts of a start relay are
   A. wired in series with the run winding.
   B. wired in series with the start winding.
   C. always normally opened.
   D. never normally opened.

733. Why are start relays *always* external to a hermetically sealed compressor?
   A. Start relays are not *always* external to a hermetically sealed compressor.
   B. Arcing could cause an explosion in the compressor.
   C. Arcing from the relay could cause acids to form in the compressor.
   D. Current will not flow in a sealed system.

734. In what position are the contacts of a centrifugal switch when in use as a start relay?
   A. closed until the motor starts, then they open
   B. opened until the motor starts, then they close
   C. closed and they remain closed until the motor stops
   D. opened and they close after the motor has reached 75% of it running speed

735. A current relay operates off the
   A. back voltage from the run winding.
   B. back voltage from the start winding.
   C. amperage from the run winding.
   D. amperage from the start winding.

736. A current relay has a set of
   A. normally closed contacts in series with the start winding.
   B. normally open contacts in parallel with the start winding.
   C. normally closed contacts in parallel with the start winding.
   D. normally open contacts in series with the start winding.

737. A potential relay has a set of
   A. normally closed contacts in series with the start winding.
   B. normally open contacts in parallel with the start winding.
   C. normally closed contacts in parallel with the start winding.
   D. normally open contacts in series with the start winding.

738. A potential relay operates off the
   A. back voltage from the run winding.
   B. back voltage from the start winding.
   C. amperage from the run winding.
   D. amperage from the start winding.

739. What terminals connect to the contacts on a potential relay?
   A. Terminals 1 and 2
   B. Terminals 1 and 5
   C. Terminals 2 and 5
   D. Terminals 4 and 5

740. What terminals connect to the coil on a potential relay?
   A. Terminals 1 and 2
   B. Terminals 1 and 5
   C. Terminals 2 and 5
   D. Terminals 4 and 5

741. What is connected to terminal # 1 on a potential relay?
   A. the start winding
   B. the run capacitor
   C. the start capacitor
   D. the common terminal

742. What is connected to terminal # 2 on a potential relay?
   A. the start winding
   B. the run capacitor
   C. the start capacitor
   D. the common terminal

743. What is connected to terminal # 5 on a potential relay?
   A. the start winding
   B. the run capacitor
   C. the start capacitor
   D. the common terminal

744. A solid state relay uses a
   A. PTC thermistor in parallel with the start winding.
   B. PTC thermistor in series with the start winding.
   C. NTC thermistor in parallel with the start winding.
   D. NTC thermistor in series with the start winding.

745. What does FLA mean?
   A. frequency loaded amperage
   B. frequency locked amperage
   C. fully locked amperage
   D. full load amperage

746. What does LRA mean?
   A. loaded rotor amperage
   B. locked rotor amperage
   C. loaded running amperage
   D. lagged running amperage

747. On a single phase compressor, external overload protectors are commonly located
   A. on the Run terminal.
   B. on the Start terminal.
   C. on the Common terminal.
   D. across the Run and Common terminals.

748. Overload protectors are designed to sense
   A. heat only.
   B. current and voltage.
   C. heat and current.
   D. heat and voltage.

749. The type of overload protector that is located in the windings of a motor and will break the circuit to a contactor coil is a
   A. line break external overload.
   B. pilot duty external overload.
   C. line break internal overload.
   D. pilot duty internal overload.

750. The type of overload protector that is located in the windings of a motor and will break the circuit directly to the compressor is a
   A. line break external overload.
   B. pilot duty external overload.
   C. line break internal overload.
   D. pilot duty internal overload.

751. The type of overload protector that is located outside of a motor and will break the circuit directly to the contactor coil is a
   A. line break external overload.
   B. pilot duty external overload.
   C. line break internal overload.
   D. pilot duty internal overload.

752. When testing a compressor motor with an ohmmeter, what reading would be obtained if its internal overload protector is in the open position?
   A. Run to Start—zero ohms
   B. Run to Common—zero ohms
   C. Run to Start—infinite ohms
   D. Run to Common—infinite ohms

Section 2 Electrical Theory and Application   153

753. Capacitors are commonly rated in
   A. microfarads.
   B. microamps.
   C. milliamps.
   D. microhertz.

754. The VAC of a capacitor
   A. can be over by 20% of its rating.
   B. should never be over its rating.
   C. is additive when wired in parallel.
   D. is additive when wired in series.

755. What is the purpose of the 20,000 ohm 2 watt resistor located on a start capacitor?
   A. to discharge the capacitor during motor operation
   B. to increase the microfarad rating by 20%
   C. to level off the sin wave for a much smoother start
   D. to discharge the capacitor after motor operation

756. The capacitor analyzer is used to measure
   A. shorted or open capacitors and microfarad ratings.
   B. microfarad ratings and the capacitors power factor.
   C. open capacitors, microfarad ratings, and the capacitors power factor.
   D. open and shorted capacitors, microfarad ratings, and the capacitor's power factor.

757. What type of relay would remove the start capacitor from the circuit?
   A. a switching relay
   B. a timed delay relay
   C. a potential relay
   D. a contactor

758. Which capacitor uses an oil to dissipate heat?
   A. the run capacitor
   B. the start capacitor
   C. the paper type capacitor
   D. the flux capacitor

759. Which capacitor is taken out of the circuit once the motor is running?
   A. the start capacitor
   B. the run capacitor
   C. the dual capacitor
   D. a run and start capacitor

760. A run capacitor is in the circuit
   A. until the motor reaches 75% of its operating speed.
   B. for 2 seconds.
   C. for the full cycle of the motor.
   D. until the capacitor is discharged.

761. When wiring two 30 μ run capacitors in parallel, what is the microfarad rating on the complete circuit?
   A. 60 μ
   B. 15 μ
   C. 90 μ
   D. 25 μ

762. When wiring two 30 μ run capacitors in series, what is the microfarad rating on the complete circuit?
   A. 60 μ
   B. 15 μ
   C. 90 μ
   D. 25 μ

763. When testing a good capacitor with an analog ohmmeter, what reading should be obtained?
   A. The needle should move toward zero and stay there.
   B. The needle should stay on infinite resistance.
   C. The needle should move toward zero and fall back toward infinite resistance.
   D. The needle should move toward zero and fall back to a readable resistance and stay.

764. When testing a shorted capacitor with an analog ohmmeter, what reading should be obtained?
   A. The needle should move toward zero and stay there.
   B. The needle should stay on infinite resistance.
   C. The needle should move toward zero and fall back toward infinite resistance.
   D. The needle should move toward zero and fall back to a readable resistance and stay.

765. When testing an open capacitor with an analog ohmmeter, what reading should be obtained?
   A. The needle should move toward zero and stay there.
   B. The needle should stay on infinite resistance.
   C. The needle should move toward zero and fall back toward infinite resistance.
   D. The needle should move toward zero and fall back to a readable resistance and stay.

766. Which motor uses no start or run capacitor, but needs a switch to remove the start windings after start up?
   A. split phase
   B. C.S. motor
   C. C.S.R. motor
   D. P.S.C. motor

767. Which motor uses a run capacitor, but has no start capacitor for operation?
   A. split phase
   B. C.S. motor
   C. C.S.R. motor
   D. P.S.C. motor

768. Which motor uses a start capacitor, but has no run capacitor for operation?
   A. split phase
   B. C.S. motor
   C. C.S.R. motor
   D. P.S.C. motor

769. Which motor uses a run capacitor and a start capacitor for operation?
   A. split phase
   B. C.S. motor
   C. C.S.R. motor
   D. P.S.C. motor

770. A three-phase motor is commonly used when the system requires
   A. a motor to have a current draw higher than 15 amps.
   B. a motor higher than 5 horsepower.
   C. a motor with the voltage higher than 240 volts.
   D. a motor with the frequency higher than 50 Hz.

771. How is the direction of a three-phase motor reversed?
   A. by reversing any two hot legs
   B. by removing and repositioning the rotor
   C. by reversing the grounding tap and one hot leg
   D. A three-phase motor cannot be reversed.

772. What is commonly used to start a three phase motor?
   A. a potential relay and a start capacitor
   B. a current relay and a start capacitor
   C. a potential relay
   D. a contactor or a starter

773. How many degrees out of phase with each other are the windings of a three-phase motor?
   A. 45 degrees
   B. 90 degrees
   C. 120 degrees
   D. The windings are in phase with each other.

ELECTRICAL CONTROLS

774. A fuse is used in an electrical circuit to
   A. protect the inductive load from an overcurrent condition.
   B. protect against a ground fault condition.
   C. protect the circuit wires from an overcurrent condition.
   D. protects against open circuits.

775. How are fuses rated?
   A. in amps only
   B. in volts only
   C. in amps with a minimum voltage rating
   D. in amps with a maximum voltage rating

776. What will an ohmmeter measure when testing an open fuse?
   A. zero resistance
   B. infinite resistance
   C. measurable resistance
   D. continuity

777. What does the above symbol represent?
   A. a cooling thermostat
   B. a low-pressure switch
   C. a heating thermostat
   D. a high-pressure switch

778. What does the above symbol represent?
   A. a reverse acting pressure switch
   B. a low-pressure switch
   C. a heating thermostat
   D. a high-pressure switch

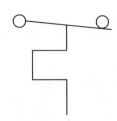

779. What does the above symbol represent?
   A. a cooling thermostat
   B. a low-pressure switch
   C. a heating thermostat
   D. a high-pressure switch

780. What does the above symbol represent?
   A. a cooling thermostat
   B. a low-pressure switch
   C. a heating thermostat
   D. a high-pressure switch

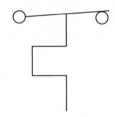

781. What does the above symbol represent?
   A. a cooling thermostat
   B. a thermal fuse
   C. a limit switch
   D. a fusible link

782. What does the above symbol represent?
   A. a normally open delay on make relay
   B. a normally open delay on break relay
   C. a normally open thermal relay
   D. a normally open timer contact

783. What does the above symbol represent?
   A. a normally open delay on make relay
   B. a normally open delay on break relay
   C. a normally open thermal relay
   D. a normally open timer contact

784. What type of sensing element is commonly found in fan and limit controls?
   A. copper
   B. helical invar
   C. bimetal
   D. mercury

785. What is used to sense space temperature in electronic temperature controls?
   A. resistors
   B. transistors
   C. capacitors
   D. thermistors

786. What is known about a liquid charged remote type temperature control?
   A. Its bulb is extremely sensitive and is completely full of liquid.
   B. Its bulb is extremely sensitive, has some gas present, and always has liquid present.
   C. Its bulb is not extremely sensitive and is completely full of liquid.
   D. Its bulb is not extremely sensitive, has some gas present, and always has liquid present.

787. When installing _____, its sensing bulb must always be the coldest part of the control.
   A. the gas charged remote temperature control
   B. the cross charged remote temperature control
   C. the liquid charged remote temperature control
   D. the bimetal type temperature control

788. A temperature control that uses a positive temperature coefficient thermistor (PTCT)
   A. decreases resistance as temperature increases.
   B. increases resistance as temperature increases.
   C. must be level for proper operation.
   D. must be installed near a supply air register for proper operation.

789. A heating thermostat
   A. opens on a drop in temperature.
   B. opens on a rise in temperature.
   C. energizes a heating system when its set point is below space temperature.
   D. de-energizes a heating system when its set point is above space temperature.

790. The heat anticipator
   A. must be energized when the system is de-energized.
   B. controls system lag.
   C. controls system overshoot.
   D. is wired in parallel with the heating thermostat.

791. When adjusting a heat anticipator on a thermostat,
   A. the heating and cooling control amperage must be measured.
   B. the heating and cooling control voltage must be measured.
   C. the cooling control amperage must be measured.
   D. the heating control amperage must be measured.

792. Choose the correct color code and its circuit for a heating and cooling thermostat.
   A. Red–heating circuit, Yellow–cooling circuit, White–power source, Green–fan relay
   B. Red–fan relay, Yellow–power source, White–heating circuit, Green–cooling circuit
   C. Red–power source, Yellow–heating circuit, White–cooling circuit, Green–fan relay
   D. Red–power source, Yellow–cooling circuit, White–heating circuit, Green–fan relay

793. How many wires does a two stage heating and two stage cooling thermostat commonly use?
   A. 4
   B. 6
   C. 8
   D. 10

794. What is the purpose of automatic changeover on a heating and cooling thermostat?
   A. It automatically changes the thermostat's set point.
   B. It automatically changes the thermostat's anticipation value.
   C. It automatically changes from heating to cooling.
   D. It automatically changes from first stage heating to second stage heating.

795. What is the definition of deadband?
   A. Deadband is the point at which the thermostat is satisfied when in the heating cycle.
   B. Deadband is the point at which the thermostat is satisfied when in the cooling cycle.
   C. Deadband is the point at which the thermostat is satisfied when in the automatic cycle.
   D. Dead band is the point at which the thermostat is satisfied between stages when using a two stage thermostat.

796. An outdoor thermostat that is used on an electric heating system is commonly used to
   A. energize electric heaters if the outdoor temperature drops below a predetermined temperature.
   B. de-energize the system if the temperature raises above a predetermined temperature.
   C. switch the system to emergency heat if the temperature drops below a predetermined temperature.
   D. energize electric heaters if the outdoor temperature rises above a predetermined temperature.

797. A cooling thermostat
   A. opens on a drop in temperature.
   B. opens on a rise in temperature.
   C. energizes a cooling system when its set point is above space temperature.
   D. de-energizes a cooling system when its set point is below space temperature.

798. A cooling anticipator
   A. must be de-energized when the system is de-energized.
   B. controls system lag.
   C. controls system overshoot.
   D. is wired in series with the cooling thermostat.

799. Outdoor thermostats are used on air conditioning systems to
   A. prevent short cycling of the compressor during the peak season.
   B. prevent short cycling of the heating system during the air conditioning season.
   C. de-energize the compressor during low ambient conditions.
   D. energize the condenser fans during low ambient conditions.

800. What type of sensing element is the most accurate?
   A. a helical bimetal
   B. a refrigerant filled sensing element
   C. a snap disc bimetal
   D. a thermistor

Section 2   Electrical Theory and Application

801. What type of sensing element is used in a programmable thermostat?
   A. a resistor
   B. a thermistor
   C. a transistor
   D. a bimetal

802. What is known about the anticipators that are located on a programmable thermostat?
   A. The heating has a fixed setting and the cooling can either be fixed or adjustable.
   B. The heating has either a fixed setting or can be adjustable and the cooling has a fixed setting.
   C. The heating has a fixed setting only and the cooling has a fixed setting.
   D. The heating has a fixed setting only and the cooling is adjustable only.

803. When using a heat pump type programmable thermostat, what cycle commonly uses an LED light?
   A. the heating cycle
   B. the cooling cycle
   C. the emergency heat cycle
   D. the 1st and 2nd stage heating and cooling cycles

804. When installing a programmable thermostat, what extra low voltage wire is commonly added to the circuit?
   A. a second power supply wire
   B. the emergency heat circuit
   C. the secondary blower circuit
   D. the common from the transformer

805. How does a two stage thermostat control a heat pump during the heating cycle?
   A. The compressor is controlled by the first stage, and the auxiliary heaters are controlled by the second stage.
   B. The auxiliary heaters are controlled by the first stage, and the compressor is controlled by the second stage.
   C. The compressor and the auxiliary heaters are controlled by the first stage, and the gas fired system is controlled by the second stage.
   D. The gas fired system is controlled by the first stage, and the compressor and auxiliary heaters are controlled by the second stage.

806. What is a battery used for on a programmable thermostat?
   A. to energize the thermostats LED lights
   B. to control the automatic changeover cycle
   C. to save the system's program if there is a loss of power
   D. to energize the heat and cool anticipators

807. When using a programmable thermostat, what is "temperature set back?"
   A. It is the temperature setting of the thermostat during an occupied time.
   B. It is the temperature setting of the thermostat during an unoccupied time.
   C. It is the point at which temperature set point of the thermostat is between the heating and cooling cycle.
   D. It is the point at which the thermostats anticipation rate is set lower than the actual system current draw.

808. What is meant by the term "intelligent recovery?"
   A. It is the feature that calculates the room temperature every 2 to 3 hours.
   B. It is the use of thermostat anticipation for the increase of energy conservation.
   C. It is the gradual return of space temperature to the systems set point after a system set back.
   D. None of the above.

809. The override feature on a programmable thermostat
   A. is used to temporarily bypass the heating control circuit.
   B. is used to temporarily bypass the cooling control circuit.
   C. is commonly accomplished by using the override or hold button.
   D. All of the above.

810. What is the purpose of the limit control on a heating system?
   A. It controls the blower motor.
   B. It limits the amount of gas that is brought to the heat exchanger.
   C. It energizes the gas valve when the thermostat calls for heat and monitors space temperature.
   D. It de-energizes the gas valve if the temperature of the heat exchanger becomes excessive.

811. The limit switch that is used on a forced air gas fired furnace is commonly
   A. a closed bimetal switch.
   B. an opened bimetal switch.
   C. a closed pressure sensitive switch.
   D. an opened pressure sensitive switch.

812. A limit switch that is used on a forced air gas fired furnace
    A. can only be an automatic type of a reset.
    B. can only be a manual type of a reset.
    C. can be an automatic or a manual type of a reset.
    D. must be replaced if the switch opens to the circuit.

813. What is the purpose of a low-pressure switch when it is used in an air conditioning system?
    A. It makes the circuit to the compressor if the low-side pressure drops too low.
    B. It breaks the circuit to the compressor if the high-side pressure drops too low.
    C. It makes the circuit to the compressor if the high-side pressure drops too low.
    D. It breaks the circuit to the compressor if the low-side pressure drops too low.

814. The low-pressure switch will
    A. close on a rise in pressure and open on a drop in pressure.
    B. open on a rise in pressure and close on a drop in pressure.
    C. close if the system has lost its refrigerant charge.
    D. open if the condenser has a loss of air flow.

815. What is the purpose of the high-pressure switch when it is used in an air conditioning system?
    A. It makes the circuit to the compressor if the low-side pressure rises too high.
    B. It breaks the circuit to the compressor if the high-side pressure rises too high.
    C. It makes the circuit to the compressor if the high-side pressure rises too high.
    D. It breaks the circuit to the compressor if the low-side pressure rises too high.

816. The high-pressure switch will
    A. close on a rise in pressure and open on a drop in pressure.
    B. open on a rise in pressure and close on a drop in pressure.
    C. close if the system has an excessive refrigerant charge.
    D. open if the air filter near the indoor coil becomes plugged.

817. A reverse acting high-pressure switch is used to
    A. energize the compressor when the pressure of the system is within operable limits.
    B. de-energize the compressor when the systems pressure becomes excessive.
    C. de-energize the condenser fans on a decrease of ambient temperature.
    D. de-energize the condenser fans on an increase of ambient temperature.

818. What is the purpose of a relay?
    A. It converts a power source from a high voltage to a low voltage.
    B. It controls a higher voltage circuit by a signal from a lower voltage circuit.
    C. It can only be used to control motor circuits.
    D. It changes an AC power supply to a DC power supply by the signal from a low voltage power supply.

819. The contacts of a relay must never exceed
    A. 5 amps.
    B. 10 amps.
    C. 20 amps.
    D. 40 amps.

820. When measuring the resistance of a relay coil and an infinite resistance is found, what is the problem with the relay?
    A. The relay is in operable condition.
    B. The relay coil is shorted.
    C. The relay coil has continuity.
    D. The relay coil is burned open.

821. What must be taken into consideration when replacing a relay?
    A. Replace the relay with the same coil voltage and switch contact configuration.
    B. Replace the relay with the same coil amperage, contact voltage rating, and switch contact configuration.
    C. Replace the relay with the same coil resistance, contact voltage rating, and mounting bracket.
    D. Replace the relay with a higher coil resistance, the same amperage rating, and always add contacts for added load demands.

822. What is the major difference between a relay and a contactor?
    A. A relay coil voltage is 24 volts and a contactor can be 24 or 120 volts.
    B. A contactor has more resistance across its coil.
    C. A contactor contact rating is higher than a relay.
    D. A contactor can be used in motor circuits and a relay cannot.

Section 2   Electrical Theory and Application

823. When testing a contactor with an ohmmeter, a good operating contactor will read
    A. zero resistance across the coil.
    B. measurable resistance across the coil.
    C. infinite resistance across the coil.
    D. measurable resistance across the contacts.

824. What is the major difference between a contactor and a starter?
    A. A starter incorporates built in overload protectors and a contactor does not.
    B. A starter is used on loads that are 460 volts, and a contactor is used on loads that are 220 volts.
    C. The starter coil voltage ranges from 220 to 460, and a contactor coil voltage is 24 volts.
    D. The contacts of a contactor will change positions immediately after the coil is energized and the changing of the starter contacts are delayed.

825. When measuring the resistance of a starter coil, zero resistance is found. What is the problem with the starter?
    A. The starter is in operable condition.
    B. The starter coil is shorted.
    C. The starter coil has no continuity.
    D. The starter coil is burned open.

826. Which type of overload protector is not incorporated in a starter?
    A. a bimetal relay
    B. a thermal relay
    C. a molten-metal relay
    D. a potential relay

827. When a time delay relay is used during the heating cycle, what part of the circuit does it commonly control?
    A. the gas valve circuit
    B. the thermostat circuit
    C. the blower circuit
    D. the accessories circuit

828. Why would a time delay relay circuit be used on an air conditioning system?
    A. to prevent short cycling of the condenser fans
    B. to prevent short cycling of the compressor
    C. to prevent short cycling of the indoor blower
    D. to delay the operation of the condenser fan until high-side pressure is established

829. What type of time delay is commonly used to de-energize a blower on a forced air heating system?
    A. a normally closed delay on make type
    B. a normally open delay on make type
    C. a normally closed delay on break type
    D. a normally open delay on break type

830. What is the purpose of a sequencer?
    A. It reduces inrush current by sequentially energizing and de-energizing loads in an HVAC system.
    B. It reduces holding current by sequentially energizing and de-energizing loads in an HVAC system.
    C. It reduces inrush voltage by sequentially energizing and de-energizing loads in an HVAC system.
    D. It reduces resistance by sequentially energizing and de-energizing loads in an HVAC system.

831. Sequencers are commonly found on
    A. residential gas fired furnaces.
    B. commercial roof top air conditioning systems.
    C. residential air conditioning systems.
    D. electric heating systems.

832. What component is often used in an electrical circuit as a sequencer?
    A. a defrost timer
    B. a magnetic switching relay
    C. a microprocessor
    D. a contactor

833. On what system is a defrost timer commonly found?
    A. a heat pump system
    B. a commercial air conditioning system
    C. a residential air conditioning system
    D. a cooling tower application

834. What are two methods that are used to control a defrost cycle?
    A. defrost timers and switching relays
    B. defrost timers and time delay relays
    C. defrost timers and sequencers
    D. defrost timers and microprocessors

835. What is the purpose of a solenoid valve?
    A. It controls the flow of current through a circuit.
    B. It controls the flow of a fluid.
    C. It allows a technician to read pressures on a manifold gauge set.
    D. It controls the amount of refrigerant in an evaporator coil.

836. An automatic pump down system uses
   A. a high-pressure switch, a thermostat, and a solenoid valve in its circuit.
   B. a high-pressure switch, a compound gauge, and a solenoid valve in its circuit.
   C. a low-pressure switch, a solenoid valve, and a thermostat in its circuit.
   D. a thermostat, a solenoid valve, and a reversing relay in its circuit.

837. What component is commonly used to control refrigerant flow in a heat pump system?
   A. a reversing valve solenoid
   B. a hot gas solenoid
   C. a liquid line solenoid
   D. a bypass solenoid

838. What component is used to protect the compressor during low outdoor ambient conditions?
   A. a subcooler
   B. a crankcase heater
   C. a thermal electric expansion valve
   D. a de-superheat coil

839. What does the above symbol represent?
   A. a fusible link
   B. a thermal overload
   C. an electric heater
   D. a time delay relay

840. What is the purpose of the crankcase heater?
   A. to increase oil viscosity during low ambient conditions
   B. to increase the efficiency of a system during low ambient conditions
   C. to prevent refrigerant migration into the compressor during low ambient conditions
   D. to decrease oil viscosity for proper lubrication of internal components

841. What is the Btu output of an electric heater that is rated at 1500 watts?
   A. 5118 Btus
   B. 4710 Btus
   C. 478 Btus
   D. 440 Btus

842. What controls the electric heating elements in an air to air heat pump system?
   A. the first stage of a two stage thermostat
   B. the second stage of a two stage thermostat
   C. the reversing valve relay
   D. the anticipator relay

843. Where are auxiliary strip heaters commonly located on an air to air heat pump?
   A. in the return air duct before the indoor blower
   B. in the return air duct after the indoor blower
   C. in the supply air plenum before the indoor coil
   D. in the supply air plenum after the indoor coil

844. What does the above symbol represent?
   A. a fuse
   B. a fusible link
   C. a thermal overload
   D. a bimetal sensor

845. What component is commonly used as a flame roll out sensor on a gas fired furnace?
   A. a diode
   B. a transistor
   C. an electric heating element
   D. a fusible link

846. How is a fusible link wired when it is in an electric heating system?
   A. in parallel with the heating element
   B. in series with the heating element
   C. in parallel with the heating element relay coil
   D. in series with the heating element relay coil

847. A step down transformer is designed to
   A. take a lower amperage and increase it to a higher amperage.
   B. take a higher amperage and decrease it to a lower amperage.
   C. take a lower voltage and increase it to a higher voltage.
   D. take a higher voltage and decrease it to a lower voltage.

848. On a step up transformer, the turn ratio from the primary side to the secondary side will be
   A. higher on the primary than on the secondary.
   B. lower on the primary than on the secondary.
   C. the same on the primary as on the secondary.
   D. the same as a step down transformer.

Section 2   Electrical Theory and Application

849. If a transformer uses a 240V primary winding that has 400 turns and a secondary that has 200 turns, what would be the voltage of the secondary winding?
   A. 200 volts
   B. 120 volts
   C. 480 volts
   D. None of the above.

850. When replacing a 24 volt control transformer that has the VA rating of 40, what is the maximum amperage of the control circuit?
   A. .06 amp.
   B. .6 amp.
   C. 1.6 amp.
   D. 9.6 amp.

851. When there is voltage at the primary side of a transformer and no voltage at the secondary, what could be the problem with the transformer?
   A. a bad secondary winding only
   B. a bad primary winding only
   C. a bad secondary or primary winding
   D. a bad fuse at the power supply

852. When measuring an open secondary winding of a transformer, what measurement will be obtained?
   A. an infinite resistance
   B. a measurable resistance
   C. a zero resistance
   D. continuity

SOLID-STATE ELECTRONICS

853. What meter must never be used when testing a circuit board?
   A. an ammeter
   B. a voltmeter
   C. an ohmmeter
   D. an oscilloscope

854. What is the purpose of a diode?
   A. It converts direct current into alternating current.
   B. It is used to increase the resistance of an electrical circuit.
   C. It stores electrons in an electrical circuit.
   D. It allows current to flow in one direction.

855. Which component is commonly used as a solid-state relay?
   A. a transistor
   B. a thermistor
   C. a diode
   D. a cathode

856. What solid-state component decreases its resistance as its temperature increases?
   A. a cathode
   B. an anode
   C. a PTC thermistor
   D. a NTC thermistor

857. Which component is used to store electrons in an electrical circuit?
   A. a capacitor
   B. a thermistor
   C. a diode
   D. a transistor

858. Which component is primarily used to drop voltage in an electrical circuit?
   A. a diode
   B. a resistor
   C. a rectifier
   D. a transistor

859. What component is used to convert alternating current to direct current?
   A. a capacitor
   B. a transistor
   C. a rectifier
   D. a thermistor

860. What component blocks DC and passes AC?
   A. a cathode
   B. an anode
   C. a rectifier
   D. a capacitor

WIRING LAYOUTS AND ELECTRICAL DIAGRAMS

861. On a forced air furnace, what loads are commonly energized with a 120 volt power supply?
   A. the indoor blower, induced draft motor, and the transformer's secondary winding
   B. the induced draft motor, the gas valve, and the indoor blower motor
   C. the electronic air cleaner, the induced draft relay, and the humidifier motor
   D. the induced draft motor, the indoor blower motor, and the transformer's primary winding

862. What is the proper method for connecting any electrical wires located in a junction box?
   A. twisting together and connecting with a wirenut
   B. twisting together and connecting with electrical tape
   C. soldering together and covering with electrical tape
   D. Wires must never be connected together in a junction box.

863. On a split system condensing unit, what loads are commonly energized with a 240 volt power supply?
   A. the compressor, the contactor, and the condenser fan
   B. the condenser fan, the reverse acting high-pressure control, and the compressor
   C. the crankcase heater, the condenser fan cycle guard, and the compressor contactor
   D. the compressor, the crankcase heater, and the condenser fan

864. On an HVAC package unit, what loads are commonly energized with a 240 volt power supply?
   A. the transformer's secondary winding, the contactor, the condenser fan, and the compressor
   B. the condenser fan, the induced draft motor, the compressor, and the indoor air blower
   C. the crankcase heater, the condenser fan relay coil, the compressor, and the thermostat
   D. the compressor, the crankcase heater, the condenser fan, and the fan relay coil

865. What is the control transformer output voltage for a residential split system?
   A. 24 volts
   B. 120 volts
   C. 208/240 volts
   D. 240 volts

866. What is the control transformer input voltage for a residential split system?
   A. 24 volts
   B. 120 volts
   C. 208/240 volts
   D. 240 volts

867. What is the control transformer input voltage for an HVAC package system?
   A. 24 volts
   B. 120 volts
   C. 208/240 volts
   D. 600 volts

868. What is the purpose of a schematic diagram?
   A. It shows the technician the physical positions of the systems components.
   B. It shows the technician the proper connections for the power supply.
   C. It shows the technician the color code of the wiring and the wire connections.
   D. It shows the technician the operational sequence of the system.

869. What do the vertical lines represent in an electrical schematic diagram?
   A. the power supply to the system
   B. the individual circuits in the electrical system
   C. all energized circuits
   D. all de-energized circuits

870. What do the horizontal lines represent in an electrical schematic diagram?
   A. the power supply to the system
   B. the individual circuits in the electrical system
   C. all energized circuits
   D. all de-energized circuits

871. A legend is used
   A. to explain the operational sequence in an electrical schematic.
   B. to identify the label of an electrical component.
   C. to locate electrical components in a wiring diagram.
   D. All of the above.

872. The top portion of an electrical schematic is commonly
   A. the low voltage circuit.
   B. the control voltage circuit.
   C. the line voltage circuit.
   D. the ignition voltage circuit.

873. The bottom portion of an electrical schematic is commonly
   A. the high voltage circuit.
   B. the control voltage circuit.
   C. the line voltage circuit.
   D. the ignition voltage circuit.

874. What component is located between the high voltage circuit and the low voltage circuit?
   A. the control relay
   B. the contactor
   C. the transformer
   D. the potential relay

Section 2   Electrical Theory and Application   163

875. A pictorial diagram is used to
   A. show the system's operational sequence.
   B. show the actual wiring and location of system components.
   C. show the wire and fuse size.
   D. show the placement of the thermostat and the power supply wiring.

876. Which diagram is also known as a label or line diagram?
   A. an installation diagram
   B. a factual diagram
   C. a schematic diagram
   D. a pictorial diagram

877. Which diagram consists of an electrical schematic and a pictorial diagram?
   A. a factual diagram
   B. an installation diagram
   C. a label diagram
   D. a systems diagram

878. Which diagram displays the system's fuse size and wire size, and shows the placement of the thermostat and power wiring?
   A. the schematic diagram
   B. the line diagram
   C. the installation diagram
   D. the ladder diagram

879. The heavy dashed lines on an installation diagram show
   A. the line voltage field wiring.
   B. the low voltage field wiring.
   C. the line voltage factory wiring.
   D. the low voltage factory wiring.

880. The heavy solid lines on an installation diagram show
   A. the line voltage field wiring.
   B. the low voltage field wiring.
   C. the line voltage factory wiring.
   D. the low voltage factory wiring.

881. The light dashed lines on an installation diagram show
   A. the line voltage field wiring.
   B. the low voltage field wiring.
   C. the line voltage factory wiring.
   D. the low voltage factory wiring.

882. The light solid lines on an installation diagram show
   A. the line voltage field wiring.
   B. the low voltage field wiring.
   C. the line voltage factory wiring.
   D. the low voltage factory wiring.

883. What is the purpose of the control labeled PR on Diagram 1–A on page 164?
   A. It directs current through the compressor's run winding.
   B. It de-energizes the compressor if there is an increase of current through the compressor circuit.
   C. It de-energizes the start capacitor after the compressor starts.
   D. It energizes the run capacitor.

884. In Diagram 1–A on page 164, what will cause control ATH to close the circuit to the low speed?
   A. An increase in ambient temperature.
   B. An increase in condenser pressure.
   C. A decrease in condenser pressure.
   D. A decrease in ambient temperature.

885. In Diagram 1–A on page 164, what will cause the compressor and condenser fan to become inoperative?
   A. When CTH and LPS are in the closed position.
   B. When either CTH or LPS are in the opened position.
   C. When the pressure in the condenser reaches an unsafe level.
   D. When FR1 and FR2 have changed positions.

886. In Diagram 1–A on page 164, what will cause switch FS to close?
   A. When terminals R and G close on the thermostat.
   B. When FR2 changes position.
   C. When the heat exchanger temperature increases.
   D. When the outdoor ambient increases.

887. In Diagram 1–A on page 164, how many thermally activated switches are present?
   A. 3
   B. 9
   C. 1
   D. 2

Diagram 1–A

Section 2 Electrical Theory and Application   165

Diagram 2–A

888. In Diagram 2–A, what switches must be in the closed position before the ignition system will energize?
   A. FR1, IDR, PS
   B. Fan Auto/On, PS
   C. Ther, PS, HL, SL
   D. Ther, Fan Auto/On, IDR

889. In Diagram 2–A, what causes PS to close?
   A. PV gas pressure
   B. IDM pressure
   C. MV gas pressure
   D. IBM pressure

890. In Diagram 2–A with the Auto/On switch closed, what will happen to the fan circuit?
   A. The FS switch changes position and operates the fan in the low speed.
   B. The IDR switch changes position and operates the IDM.
   C. The FR1 closes and operates the IBM on the high speed.
   D. The FR2 closes and operates the IBM on the low speed.

891. In Diagram 2–A, how many relay coils are being used?
   A. 7
   B. 5
   C. 3
   D. 1

892. How many switches are shown in Diagram 2–A?
   A. 8
   B. 7
   C. 6
   D. 5

893. In Diagram 3–A on page 166, what voltage should be present when testing across terminals SEC 1 and SEC 2?
   A. 240V
   B. 120V
   C. 24V
   D. 0V

Diagram 3–A

894. In Diagram 3–A, when the thermostat is calling for heat, what two terminals should read 24 volts?
    A. W and C
    B. R and W
    C. Gas 2 and C
    D. GC and C

895. In Diagram 3–A, what two terminals should be used to test for voltage to the fan motor during the heating season?
    A. PR1 and PR2
    B. COM and LO
    C. COM and HI
    D. GAS 3 and LO

896. In Diagram 3–A, the pilot is lit and the main gas valve is not energized. What action could be taken to test the operation of the main gas valve?
    A. bypass the yellow and the white wire on the sensor
    B. bypass the yellow and the green wire on the sensor
    C. bypass GAS 3 and GAS 2
    D. bypass GAS 1 and GAS 2

897. In Diagram 3–A, what terminals are used to test the main gas valve for voltage?
    A. GAS 2 and Ground
    B. GAS 1 and GAS 2
    C. GAS 1 and GAS 3
    D. GAS 2 and GAS 3

Section 2   Electrical Theory and Application

Diagram 4–A

898. In Diagram 4–A, what switches must be closed to energize the orange wire?
   A. DISC, LIM, and FS
   B. DISC, LIM, and IT
   C. DISC, TH, and LIM
   D. DISC and TH

899. In Diagram 4–A, what can cause the system not to fire on start up?
   A. If the fan switch is in the closed position.
   B. If the cad-cell is positioned in the dark.
   C. If the cad-cell is positioned in the light.
   D. If contacts R and W on TH are in the closed position.

900. In Diagram 4–A, what is the first test that should be made when troubleshooting the system?
   A. Test for resistance across the orange and white wires.
   B. Test for voltage across the FF terminals.
   C. Test for resistance across the TT terminals.
   D. Test for voltage across the black and white wires.

901. In Diagram 5–A on page 168, what is the purpose of the holding relay?
   A. It holds in the emergency heat circuit.
   B. It de-energizes the reversing valve circuit during a fault condition.
   C. It de-energizes the contactor and energizes a light at the thermostat.
   D. It de-energizes the control relay during a fault condition.

902. In Diagram 5–A on page 168, what loads are controlled by the defrost relay, DR?
   A. RV, ADR, APS, and TD1
   B. ADR, TM, RV, and TD2
   C. RV, ADR, APS, and TD1
   D. ODT, TD2, TM, and RV

903. In Diagram 5–A on page 168, what loads are energized during the emergency heating cycle if the outdoor thermostat ODT is closed from terminals 1 and 2?
   A. TD1, H1, H2, FR, IFM, and R
   B. TM, TD1, TD2, IFM, H1, H2 and H3
   C. Comp, CF, RV, FR, IFM
   D. ADR, FR, IFM, and H1

Diagram 5–A

904. In Diagram 5–A, when is the reversing valve, RV, energized?

A. during defrost only
B. during 1st stage cooling only
C. during defrost and 1st stage cooling
D. during defrost, 1st stage cooling, and second stage cooling

905. In Diagram 5–A, what does the outdoor thermostat, ODT, control?

A. TD1 and TD2
B. FR and TD1
C. TD2
D. FR, TD1, and TD2

Section 2 Electrical Theory and Application

ELECTRICAL METERS

906. The most accurate reading on an analog meter is
    A. from the middle to the lower $\frac{1}{3}$ of the scale.
    B. in the middle of the scale.
    C. below the middle of the scale.
    D. from the middle to the upper $\frac{2}{3}$ of the scale.

907. When using a digital meter that has the capability of displaying the proper scale to be used, this is called
    A. autoramping.
    B. autoranging.
    C. pre-scaling.
    D. self-scaling.

908. The meter used to measure the electrical pressure in a circuit is the
    A. ammeter.
    B. voltmeter.
    C. ohmmeter.
    D. wattmeter.

909. The meter used to measure the electrical current in a circuit is the
    A. ammeter.
    B. voltmeter.
    C. ohmmeter.
    D. wattmeter.

910. The meter used to measure the electrical resistance in a circuit is the
    A. ammeter.
    B. voltmeter.
    C. ohmmeter.
    D. wattmeter.

911. The meter used to measure the electrical power in a circuit is the
    A. ammeter.
    B. voltmeter.
    C. ohmmeter.
    D. wattmeter.

912. When measuring voltage to a load, the voltmeter must
    A. always be placed in series with the load.
    B. always be placed in parallel with the load.
    C. be in line with the switch controlling the load.
    D. be used with the load disconnected.

913. When testing for voltage across a switch located in a 120 volt circuit, what reading would be obtained if the switch were in the open position?
    A. 0 volts
    B. 60 volts
    C. 120 volts
    D. 240 volts

914. When testing for voltage across a switch located in a 120 volt circuit, what reading would be obtained if the switch were in the closed position?
    A. 0 volts
    B. 60 volts
    C. 120 volts
    D. 240 volts

915. A millivolt meter is designed to measure
    A. 1000 volts
    B. 1 volt
    C. 1/1000 of a volt
    D. 1/100,000 of a volt

916. A voltmeter should
    A. always be set to the lowest scale when working with an unknown voltage.
    B. always be set to the highest scale when working with an unknown voltage.
    C. be set to a mid-scale when working with unknown voltage.
    D. always be used with the power supply isolated from the circuit.

917. VOM stands for
    A. volt, ohm, meter.
    B. volt, ohm, mega-ohmmeter.
    C. volt, ohm, milli-ammeter.
    D. volt, ohm, micro-ammeter.

918. An ammeter must
    A. always be placed in series with the load.
    B. always be placed in parallel with the load.
    C. be in parallel with the switch controlling the load.
    D. be used with the load disconnected.

919. When using a clamp on ammeter, a wire is coiled 10 times and the clamp on ammeter is placed around the coil of wire. What effect, if any, will this have on the amp reading?
    A. The amperage will not be effected.
    B. The amperage will multiply by 10.
    C. The amperage will divide by 10.
    D. The amperage will change depending where the coil of wire is placed.

920. Which meter is more accurate?
    A. the clamp on digital ammeter
    B. the clamp on analog ammeter
    C. The accuracy of the analog is the same as the digital.
    D. the in-line analog ammeter

921. What is the purpose of the hold button located on a clamp-on ammeter?
   A. It automatically adjusts the meter scale.
   B. It retains the amperage reading.
   C. It keeps the meter on the amperage scale.
   D. It allows the meter to be suspended from the circuit being tested.

922. When using an ammeter,
   A. the circuit must always be disconnected.
   B. the circuit must always be energized.
   C. the circuit must always be isolated.
   D. the component must always be isolated.

923. An ammeter should
   A. always be set to the lowest scale when working with an unknown amperage.
   B. always be set to the highest scale when working with an unknown amperage.
   C. never be used when working with unknown amperage.
   D. always be used after testing the voltage.

924. A clamp-on ammeter must always measure the amperage by
   A. clamping around one wire.
   B. clamping around two wires.
   C. clamping around one wire and the ground.
   D. clamping around the ground wire.

925. When measuring resistance of a load, the ohmmeter must always
   A. be placed in series with the load.
   B. be placed in parallel with the load.
   C. be used with the load disconnected.
   D. Both B and C.

926. When measuring the resistance across a switch, what reading would be obtained if the switch were closed?
   A. 0 ohms
   B. measurable resistance
   C. infinite resistance
   D. 1 megaohm

927. When measuring the resistance across a switch, what reading would be obtained if the switch were open?
   A. 0 ohms
   B. measurable resistance
   C. infinite resistance
   D. 1 megaohm

928. Which meter has its own built in power supply?
   A. the voltmeter
   B. the ammeter
   C. the ohmmeter
   D. the wattmeter

929. Which meter must always be used with the power supply disconnected from the load?
   A. the voltmeter
   B. the ammeter
   C. the ohmmeter
   D. the wattmeter

930. A wattmeter is used to measure
   A. the input power rating of the load.
   B. the output power rating of the load.
   C. the power factor input rating.
   D. the power factor output rating.

931. If a motor has a power factor of 85% and the voltage of 120V with the amperage at 10 amps, the power rating would be
   A. 1.1 kW.
   B. 1000 watts.
   C. 1.2 kW.
   D. 1020 watts.

ELECTRICAL TROUBLESHOOTING

932. When testing an electrical circuit that is inoperable, what is commonly the first test that is performed?
   A. an amperage test
   B. a resistance test
   C. a voltage test
   D. a power test

933. When a relay or contactor coil is energized, what type of reading should be obtained across a normally closed set of contacts that are in an operating circuit?
   A. 0 volts
   B. source voltage
   C. 0 ohms
   D. circuit amperage

934. If source voltage is applied to an electrical load and the load is inoperable, what can be said about the load?
   A. It has a short circuit.
   B. It has a defective control circuit.
   C. It has a grounded circuit.
   D. It has an open circuit.

Section 2   Electrical Theory and Application

935. If a measurable voltage reading is obtained when testing across a normally open set of contacts on a relay that is energized, what is the problem with the relay?
   A. There is no problem.
   B. The contacts are burned.
   C. The contacts are closed.
   D. The contacts are shorted.

936. When troubleshooting an electrical circuit using an ohmmeter,
   A. power must be supplied to the circuit.
   B. the meter must always be set to the R x 1 scale.
   C. the component being tested must be removed from the circuit.
   D. the circuit being tested must have current flow.

937. If a measurable resistance is obtained when testing across a normally closed set of contacts on a relay when using an ohmmeter, what is the problem with the relay?
   A. There is no problem.
   B. The contacts are burned.
   C. The contacts are opened.
   D. The contacts are shorted.

938. When measuring the resistance of a load,
   A. start by setting the ohmmeter to the R x 100,000 scale.
   B. start by setting the ohmmeter to the R x 10,000 scale.
   C. start by setting the ohmmeter to the R x 1,000 scale.
   D. start by setting the ohmmeter to the R x 1 scale.

939. When measuring the resistance of a load, what type of resistance should be obtained if the load is operable?
   A. infinite resistance
   B. measurable resistance
   C. zero resistance
   D. no continuity

940. Which of the following resistance readings best display the term continuity?
   A. an infinite and a zero resistance reading
   B. a measurable and an infinite resistance reading
   C. a measurable and a zero resistance reading
   D. an infinite resistance reading

941. When measuring the resistance of a shorted electrical load, what reading will be obtained?
   A. infinite resistance
   B. zero or very low measurable resistance
   C. 0 ohms only
   D. high measurable resistance

942. When measuring the resistance of an electrical load that has burned open, what reading will be obtained?
   A. infinite resistance
   B. zero or very low measurable resistance
   C. 0 ohms only
   D. high measurable resistance

943. When measuring the resistance of a load that has a short to ground, it should be tested with the
   A. R x 1 scale.
   B. R x 10 scale.
   C. R x 1000 scale.
   D. R x 1,000,000 scale.

944. When troubleshooting an electrical circuit using a clamp-on ammeter,
   A. the circuit must be de-energized and the ammeter must be clamped around two wires.
   B. the circuit must be energized and the ammeter must be clamped around two wires.
   C. the circuit must be energized and the ammeter must be clamped around one wire.
   D. the circuit must be de-energized and the ammeter must be clamped around one wire.

945. When testing a 24 volt circuit using an ammeter, what must be used to increase the amperage reading to a measurable level?
   A. a current multiplier
   B. a voltage rectifier
   C. a step up transformer
   D. a potentiometer

946. What reading would be obtained when measuring the amperage of a short circuit?
   A. 0 amps
   B. steady measurable amperage
   C. excessive amperage
   D. low amperage

947. What reading would be obtained when measuring the amperage of an open circuit?
   A. 0 amps
   B. steady measurable amperage
   C. excessive amperage
   D. low amperage

Section 3

Air Flow Components and Duct Fabrication

AIR FLOW TOOLS AND MEASUREMENTS

948. A strobe tachometer is used to measure
   A. air speed.
   B. blower speed.
   C. air volume.
   D. static pressure.

949. A strobe tachometer's measurements are in
   A. FPM.
   B. RPM.
   C. CFM.
   D. inches WC.

950. A Pitot-tube has the capability to measure
   A. static pressure.
   B. velocity pressure.
   C. velocity and static pressure.
   D. CFM.

951. The holes on the side of a Pitot-tube are used to measure
   A. static pressure.
   B. velocity pressure.
   C. velocity and static pressure.
   D. total pressure.

952. The hole on the end of a Pitot-tube is used to measure
   A. static pressure.
   B. velocity pressure.
   C. velocity and static pressure.
   D. total pressure.

953. The speed of air in a duct system is measured in
   A. CFM, Cubic Feet per Minute.
   B. FPM, Feet per Minute.
   C. GPH, Gallons per Hour.
   D. CFH, Cubic Feet per Hour.

954. A Pitot-tube traverse is commonly used to
   A. take an average air velocity pressure reading in an air distribution system for the calculation of air volume.
   B. take an average static pressure reading in an air distribution system for the calculation of air velocity.
   C. take an average air volume reading in an air distribution system for the calculation of air velocity.
   D. None of the above.

955. A Pitot-tube traverse
   A. must be taken in the main supply and return duct system only.
   B. must be taken in the individual branch runs only.
   C. must be taken in the main supply duct only.
   D. can be taken in the main supply, return, and individual branch runs.

956. A velometer measures
    A. the quantity of air in a duct system.
    B. the speed of the air in a duct system or at a diffuser.
    C. the static pressure of the duct system.
    D. None of the above.

957. What instrument is commonly used with a velometer to measure the air velocity at a supply air register?
    A. a Pitot-tube
    B. a diffuser probe
    C. an air scoop
    D. a static ring

958. An anenometer measures
    A. static pressure.
    B. velocity in inches of water column.
    C. velocity in feet per minute.
    D. quantity of air in cubic feet per minute.

959. What are two common types of anenometers used for air flow testing?
    A. a vane type and a Pitot type
    B. a hot wire and a vane type
    C. a hot wire type and a Pitot type
    D. None of the above.

960. What is considered static pressure in a duct system?
    A. the speed of the air in the duct system
    B. the pressure of the air flowing down the duct system
    C. the pressure of the air pushing equally in all directions in the duct system
    D. the total pressure in the duct system

961. What instrument must be used with a manometer to measure static, velocity, and total pressure?
    A. a Pitot-tube
    B. an incline gauge
    C. an anenometer
    D. a static disc

962. Which of the following formulas are correct?
    A. Static Pressure / Velocity Pressure = Total Pressure
    B. Velocity Pressure − Static Pressure = Total Pressure
    C. Velocity Pressure × Static Pressure = Total Pressure
    D. Static Pressure + Velocity Pressure = Total Pressure

963. A U-Tube manometer is commonly used to measure pressures
    A. below 1" WC.
    B. above 1" WC.
    C. up to 40" WC.
    D. None of the above.

964. An incline manometer is commonly used to measure pressures
    A. below 1" WC.
    B. above 1" WC.
    C. up to 40" WC.
    D. None of the above.

965. What type of gauge uses a high and low pressure port to measure a pressure difference between two different pressure locations?
    A. a high-pressure gauge
    B. a low-pressure gauge
    C. a differential pressure gauge
    D. a high/low pressure gauge

966. Air volume is measured in
    A. CFM.
    B. FPM.
    C. inches WC.
    D. inches Hg.

967. Which one of the following formulas is correct?
    A. Area + Velocity = CFM
    B. Velocity × Area = CFM
    C. Velocity / Area = CFM
    D. Area − Velocity = CFM

968. If an 8" x 22" supply air duct has air that is traveling at the speed of 1000 FPM, what is the quantity of air that is moving through the system?
    A. 176,000 CFM
    B. 1760 CFM
    C. 1222 CFM
    D. 122 CFM

969. An 8" x 12" register has the velocity of 300 FPM with the free area of 75%. What is the quantity of air exiting the register?
    A. 1800 CFM
    B. 2880 CFM
    C. 200 CFM
    D. 150 CFM

970. An airhood is commonly used for measuring
    A. CFM.
    B. FPM.
    C. static pressure.
    D. velocity pressure.

971. Which of the following formulas are correct?
    A. CFM = Input (Btu/h) / 1.08 × temperature difference
    B. CFM = Output (Btu/h) × 1.08 / temperature difference
    C. CFM = Input (Btu/h) / 1.08 × temperature difference
    D. CFM = Output (Btu/h) / 1.08 × temperature difference

Section 3   Air Flow Components and Duct Fabrication   175

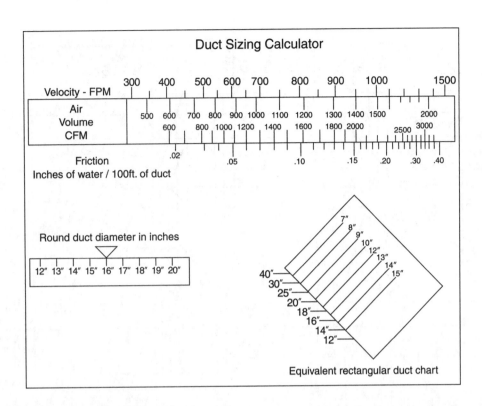

AIR FLOW CALCULATIONS

Air Flow Through a Duct System

V = velocity in FPM (Feet Per Minute)
A = area of a duct in sq. ft. (Square Feet)
Q = quantity of air in CFM (Cubic Feet Per Minute)

Q = AV or (CFM = sq. ft ¥ FPM)
V = Q/A or (FPM = CFM / sq. ft.)
A = Q/V or (sq. ft. = CFM / FPM)

Calculation for Furnace Blower Output

CFM = Output (Btu/h) / 1.08 ¥ temperature difference

Calculation for Weight of Air

Weight of air = W   Volume = V   Density = D
W = DV   or   W/D = V   or   W/V = D

972. Using the Duct Sizing Calculator, if the volume of air in a system is 1600 CFM and the height of a rectangular duct is 8 inches, what should be the width of the duct?
   A. 40 inches
   B. 30 inches
   C. 20 inches
   D. 16 inches

973. Using the Duct Sizing Calculator, if the volume of air in a duct system is 1600 CFM, what would the speed of the air be in the main trunk?
   A. .10" WC
   B. 1.6" WC
   C. 1100 FPM
   D. 1300 FPM

974. Using the Duct Sizing Calculator, if the quantity of air in a duct system is 1600 CFM, what would be the static pressure of the duct per 100 feet?
   A. .10" WC
   B. 1.6" WC
   C. 1100 FPM
   D. 1300 FPM

975. Using the Duct Sizing Calculator, a 16 inch round duct has the same face area as a
   A. 14" x 20" rectangular duct.
   B. 12" x 19" rectangular duct.
   C. 10" x 30" rectangular duct.
   D. 8" x 40" rectangular duct.

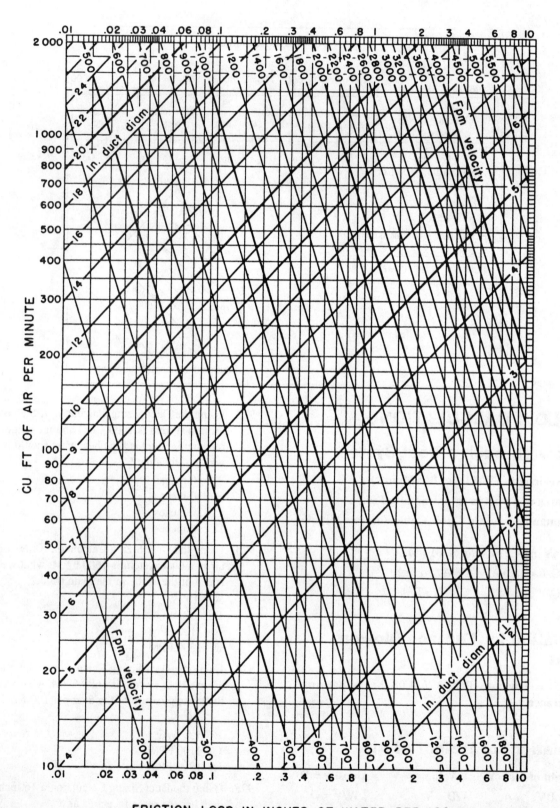

FRICTION LOSS IN INCHES OF WATER PER 100 FT

(Based on Standard Air of 0.075 lb per cu ft density flowing through average, clean, round, galvanized metal ducts having approximately 40 joints per 100 ft.) Caution: Do not extrapolate below chart.

Section 3 Air Flow Components and Duct Fabrication   177

976. If the temperature difference across a heat exchanger was 70° F and the output of a furnace was 82,000 Btu/h, what would be the CFM of the blower?
A. 950 CFM
B. 1045 CFM
C. 1085 CFM
D. 1125 CFM

977. What do the vertical lines on the Friction Loss Chart on page 176 represent?
A. CFM
B. FPM
C. inches WC
D. inches duct diameter

978. What do the horizontal lines on the Friction Loss Chart on page 176 represent?
A. CFM
B. FPM
C. inches WC
D. inches duct diameter

979. Using the Friction Loss Chart on page 176, what would the velocity of the air be at a 0.1 friction loss per 100 feet and 200 CFM of air?
A. 85 CFM
B. 550 CFM
C. 550 FPM
D. 650 FPM

980. Using the Friction Loss Chart on page 176, what would the diameter of the duct have to be to move 120 CFM at the velocity of 600 FPM?
A. 5 inches
B. 6 inches
C. 7 inches
D. 8 inches

981. Using the Friction Loss Chart on page 176, what would be the friction loss per 100 feet if the speed of the air down a 5 inch duct were 600 FPM?
A. .05
B. .08
C. 1.0
D. .15

BLOWERS AND FANS

982. Dry bearings can cause
A. excessive voltage to a motor.
B. an increase in efficiency.
C. an increase in amperage.
D. a decrease in amperage.

983. The sleeve in a sleeve bearing is commonly made of
A. copper.
B. bronze.
C. silver.
D. iron.

984. An oil type motor typically requires oiling every
A. 1 to 3 months.
B. 3 to 6 months.
C. 12 months.
D. 18 months.

985. What type of bearing can handle a larger load?
A. sleeve bearing
B. bushing
C. ball bearing
D. axial bearings

986. Increasing the pitch of an axial fan blade will
A. increase the amperage of a fan motor.
B. decrease the amperage of a fan motor.
C. increase the torque of a fan motor.
D. increase the RPMs of a fan motor.

987. Changing a condenser fan blade from a 4 to a 5 blade will
A. increase the amperage of a fan motor.
B. decrease the amperage of a fan motor.
C. increase the torque of a fan motor.
D. increase the RPMs of a fan motor.

988. As the air volume is decreased over a fan the motor amperage will
A. increase.
B. decrease.
C. stay the same.
D. Air volume cannot be changed.

989. Direct drive condenser fans that are lower than 5 horsepower are commonly a
A. three-phase type motor.
B. PSC type motor.
C. CSIR type motor.
D. CSCR type motor.

990. How are condenser fan motors cycled during low ambient conditions?
A. by the use of a high ambient pressure switch
B. by the use of a low pressure switch
C. by the use of a high pressure switch
D. by the use of a reverse acting high pressure switch

991. Why is it necessary to cycle the condenser fans during low ambient conditions?
   A. to prevent an increase in high-side pressure and the flooding of the evaporator coil
   B. to prevent a decrease in high-side pressure and the flooding of the evaporator coil
   C. to prevent an increase in high-side pressure and the starving of the evaporator coil
   D. to prevent an decrease in high-side pressure and the starving of the evaporator coil

992. Belt drive blower motors lower than 5 horsepower commonly use a
   A. three-phase type motor.
   B. split phase type motor.
   C. CSIR type motor.
   D. PSC type motor.

993. What is the color of the low speed on a 120V three speed motor?
   A. white
   B. red
   C. black
   D. yellow

994. What is the color of the high speed on a 120V three speed motor?
   A. white
   B. red
   C. black
   D. yellow

995. What is the color of the common lead on a 120V three speed motor?
   A. green
   B. white
   C. orange
   D. black

996. A fan motor that operates at 1050 RPM and has a pulley size of 6 inches operates a blower that has a pulley size of 9 inches. What is the RPM of the blower?
   A. 1500
   B. 1575
   C. 775
   D. 700

997. How many CFMs of air must a blower provide per ton of refrigeration effect?
   A. 300
   B. 400
   C. 600
   D. 800

998. How does an increased static pressure in a duct system effect the blower motor?
   A. the amperage increases
   B. the amperage decreases
   C. the air quantity increases
   D. the air quantity decreases

999. Name two common fans that are used on residential and light commercial heating and air conditioning systems.
   A. a centrifugal and rotary type
   B. an axial and reverse rotary
   C. a squirrel cage and a centrifugal type
   D. an axial and a squirrel cage type

1000. What is the FLA of a motor?
   A. the starting amperage
   B. the running amperage
   C. the counter electromotive force amperage
   D. the fully locked amperage

1001. What is the LRA of a motor?
   A. the starting amperage
   B. the running amperage
   C. the counter electromotive force amperage
   D. the locked resistance amperage

1002. Common external static pressure for a residential furnace blower is
   A. .5" WC.
   B. 1" WC.
   C. 1.5" WC.
   D. 2" WC.

AIR DUCT SYSTEMS AND FABRICATION

1003. How often must a horizontal rectangular duct under 60 inches wide be supported?
   A. not less than every 5 feet
   B. not less than every 8 feet
   C. not exceeding every 10 feet
   D. not exceeding every 12 feet

1004. What is the minimum width of the galvanized steel hanger that supports a horizontal duct of 60 inches or less?
   A. $\frac{3}{4}$ inches
   B. 1 inches
   C. $1\frac{1}{4}$ inches
   D. $1\frac{1}{2}$ inches

Section 3 Air Flow Components and Duct Fabrication 179

1005. Duct joints must always
A. be sealed with a pressure sensitive tape.
B. be sealed with mastics.
C. be sealed with gasket material.
D. be made air tight.

1006. The crimp of a round duct joint must have a contact lap of at least
A. 1½ inches.
B. 1 inch.
C. 2 inches.
D. 1¾ inches.

1007. When connecting two sections of round duct,
A. the air flows from the female section of the duct to the male section.
B. a minimum of three sheet metal screws must be used per section.
C. screws or tape are sufficient to hold the sections together.
D. a minimum of two sheet metal screws must be used per section.

1008. Vibration isolators that are installed between metal ducts and mechanical equipment
A. must not exceed 10 inches in length.
B. must not exceed 5 inches in length.
C. must not exceed 3 inches in length.
D. must be at least 2 inches in length.

1009. What is the minimum distance that a duct system is allowed to be installed from the ground?
A. flush
B. 2 inches
C. 1 inch
D. 4 inches

1010. When a metal duct system is installed under a concrete slab,
A. the duct must be encased with at least two inches of concrete.
B. the duct must be fabricated of stainless steel.
C. the duct must be aerated to prevent moisture build up.
D. a duct is not allowed to be installed under a concrete slab.

1011. When circular bands are used to connect sections of round duct,
A. the bands must be at least 4 inches wide and use the same gauge metal as the duct system.
B. the bands must be at least 3 inches wide and use the same gauge metal as the duct system.
C. the bands must be at least 2 inches wide and use the same gauge metal as the duct system.
D. the bands must be at least 1 inch wide and use the same gauge metal as the duct system.

1012. The minimum total area of a supply air duct used on a forced warm air furnace
A. must be at least 1 square inch per 1000 Btu/h approved output rating.
B. must be at least 2 square inches per 1000 Btu/h approved output rating.
C. must be at least 4 square inches per 1000 Btu/h approved output rating.
D. must be at least 6 square inches per 1000 Btu/h approved output rating.

1013. The minimum total area of a supply air duct used on a heat pump
A. must be at least 1 square inch per 1000 Btu/h approved output rating.
B. must be at least 2 square inches per 1000 Btu/h approved output rating.
C. must be at least 4 square inches per 1000 Btu/h approved output rating.
D. must be at least 6 square inches per 1000 Btu/h approved output rating.

1014. When installing a metal air duct system on a roof,
A. the duct system should have at least 1 inch of insulation and be sealed with an approved weatherproof barrier.
B. the duct system should have at least 3 inches of insulation and be sealed with an approved weatherproof barrier.
C. the maximum length of the duct system must not exceed 10 feet.
D. the duct must be at least 28 gauge.

1015. When installing a metal air duct in a crawl space or attic,
A. the duct system should have at least 1 inch of insulation and be sealed with an approved vapor barrier.
B. the duct system should have at least 3 inches of insulation and be sealed with an approved vapor barrier.
C. the duct must have fire dampers installed.
D. the duct must always be pitched toward the furnace.

1016. In a single dwelling unit, exposed rectangular ducts or plenums 14 inches or less must be fabricated with a minimum of
A. 24 gauge galvanized steel.
B. 26 gauge galvanized steel.
C. 28 gauge galvanized steel.
D. 30 gauge galvanized steel.

1017. In a single dwelling unit, exposed rectangular ducts or plenums of more than 14 inches must be fabricated with a minimum of
   A. 24 gauge galvanized steel.
   B. 26 gauge galvanized steel.
   C. 28 gauge galvanized steel.
   D. 30 gauge galvanized steel.

1018. What is the purpose of a cross break in a section of duct?
   A. to increase air volume
   B. to increase air velocity
   C. to decrease air currents caused by air velocity
   D. to increase its rigidity and decrease noise

1019. Low pressure duct systems are designed for a maximum static pressure of
   A. .5" WC.
   B. 1" WC.
   C. 2" WC.
   D. 2.5" WC.

1020. When air ducts are run through floors, what special provision must be made to meet code requirements?
   A. The air duct must be at least 5 inches in diameter.
   B. There must be at least 2 inches from any combustible surface.
   C. Shut off dampers must be within 4 inches of the floor.
   D. A fire stop must be placed around the duct.

1021. What type of duct system is one where the furnace is in a central location of the home and the trunk is branched from two sides of the furnace?
   A. an extended plenum
   B. a reducing extended plenum
   C. a perimeter loop
   D. a radial system

1022. What type of duct system is commonly used in homes with concrete slab floors where a single pipe runs on the outside wall?
   A. an extended plenum
   B. a reducing extended plenum
   C. a perimeter loop
   D. a radial system

1023. What type of duct system is commonly installed on the first floor where several ducts run from the furnace to the outside walls?
   A. an extended plenum
   B. a reducing extended plenum
   C. a perimeter loop
   D. a radial system

1024. What type of duct system has the furnace located at one end of the house and the trunk size is reduced as it increases in distance from the furnace?
   A. an extended plenum
   B. a reducing extended plenum
   C. a perimeter loop
   D. a radial system

1025. On an extended plenum, what is the maximum distance that a duct can extend without having a reduction in size?
   A. 8 feet
   B. 12 feet
   C. 16 feet
   D. 24 feet

1026. Decreasing the size on a reducing extended plenum will
   A. increase air velocity.
   B. increase air volume.
   C. decrease air volume.
   D. reduce air turbulence.

1027. What is the name of the section of duct that connects two different sized ducts together?
   A. a duct intersection
   B. a duct transverse
   C. a duct transition
   D. a duct collar

1028. Which tool is commonly used to decrease the diameter of the male fitting on a round duct?
   A. the duct seamer
   B. the crimping tool
   C. the bar folder
   D. the reducing tool

1029. Where should a balancing damper be located on a branch run?
   A. at the center point between the plenum and the register
   B. as close the register as possible
   C. as close to the supply air plenum as possible
   D. If the system is sized properly, a balancing damper is not required.

1030. When adjusting a balancing damper located in a branch run, the damper is closed
   A. when the adjusting handle is perpendicular to the duct.
   B. when the adjusting handle is parallel to the duct.
   C. when the adjusting handle is pointed 45 degrees above the duct.
   D. when the adjusting handle is pointed 45 degrees below the duct.

Section 3 Air Flow Components and Duct Fabrication

1031. A main supply air plenum must be located
A. at least 3 inches from a combustible surface.
B. at least 2½ inches from a combustible surface.
C. at least 2 inches from a combustible surface.
D. at least 1 inch from a combustible surface.

1032. What is the major difference between a register and a grill?
A. A register is adjustable and is used on the supply air side of a system. A grill is not adjustable and is used on the return air side of a system.
B. A register is not adjustable and is used on the supply air side of a system. A grill is adjustable and is used on the return air side of a system.
C. A grill is not adjustable and is used on the supply air side of a system. A register is adjustable and is used on the return air side of a system.
D. A grill is adjustable and is used on the supply air side of a system. A register is not adjustable and is used on the return air side of a system.

1033. What type of supply air register is required for bathrooms, kitchens, and utility rooms?
A. a floor register
B. a floor register with a return air grill
C. a toe kick with a return air grill
D. a toe kick or side wall register

1034. In homes, what is most commonly used to return the conditioned space air back to the furnace?
A. prefabricated sheet metal return air channels
B. wall studs and gypsum boar
C. individual chases
D. 26 gauge return air runs

1035. What is the name for the above metal duct connection?
A. Hemmed "S" slip
B. Drive slip
C. Plain "S" slip
D. Standing drive slip

1036. What is the name for the above metal duct connection?
A. Hemmed "S" slip
B. Drive slip
C. Plain "S" slip
D. Standing drive slip

1037. What is the name for the above metal duct connection?
A. Hemmed "S" slip
B. Inside slip joint
C. Double "S" slip
D. Standing "S" slip

1038. What is the name for the above metal duct connection?
A. Drive slip
B. Inside slip joint
C. Hemmed "S" slip
D. Standing drive slip

1039. What is the name for the above metal duct connection?
A. Hemmed "S" slip
B. Inside slip joint
C. Drive slip
D. Standing drive slip

1040. What is the name for the above metal duct connection?
A. Hemmed "S" slip
B. Inside slip joint
C. Double "S" slip
D. Standing "S" slip

1041. What is the name for the above metal duct connection?
   A. Hemmed "S" slip
   B. Inside slip joint
   C. Double "S" slip
   D. Standing "S" slip

1042. "S" slips are commonly connected
   A. horizontally on the duct sections.
   B. vertically on the duct sections.
   C. by folding its end tabs.
   D. using rivets every 2 feet.

1043. Drive slips are commonly connected
   A. horizontally on the duct sections.
   B. vertically on the duct sections.
   C. using sheetmetal screws.
   D. using rivets every 2 feet.

FLEXIBLE (FLEX) DUCT

1044. When installing a flexible duct, it is important to always
   A. support the duct every 10 feet.
   B. fully extend the duct section to prevent the reduction of air flow.
   C. use a supporting saddle material greater than 3 inches.
   D. decrease the duct sag down to 2 inches per foot of section.

1045. The radius of a flexible duct must not be less than
   A. half of the diameter of duct being used.
   B. one diameter of the duct being used.
   C. two diameters of the duct being used.
   D. three diameters of the duct being used.

1046. Sheetmetal collars that are connected to a flexible duct must be a minimum of
   A. $1\frac{1}{2}$ inches in length.
   B. 3 inches in length.
   C. 1 inch in length.
   D. 2 inches in length.

1047. Sheetmetal sleeves used for joining two sections of flexible duct must be a minimum of
   A. 3 inches in length.
   B. 6 inches in length.
   C. 2 inches in length.
   D. 4 inches in length.

1048. A horizontal flexible duct must be supported
   A. every 8 feet.
   B. every 2 feet.
   C. every 4 feet.
   D. every 6 feet.

1049. A vertical flexible duct must be supported
   A. every 8 feet.
   B. every 2 feet.
   C. every 4 feet.
   D. every 6 feet.

1050. What is the minimum width of saddle material used to support a flexible duct?
   A. 1 inch
   B. $1\frac{1}{2}$ inches
   C. 2 inches
   D. 3 inches

1051. What is the maximum sag that is allowed for flexible duct between supports?
   A. $\frac{1}{2}$ inch per foot
   B. 1 inch per foot
   C. $1\frac{1}{2}$ inches per foot
   D. 2 inches per foot

FIBERGLASS DUCTBOARD

1052. Fiberglass ductboard is available in thicknesses of
   A. $\frac{1}{2}$" and 1".
   B. $\frac{1}{2}$" and $1\frac{1}{2}$".
   C. 1" and $1\frac{1}{2}$".
   D. $1\frac{1}{2}$" and $2\frac{1}{2}$".

1053. What is the maximum static pressure that can be obtained using fiberglass ductboard on the positive side of the system?
   A. $\frac{1}{2}$" WC
   B. 1" WC
   C. $1\frac{1}{2}$" WC
   D. 2" WC

1054. What is the maximum static pressure that can be obtained using fiberglass ductboard on the negative side of the system?
   A. $\frac{1}{2}$" WC
   B. 1" WC
   C. $1\frac{1}{2}$" WC
   D. 2" WC

1055. What is the maximum air velocity that can be used in a fiberglass ductboard system?
   A. 1200 FPM
   B. 1800 FPM
   C. 2400 FPM
   D. 3600 FPM

Section 3 Air Flow Components and Duct Fabrication

1056. What is the maximum temperature that is permitted inside a fiberglass ductboard system?
   A. 200° F
   B. 250° F
   C. 300° F
   D. 350° F

1057. What is the maximum temperature that is permitted on the outside surface of a fiberglass ductboard system?
   A. 150° F
   B. 200° F
   C. 100° F
   D. 250° F

1058. When screws are used to fasten ductboard to a metal surface, which is the proper procedure?
   A. use a # 8 sheet metal screw with a 2 inch round washer
   B. use a # 10 sheet metal screw with a $2\frac{1}{2}$ inch square washer
   C. use a # 8 sheet metal screw with a $2\frac{1}{2}$ inch square washer
   D. use a # 10 sheet metal screw with a 2 inch round washer

1059. Screws used to fasten ductboard material must be
   A. a minimum of $\frac{1}{2}$ inch longer than the board thickness.
   B. a minimum of $\frac{1}{4}$ inch longer than the board thickness.
   C. a minimum of $\frac{3}{4}$ inch longer than the board thickness.
   D. a minimum of 1 inch longer than the board thickness.

1060. A shiplap joint flap must be stapled every
   A. 2 inches on center.
   B. 3 inches on center.
   C. 4 inches on center.
   D. 5 inches on center.

1061. What is commonly used to cover a stapled shiplap joint flap?
   A. 2% silver tape
   B. duct tape
   C. aluminum foil tape
   D. mastic used with masking tape

1062. Crosstabs are used to hold seams in position when staples are not used. A crosstab is made of
   A. a 5 inch minimum length of aluminum foil tape that is placed across a taped joint every 15 inches.
   B. an 8 inch minimum length of aluminum foil tape that is placed across a taped joint every 12 inches.
   C. a 10 inch minimum length of aluminum foil tape that is placed across a taped joint every 15 inches.
   D. a 12 inch minimum length of aluminum foil tape that is placed across a taped joint every 12 inches.

1063. The minimum width of pressure sensitive aluminum foil tape that is used on fiberglass ductboard is
   A. $1\frac{1}{2}$ inches.
   B. 2 inches.
   C. $2\frac{1}{2}$ inches.
   D. 3 inches.

1064. Aluminum foil tape must overlap each side of a fiberglass ductboard joint by a minimum of
   A. 1 inch.
   B. $1\frac{1}{2}$ inches.
   C. 2 inches.
   D. $2\frac{1}{2}$ inches.

1065. If aluminum foil tape is stored in an area less than 50° F,
   A. the tape is permanently damaged and should not be used.
   B. the tape can be used only with a mastic material.
   C. the tape must be preheated before it is used.
   D. additional glue must be added to the tape before use.

1066. When using heat activated tape, what temperature must the tape be heated to after being applied to the fiberglass ductboard?
   A. 300°–400° F
   B. 400°–500° F
   C. 450°–550° F
   D. 550°–600° F

1067. What is the minimum width of heat activated tape that is used on fiberglass ductboard?
   A. 2 inches
   B. $2\frac{1}{2}$ inches
   C. 3 inches
   D. $3\frac{1}{2}$ inches

1068. What is the purpose of a tie rod in a fiberglass ductboard system?
   A. It is used to hang the duct from a ceiling.
   B. It is used to raise the duct off of the ceiling joists when placed in an attic.
   C. It is used to connect two sections of duct together.
   D. It is used to increase the ducts rigidity.

1069. A tie rod used for fiberglass ductboard must be made of
   A. 18 gauge stainless steel wire.
   B. 16 gauge galvanized steel wire.
   C. 14 gauge stainless steel wire.
   D. 12 gauge galvanized steel wire.

1070. A tie rod
   A. is commonly placed horizontally in a ductboard system.
   B. is commonly placed vertically in a ductboard system.
   C. can only be used with 1 inch thick fiberglass ductboard systems.
   D. can only be used with $1\frac{1}{2}$ inches thick fiberglass ductboard systems.

1071. When a transverse joint is used in a fiberglass ductboard system, the tie rod must be placed
   A. 4 inches from the end of the male joint.
   B. 4 inches from the end of the female joint.
   C. 5 inches from the end of the male joint.
   D. 5 inches from the end of the female joint.

1072. When a butt joint is used in a fiberglass ductboard system, the tie rod must be placed
   A. 3 inches on either side of the joint.
   B. 4 inches on either side of the joint.
   C. 5 inches on either side of the joint.
   D. 6 inches on either side of the joint.

1073. When using a male and female shiplap joint in a fiberglass ductboard system,
   A. air must always flow from the female to the male end of the joint.
   B. air must always flow from the male to the female end of the joint.
   C. air can flow in either direction.
   D. Male shiplap joints are never used with female joints.

1074. When are V-Grooves most commonly used in fiberglass ductboard systems?
   A. when connecting two ductboard sections together
   B. on ductboard corners as a seam
   C. on single flat board sections to create a rectangular duct
   D. to connect supports to the ductboard system

1075. What type of joint is not permitted on corners of a ductboard system?
   A. male to female shiplap joints
   B. square cut to square cut joints
   C. female shiplap to square cut joints
   D. V-grooves

1076. In the above diagram, what type of corner is being used?
   A. male shiplap with a square cut
   B. male shiplap with a female shiplap
   C. female shiplap with a square cut
   D. square cut to square cut

1077. In the above diagram, what type of corner is being used?
   A. male shiplap with a square cut
   B. shiplap corner fold
   C. female shiplap with a square cut
   D. male shiplap with a female shiplap

Section 3 Air Flow Components and Duct Fabrication   185

1078. In the above diagram, what type of corner is being used?
   A. male shiplap and square cut with a flap
   B. shiplap corner fold with a flap
   C. female shiplap and square cut with a flap
   D. male shiplap and a female shiplap with a flap

1079. In the above diagram, what type of corner is being used?
   A. male to female shiplap joint
   B. square cut to square cut joint
   C. female shiplap to square cut joint
   D. V-groove corner fold

AIR SIDE COMPONENTS

1080. Two general types of damper blades are
   A. adjustable and non-adjustable.
   B. parallel blade and opposing blade.
   C. eccentric blade and opposing blade.
   D. direct and indirect.

1081. Which type of damper is known to give more uniformity of air flow through a system?
   A. adjustable direct damper
   B. parallel blade damper
   C. eccentric damper
   D. opposing blade damper

1082. A mixed air sensor is commonly used in conjunction with
   A. an economizer.
   B. an ignition control system.
   C. a compressor circuit.
   D. an electric heating coil.

1083. The main purpose of an economizer circuit is to
   A. add a heat load to an air conditioning system during low load conditions.
   B. introduce outside air for ventilation and free cooling.
   C. ventilate unwanted air from a building.
   D. bypass cooling coils during low load conditions.

1084. What controls are commonly found in economizer circuits?
   A. enthalpy, air volume, and air velocity sensors
   B. humidistats, thermostats, and discharge air sensors
   C. pressure and temperature sensors
   D. enthalpy, discharge air, and mixed air sensors

1085. What can be used to prevent the pressurization of a building?
   A. an economizer system
   B. relief dampers
   C. opposing blade type dampers
   D. face and bypass dampers

1086. What is used together to control the amount of air over a heating or cooling coil in a constant air volume system?
   A. an economizer system
   B. relief dampers
   C. opposing blade type dampers
   D. face and bypass dampers

1087. How does a variable air volume (VAV) system operate?
   A. air volume is reduced as the load is increased
   B. air volume is increased as the load increases
   C. air volume is increased as the room temperature decreases during cooling
   D. air volume is decreased as the room temperature increases during heating

1088. What is the primary purpose of a damper?
   A. to increase the velocity in a branch run
   B. to decrease velocity in a branch run
   C. to isolate a duct section
   D. to balance air flow in a duct system

1089. What type of damper is required in a commercial building where a section of duct extends through two different rooms?
   A. a balancing damper
   B. a partitioning damper
   C. a remote damper
   D. a fire damper

1090. What is used to open and close a fire damper?
   A. an electrical cam and switching mechanism
   B. a bimetal rod
   C. a fusible link
   D. an expansion type joint

1091. What is commonly used to open and close a smoke damper?
   A. a photoelectric sensor
   B. a cadmium sensor
   C. a fusible link
   D. an expansion joint

1092. What is the purpose for proper ventilation?
   A. to bring the proper combustion air into a heating system
   B. to bring the proper primary and secondary air into a heating system
   C. so a system can extract unwanted dust particles
   D. to increase oxygen levels in the air

1093. The temperature difference between conditioned space air and ventilation air must not exceed
   A. 5° F.
   B. 10° F.
   C. 20° F.
   D. 30° F.

AIR FILTERS

1094. Where should a filter be placed in an HVAC system?
   A. in the return air system upstream from a heat exchanger or a cooling coil
   B. in the return air system downstream from a heat exchanger or a cooling coil
   C. in the supply air system upstream from a heat exchanger or a cooling coil
   D. in the supply air system downstream from a heat exchanger or a cooling coil

1095. Air filters must be installed
   A. so that they are readily removable.
   B. so that they are always next to a return air boot.
   C. within 6 inches maximum from the furnace return air opening.
   D. within 10 inches maximum from the furnace return air opening.

1096. What aids an air distribution system in evenly disbursing air across a filter media?
   A. an accordion type air filter
   B. a non-adjustable damper
   C. turning vanes
   D. an adjustable damper

1097. What is meant by the term ionizing?
   A. a process of removing heat
   B. charging a particle with a negative or positive charge
   C. adding resistance to a circuit
   D. coating a material with a solution

1098. What are three common types of electronic air filters?
   A. removable plate type, ionizing plate type, and charged media ionizing type
   B. retractable static type, ionizing type, and non-ionizing type
   C. ionizing plate type, charged media non-ionizing type, and charged media ionizing type
   D. removable plate type, retractable static type, and ionizing plate type

1099. What type of electronic filter generates positive ions on wires and uses a high voltage panel that holds the dust particles?
   A. ionizing plate type
   B. retractable static type
   C. charged media non-ionizing type
   D. removable plate type

1100. What type of electronic filter uses a dielectric medium that is supported by an electrically charged grid?
   A. ionizing plate type
   B. retractable static type
   C. charged media ionizing type
   D. charged media non-ionizing type

1101. What type of electronic filter generates ions and uses a dielectric medium that is supported by an electrically charged grid?
   A. ionizing plate type
   B. retractable static type
   C. charged media ionizing type
   D. charged media non-ionizing type

1102. An electronic air filter removes particles as small as
   A. 0.01 microns.
   B. 0.001 microns.
   C. 0.0001 microns.
   D. 0.00001 microns.

Section 3   Air Flow Components and Duct Fabrication   187

1103. Where is an electronic air filter commonly installed on a furnace?
A. ahead of the heat exchanger
B. down stream from the blower
C. in the supply side of the furnace
D. in the return side of the furnace

1104. What is commonly used with electronic air filters to trap large dust particles?
A. a mesh screen after the electronic element
B. a pre-filter before the electronic element
C. a pre-filter after the electronic element
D. an electrostatic media before the element

1105. The average that the cells of an electronic air cleaner must be cleaned are
A. once a week.
B. once every 2 months.
C. once every 6 months.
D. once every year.

1106. What is used to measure the pressure drop across a filter?
A. a water gauge manometer
B. a mercury gauge manometer
C. an anenometer
D. a velometer

1107. Disposable filters should be changed at least
A. once a week.
B. once every 2 months.
C. once every 6 months.
D. once every year.

HUMIDIFIERS

1108. A plate type humidifier, a rotating drum type humidifier, and a fan powered humidifier are all considered
A. evaporative type humidifiers.
B. atomizing type humidifiers.
C. vaporizing type humidifiers.
D. Both A and C.

1109. Which two of the following humidifiers use a bypass duct?
A. the plate type and rotating plate type humidifiers
B. the drum type and plate type humidifiers
C. the flow through type and the fan powered type
D. the drum type and the flow through type

1110. What is the purpose of the bypass duct that is used on some humidifiers?
A. It allows the humidifier to be closed during the summer months.
B. It allows for the humidifier to be used only in the winter months.
C. It increases the efficiency of the humidifier.
D. All of the above.

1111. Which type of humidifier uses a solenoid valve to control water flow?
A. a flow through type
B. a fan powered type
C. a rotating plate type
D. a rotating drum type

1112. Which type of humidifier uses a float valve to control water flow?
A. a plate type
B. a rotating plate type
C. a rotating drum type
D. All of the above.

1113. Which two of the following humidifiers are commonly mounted on the supply duct without the use of a bypass duct?
A. the plate type and fan powered type
B. the drum type and the flow through type
C. the flow through type and the fan powered type
D. the drum type and plate type

1114. Which humidifier commonly uses a heating element?
A. the vaporizing type
B. the atomizing type
C. the evaporative type
D. the flow through type

1115. What is the average recommended relative humidity for a home during the winter months?
A. 10% to 20%
B. 35% to 45%
C. 65% to 75%
D. 85% to 95%

1116. Why must the humidity level in a home be increased during the winter months?
A. This will cause the moisture on the skin to evaporate more rapidly, improving the comfort level.
B. This will cause the moisture on the skin to evaporate more slowly, improving the comfort level.
C. This will not affect the rate of the evaporation of moisture on the skin, it mainly has to do with a decrease of static electricity.
D. None of the above.

1117. The most common type of humidifier circuit for residential applications is
   A. 120V / 15A.
   B. 24V / 10A.
   C. 24V / 20 VA.
   D. 120V / 40 VA.

1118. What is the purpose of the humidistat when used to control a humidifier?
   A. It senses humidity and closes on a decrease.
   B. It senses temperature and humidity and closes on a decrease.
   C. It senses temperature and humidity and opens on a decrease.
   D. It senses humidity and opens on a decrease.

1119. What load must be energized before the humidifier circuit is energized?
   A. the gas valve
   B. the induced draft blower
   C. the indoor blower
   D. All of the above.

1120. What is the purpose of a current sensing relay?
   A. It senses the high speed of the blower and energizes the humidifier transformer.
   B. It always senses the low speed of the blower and energizes the 24V humidifier circuit.
   C. It senses the heating speed of the blower and energizes the humidifier transformer or the 24V humidifier circuit.
   D. A current sensing relay is only used on the cooling circuit.

1121. On modern furnace controls, the terminals marked H or HUM
   A. energize the 120V primary side of the humidifier transformer only.
   B. energize the 24V humidifier control circuit only.
   C. can energize the 120V primary side of the transformer or the 24V humidifier control circuit.
   D. None of the above.

TEMPERATURE AND HUMIDITY

1122. What is commonly used as a sensing element for an electronic thermometer?
   A. a resistor
   B. a transistor
   C. a thermistor
   D. a diode

1123. A dry bulb thermometer
   A. measures sensible temperature of a substance.
   B. senses the point at which moisture condenses in the air.
   C. uses a saturated wick over the end of a thermometer.
   D. senses the moisture content of the air.

1124. Which instrument contains both the wet-bulb and the dry-bulb thermometer?
   A. a sling psychrometer
   B. a dial thermometer
   C. a recording thermometer
   D. an electronic thermometer

1125. A sling psychrometer is used to determine
   A. the absolute humidity of the air.
   B. the relative humidity of the air.
   C. the dew point temperature of the air.
   D. the apparatus dew point temperature.

1126. When using a wet-bulb thermometer, as the moisture content of the air decreases
   A. the temperature of the bulb increases.
   B. the temperature of the bulb decreases.
   C. the temperature of the bulb stays the same.
   D. moisture content does not effect the wetbulb temperature.

1127. A Psychrometric chart is used to
   A. measure air content.
   B. measure air currents.
   C. measure air properties.
   D. measure air in SI units.

Section 3 Air Flow Components and Duct Fabrication

1128. On a Psychrometric chart, which lines are vertical?
   A. the dry-bulb temperature lines
   B. the wet-bulb temperature lines
   C. the dewpoint temperature lines
   D. the relative humidity lines

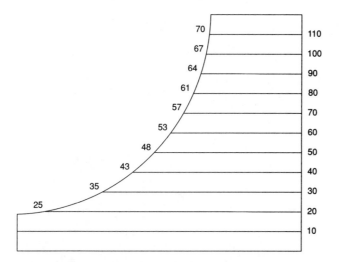

1129. On a Psychrometric chart, which lines are horizontal?
   A. the dry-bulb temperature lines
   B. the wet-bulb temperature lines
   C. the dewpoint temperature lines
   D. the relative humidity lines

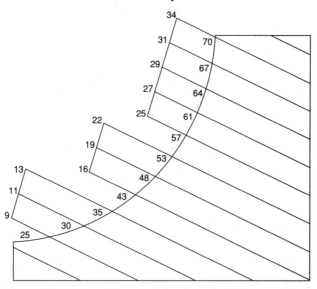

1130. On a Psychrometric chart, which lines are diagonal?
   A. the dry-bulb temperature lines
   B. the wet-bulb temperature lines
   C. the dewpoint temperature lines
   D. the relative humidity lines

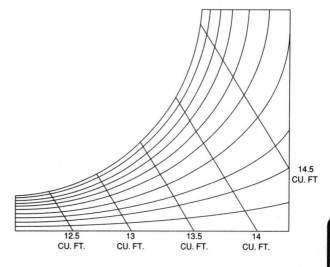

1131. On a Psychrometric chart, which lines are curved?
   A. the dry-bulb temperature lines
   B. the wet-bulb temperature lines
   C. the dewpoint temperature lines
   D. the relative humidity lines

1132. Complete air conditioning means to
   A. cool air within a space.
   B. dehumidify air within a space.
   C. cool and dehumidify air within a space.
   D. heat, humidify, cool, dehumidify, and filter air within a space.

1133. A typical comfort goal for the HVAC industry is maintaining air
   A. to an 80° F wet-bulb temperature at 70% relative humidity.
   B. to a 75° F dry-bulb temperature at 50% relative humidity.
   C. to a 65° F dry-bulb temperature at a 40% relative humidity.
   D. to a 60° F wet-bulb temperature at 35% relative humidity.

1134. What two factors decrease the rate of evaporation of moisture from a person to the surrounding air?
   A. a high dry-bulb temperature and a low relative humidity
   B. a low dry-bulb temperature and a high relative humidity
   C. a low dry-bulb temperature and a low relative humidity
   D. a high dry-bulb temperature and a high relative humidity

Section 4

Indoor Air Quality (IAQ) and Safety

INDOOR AIR QUALITY (IAQ)

1135. What's the difference between a contaminant and a pollutant?
   A. A contaminant may or may not be a health risk and a pollutant is always a health risk.
   B. A pollutant may or may not be a health risk and a contaminant is always a health risk.
   C. There is no difference. Contaminants and a pollutants are always health risks.
   D. There is no difference. Contaminants and a pollutants are never health risks.

1136. Respirable particles that are breathed deep into the lung are commonly
   A. 100 microns or less.
   B. 50 microns or less.
   C. 40 microns or less.
   D. 10 microns or less.

1137. Asbestos is made from a mineral called chrysolite. If it is breathed into the lungs, it could cause
   A. scarring of the lung tissue.
   B. lung cancer.
   C. damaging to the abdomen.
   D. All of the above.

1138. The first task a technician should do to improve air quality is to
   A. clean the ventilation system.
   B. close off return air then clean ductwork.
   C. change the air filters.
   D. open all dampers to supplies.

1139. Most inexpensive filters are
   A. approximately 7% efficient.
   B. approximately 50% efficient.
   C. approximately 75% efficient.
   D. approximately 90% efficient.

1140. What is the efficiency of a common pleated filter?
   A. 5%
   B. 30%
   C. 80%
   D. 90%

1141. What type of air filter would remove tobacco smoke particles efficiently?
   A. disposable filters
   B. electronic filters
   C. pleated filters
   D. carbon type filters

1142. Stratification refers to air that
   A. has been over heated.
   B. has been over cooled.
   C. settles in different temperature layers.
   D. has an extreme temperature rise.

1143. Air must be kept moving
   A. to eliminate stratification.
   B. to prevent a pressure drop.
   C. to increase a positive building pressure.
   D. to decrease a positive building pressure.

1144. Into which four groups are air contaminants classified?
   A. gasses / vapors, inert particles, micro-organisms, pollen.
   B. gasses / vapors, pollen, radon, organisms.
   C. radon, gasses / vapors, molds, pollen
   D. pollen, molds, inert particles, micro-organisms

1145. Manmade fibers, dust, and cigarette smoke are in which of the following groups?
   A. vapors
   B. gasses
   C. inert particles
   D. organisms

1146. Microorganisms include
   A. gasses and bacteria.
   B. fungi, mold, and bacteria.
   C. vapors and mold.
   D. mold and fibers.

1147. Which of the following contaminants greatly effect HVAC/R air distribution equipment?
   A. microorganisms
   B. inert particles
   C. gasses and vapors
   D. pollens

1148. Which of the following are considered to be pollen type contaminants?
   A. molds and bacteria
   B. ragweed and molds
   C. fungi and grass
   D. grass and seasonal plants

1149. Symptoms such as eye, nose, and throat irritations; dizziness; headaches; drowsiness; fatigue; and burning eyes are commonly caused by
   A. influenza symptoms.
   B. cold symptoms.
   C. smoke inhalation symptoms.
   D. sick building syndrome (SBS).

1150. A satisfactory CO_2 measurement in an indoor environment would be
   A. a maximum of 1400 PPM.
   B. a maximum of 1200 PPM.
   C. a maximum of 1000 PPM.
   D. a maximum of 800 PPM.

1151. Indoor air contaminants are
   A. commonly 2–5 times higher in quantity than outdoor air contaminants.
   B. commonly 8–10 times higher in quantity than outdoor air contaminants.
   C. commonly 12–14 times higher in quantity than outdoor air contaminants.
   D. commonly the same quantity as outdoor air contaminants.

1152. Which illness is commonly caused by untreated cooling tower water?
   A. influenza virus
   B. Parkinson's disease
   C. Legionnaires disease
   D. respiratory symptoms

1153. When cleaning non-porous materials, what is commonly used to prevent the growth of mold?
   A. a non-oil base detergent
   B. a solution of 90% water and 10% bleach
   C. 50% bleach and 50% soap
   D. a toxic solution of 50% bleach and 50% ammonia

1154. Some air leakage into a building will help
   A. decrease the temperature rise.
   B. increase return air temperatures.
   C. reduce indoor air pollution.
   D. increase indoor air pollution.

1155. A high relative humidity within a building can
   A. cause static electricity.
   B. cause a heating system to overshoot.
   C. decrease the supply air temperature of a heating system.
   D. produce mold.

1156. A ventilation system will exhaust stale indoor air and
   A. eliminate air pollution.
   B. maintain a low relative humidity.
   C. introduce fresh outside air.
   D. maintain a high relative humidity.

1157. Common types of ventilation systems are
   A. passive and exhaust only types.
   B. non-heat recovery units and passive types.
   C. balanced mechanical and heat recovery types.
   D. All of the above.

1158. What must a public building have for proper ventilation?
   A. an outside air supply and an exhaust system
   B. closed dampers to the interior room
   C. at least fifty percent outside air at all times
   D. at least forty percent outside air at all times

Section 4   Indoor Air Quality (IAQ) and Safety

1159. What is a VOC (volatile organic compound)?
   A. a highly evaporative chemical that gives off gasses
   B. a chemical that stays a liquid and has a high boiling point
   C. a solid compound that is commonly found in the ground
   D. None of the above.

1160. What symptoms can VOCs cause on a human being?
   A. rashes and nausea
   B. sinus congestion and drowsiness
   C. headaches and respiratory problems
   D. All of the above.

1161. The radioactive gas that comes from the Earth's crust is known as
   A. phosgene.
   B. radon.
   C. VOC compound.
   D. formaldehyde.

1162. Radon enters the home through
   A. cracks, joints, and non-trapped drains.
   B. sump pumps and water supplies.
   C. underground service openings and crawlspaces.
   D. All of the above.

1163. Why does radon commonly stay near ground level?
   A. because it's lighter that air
   B. because it's nine times heavier than air
   C. because radioactive particles are commonly attracted to the Earth's crust
   D. None of the above.

1164. Radon is not considered harmful to humans until it
   A. decays.
   B. mixes with mold bacteria.
   C. mixes with oxygen.
   D. is breathed into the lungs.

1165. Radon levels inside a building should be considered harmful when the average annual concentration in the normal living areas exceeds
   A. 300 beccqirels (Bq) per cubic meter.
   B. 600 beccqirels (Bq) per cubic meter.
   C. 800 becquerels (Bq) per cubic meter.
   D. 400 beccqirels (Bq) per cubic meter.

1166. What organic compounds are produced when materials containing carbon and hydrogen are burned?
   A. hydrogen oxide
   B. polynuclear aromatic hydrocarbons
   C. phosgene gasses
   D. halogen gasses

1167. Outdoor sources of "PAHs" are commonly caused by
   A. wood burning appliances.
   B. coal burning equipment.
   C. internal combustion engines.
   D. All of the above.

1168. Indoor sources of "PAHs" are commonly caused by
   A. cigarette smoke.
   B. burning of food.
   C. fireplaces.
   D. All of the above.

1169. What is ozone?
   A. a stable oxygen molecule
   B. 1 oxygen atom and 1 hydrogen atom
   C. a makeup of oxygen and nitrogen molecules
   D. 3 oxygen atoms

1170. How is ozone produced?
   A. when air is heated by the sun and the humidity is increased
   B. when chlorine is released to the atmosphere
   C. during an electrical storm
   D. Both A and B.

1171. The level of ozone inside the home
   A. is directly correlated with the levels outside the home.
   B. is commonly higher than the levels outside the home.
   C. is caused by heat producing appliances.
   D. None of the above.

1172. Nitric Oxide (NO) is a poisonous gas which is produced
   A. during low temperature combustion.
   B. when there is a lack of primary air.
   C. when the combustion process has been delayed too long.
   D. during high temperature combustion.

1173. Nitric oxide is
   A. highly toxic.
   B. non-toxic.
   C. flammable.
   D. explosive.

1174. Common indoor levels of nitrogen oxide are
   A. 0.25 PPM or less.
   B. 5 PPM or less.
   C. 20 PPM or less.
   D. 50 PPM or less.

1175. Nitric oxide and nitrogen dioxide are produced when
   A. fuel burning appliances are not vented.
   B. nitrogen and oxygen are mixed.
   C. carbon dioxide and oxygen are combined.
   D. not enough air is infiltrated into a home.

1176. Once nitric oxide is released into the air, it quickly changes to
   A. carbon dioxide.
   B. carbon monoxide.
   C. nitrogen.
   D. nitrogen dioxide.

1177. Carbon monoxide (CO) is a poisonous gas that is produced
   A. during complete combustion of gas products.
   B. during incomplete combustion of fossil fuels.
   C. during any burning of fuels, as it's a product of combustion.
   D. All of the above.

1178. Indoor sources of carbon monoxide are
   A. cracked heat exchangers and chimneys.
   B. unvented appliances.
   C. cigarette smoke.
   D. All of the above.

1179. Average indoor levels of carbon monoxide range from
   A. 0.5 to 5 PPM.
   B. 8 to 10 PPM.
   C. 10 to 15 PPM.
   D. 20 to 50 PPM

1180. If a home has excessive negative pressure, the combustion gasses of an appliance could
   A. enter back into the home.
   B. produce carbon monoxide.
   C. produce ozone.
   D. Both A and B.

1181. What type of contaminant is formaldehyde considered?
   A. a VOC
   B. an inert particle
   C. a micro-organism
   D. None of the above.

1182. Medium density fiberboard (MDF) and particle board inside a home is a common source of which indoor contaminant?
   A. formaldehyde
   B. nitric acid
   C. carbon monoxide
   D. excessive CO_2

1183. The rate that formaldehyde gas is released in a home is directly proportional to
   A. an increase of humidity and a decrease of temperature.
   B. a decrease of humidity and an increase of temperature.
   C. a decrease of humidity and a decrease of temperature.
   D. an increase of humidity and an increase of temperature.

PERSONAL SAFETY

1184. What type of safety equipment is commonly used to protect against falling objects?
   A. safety glasses
   B. ear plugs
   C. gloves
   D. hard hats

1185. What type of safety equipment is commonly used to protect against flying objects?
   A. safety glasses
   B. respirators
   C. gloves
   D. a safety harness

1186. What type of safety equipment is commonly used to protect against scrapes, cuts, and burns?
   A. safety glasses
   B. respirators
   C. gloves
   D. a safety harness

1187. What type of safety equipment is commonly used to protect against falling objects and punctures?
   A. ear plugs
   B. respirators
   C. steel toe boots
   D. a safety harness

1188. What type of safety equipment is commonly used for protection in high noise level areas?
   A. ear plugs
   B. respirators
   C. steel toe boots
   D. hard hats

1189. What type of safety equipment is commonly used for protection in a highly contaminated area?
   A. ear plugs
   B. respirators
   C. safety glasses
   D. hard hats

Section 4   Indoor Air Quality (IAQ) and Safety

1190. What type of safety equipment is commonly used for protection against falls when working in high areas?
   A. ear plugs
   B. respirators
   C. steel toe boots
   D. safety harness

1191. A safety harness should be used when working on objects
   A. more than 4 feet above the ground.
   B. more than 10 feet above the ground.
   C. more than 6 feet above the ground.
   D. more than 12 feet above the ground.

1192. When altering any safety equipment, what should be done after use?
   A. Always change it back to it original working order after use.
   B. Place a note near the equipment explaining the alteration in detail.
   C. If the alteration worked, nothing is needed.
   D. Safety equipment must never be altered in any way.

1193. To prevent an injury when using a wrench, it is safer to
   A. pull the wrench towards you.
   B. push the wrench away from you.
   C. pull or push the wrench.
   D. use maximum force at all times.

1194. When lifting objects,
   A. it is best to use your back muscles only.
   B. bend your back and use leg muscles.
   C. keep your back straight and use leg muscles.
   D. always use your arm and chest muscles.

1195. When lifting heavy objects,
   A. always use your back muscles.
   B. always use your arm muscles.
   C. always use your arm and back muscles.
   D. always ask for assistance.

1196. To prevent a personal injury when working with power tools, always
   A. wear hand protection.
   B. wear eye protection.
   C. wear long sleeves to protect arms.
   D. use AGA approved power tools.

1197. To prevent a personal injury when handling sheetmetal, always
   A. wear gloves that are a non-slip type.
   B. wear safety glasses.
   C. ask for assistance when handling large or heavy sections.
   D. All of the above.

1198. When using extension ladder, you should
   A. keep a three-point contact with the ladder.
   B. extend the ladder all the way.
   C. extend the ladder only one half the way.
   D. always get help extending the ladder.

1199. Never use a hammer and chisel without
   A. hand protection.
   B. ear protection.
   C. eye protection.
   D. foot protection.

1200. Gasoline should never be used to clean tools because
   A. the vapors are flammable.
   B. it will dissolve the tool being cleaned.
   C. vapors will remain with the equipment being cleaned.
   D. it does not cut grease well enough.

1201. To prevent a personal injury,
   A. never use a hammer with a rounded head.
   B. never use a worn chisel.
   C. use only well oiled hand tools.
   D. hold a screw driver only by its handle when walking.

REFRIGERANT SAFETY

1202. To prevent an injury when working with refrigerants, what safety precautions are necessary?
   A. always have a fire extinguisher nearby
   B. always have a phosgene sensor near the equipment being serviced
   C. always use safety glasses and gloves
   D. always use ear protection

1203. What two types of injuries can CFC, HCFC, and HFC refrigerants cause?
   A. lung cancer and heart attacks
   B. heart attacks and frost bite burns
   C. carbon monoxide poisoning and frost bite burns
   D. frost bite burns and suffocation

1204. Which of the following may cause suffocation?
   A. using a torch on a refrigeration system under pressure
   B. venting refrigerant in a confined space
   C. using a vacuum pump in a confined space
   D. using a torch while in an enclosed walk-in cooler

1205. Which one of the following statements is true?
    A. A system should be evacuated while under pressure.
    B. It is safe to heat a refrigerant cylinder above 125° F.
    C. Refrigerant is lighter than air.
    D. Air is lighter than refrigerant.

1206. What is required in equipment rooms that use systems containing large quantities of refrigerant?
    A. phosgene and refrigerant alarms and sensors
    B. refrigerant and carbon monoxide alarms and sensors
    C. carbon monoxide and oxygen alarms and sensors
    D. oxygen and refrigerant alarms and sensors

1207. When halogens are in contact with an open flame,
    A. an explosion will occur.
    B. phosgene gas is produced.
    C. it is noticeable from a green color.
    D. it is noticeable from a purple color.

1208. What care must be taken when working around toxic fumes?
    A. open a window if possible for ventilation
    B. hold your breath as long as possible
    C. a breathing apparatus must be used
    D. no extra care must be taken

1209. What must always be available when working with large amounts of refrigerants in equipment rooms?
    A. a self contained breathing apparatus
    B. a fire extinguisher
    C. a carbon monoxide detector
    D. a smoke detector

1210. When leak testing systems using nitrogen, the technician should
    A. not mix refrigerant with the nitrogen.
    B. leak test through the high-side only.
    C. always use a pressure regulator and safety relief valve at the nitrogen tank.
    D. leak test through the low-side only.

1211. When using nitrogen in a system for leak detection, a technician
    A. must use two separate gauges when connecting the nitrogen and a trace refrigerant to the system at the same time.
    B. should add mostly refrigerant when testing a system for leaks.
    C. does not have to use a pressure regulator on the nitrogen tank.
    D. must never connect nitrogen and refrigerant to the system at the same time.

1212. When pressure testing a refrigeration system,
    A. the system must always be in operation.
    B. the pressures must be at least 50 PSIG.
    C. the pressures must not exceed system nameplate pressures.
    D. the system must always be de-energized.

1213. What mixture is the most explosive?
    A. air and refrigerant
    B. air and nitrogen
    C. nitrogen and oil
    D. oxygen and oil

1214. To prevent an explosion, never leak test a system using
    A. CO_2.
    B. oxygen.
    C. nitrogen.
    D. R-22.

1215. When charging refrigerant into an air conditioning system, the technician should
    A. never charge in a liquid state.
    B. never charge in a vapor state.
    C. always charge by superheat method only.
    D. always charge by the manufacturer's recommendation.

1216. To prevent an explosion, the discharge service valve on a compressor must never be
    A. in the mid-seated position during operation.
    B. in the back-seated position during operation.
    C. in the front-seated position during operation.
    D. in the back-seated during a system pump down.

1217. Which refrigerant lines might cause a serious burn?
    A. the discharge line
    B. the suction line
    C. the liquid line
    D. the inlet line

1218. When handling large air conditioning fin and tube coils,
    A. always use the proper safety harness.
    B. always evacuate the coil.
    C. always use hand protection and ask for assistance.
    D. always remove the coil from the package system.

1219. Acidic oil from compressors is harmful because
    A. it causes skin burns.
    B. it contains oil.
    C. it contains oil and refrigerant.
    D. it contains refrigerant properties.

Section 4 Indoor Air Quality (IAQ) and Safety

1220. When testing for natural gas leaks, never use
   A. a soap bubble solution.
   B. an open flame.
   C. electronic detectors.
   D. a heavy soap solution containing chlorine.

1221. To prevent danger caused by excessive pressures,
   A. never overcharge an air conditioning system.
   B. never undercharge an air conditioning system.
   C. never allow an air conditioning system to operate above 200 PSIG.
   D. never allow an air conditioning system to operate above 150 PSIG.

1222. Before opening a refrigeration system to the atmosphere,
   A. always purge air from the manifold gauge set.
   B. connect a vacuum pump and evacuate the system to 500 microns.
   C. remove the refrigerant until the system's pressure is 0 PSIG.
   D. remove the refrigerant until the system's pressure is 30 PSIG.

1223. To prevent the refrigerant from slugging the compressor,
   A. never charge vapor into the high side while under a vacuum.
   B. never charge liquid into the high side while under a vacuum.
   C. never charge vapor into the low side.
   D. never charge liquid into the low side.

1224. If refrigerant vapor is vented into the atmosphere in a confined condition, it will
   A. cause an explosive mixture.
   B. be very flammable.
   C. displace oxygen.
   D. create phosgene gas.

1225. Disposable refrigerant cylinders
   A. are rarely color-coded.
   B. can be refilled with the proper equipment.
   C. can be filled up to 80% of tank capacity.
   D. should never be refilled.

1226. A rechargeable refrigerant cylinder must
   A. not be used as a storage tank.
   B. be over 45 pounds.
   C. never be filled over 80% capacity.
   D. be heated before it can be refilled with refrigerant.

1227. When charging a rechargeable cylinder, always take into account the cylinder weight known as the
   A. tare lbs.
   B. tank lbs.
   C. cylinder lbs.
   D. None of the above.

1228. Refrigerant cylinders should never be
   A. heated with warm water.
   B. heated with a torch.
   C. stored at temperatures below 30° F.
   D. refilled.

1229. Any stored refrigerant cylinder must always be
   A. properly labeled and placed in an upright position.
   B. kept at temperatures below 40° F to prevent expansion.
   C. no more than 30 pounds.
   D. at least 20% full at all times.

1230. Refrigerant recovery cylinders are commonly color coded
   A. as yellow.
   B. as gray.
   C. with a gray bottom and a yellow top.
   D. with a yellow bottom and a gray top.

1231. When heating a refrigerant cylinder using warm water, the temperature of the cylinder must never exceed
   A. 60° F, or the rupture disc could open.
   B. 90° F, or the rupture disc could open.
   C. 125° F, or the rupture disc could open.
   D. Cylinders must never be heated with warm water.

1232. Refrigerant cylinders must be approved by
   A. GAMA.
   B. AGA.
   C. DOT.
   D. ACCA.

1233. Which of the following are common types of pressure relief devices?
   A. fusible plug, rupture disc, and pressure relief valve
   B. fusible plug, vacuum break, and rupture disc
   C. thermal disc, fusible disc, and rupture plug
   D. fusible link, vacuum plug, and pressure relief valve

1234. Which of the following valves must never be capped?
   A. the discharge service valve
   B. the suction service valve
   C. the king valve
   D. the pressure relief valve

1235. Relief devices must
   A. be installed in series with one another.
   B. not be installed in parallel with one another.
   C. not be installed in series with one another.
   D. be inspected annually by DOT.

1236. When pressurizing a low pressure chiller, never exceed
   A. 25 PSIG on the low side.
   B. 25 PSIG on the high side.
   C. 10 PSIG on the low side.
   D. 10 PSIG on the high side.

1237. When charging a liquid chiller,
   A. refrigerant liquid is commonly added to the low side of the system.
   B. refrigerant vapor is commonly added to the low side of the system.
   C. refrigerant liquid is commonly added to the high side of the system.
   D. refrigerant vapor is commonly added to the high side of the system.

1238. A fill limit device used on a recovery system
   A. shuts down the machine when a container reaches 50% full.
   B. shuts down the machine when a container reaches 80% full.
   C. shuts down the machine when a container reaches 90% full.
   D. shuts down the machine when a container reaches 100% full.

1239. According to EPA regulations, you must never
   A. recover all refrigerants.
   B. vent refrigerant to the atmosphere.
   C. re-use refrigerant.
   D. sell reclaimed refrigerant.

1240. More caution should be given when handling class II refrigerants because
   A. they are more expensive.
   B. they are mostly used on industrial applications.
   C. they are more toxic and flammable.
   D. they are heavier.

1241. Refrigerant cylinders should never be
   A. subjected to extreme ambient temperatures.
   B. labeled.
   C. DOT or UL approved.
   D. properly color-coded.

1242. When working with refrigerants, what is most likely to describe the flammability and the toxicity of a refrigerant?
   A. the BOCA code
   B. the EPA
   C. the MSDS
   D. None of the above.

1243. What precautions must be made when storing or transporting cylinders?
   A. The cylinders must be kept under 80° F.
   B. The valve and hood caps must always be secured.
   C. The tanks must be inspected by the local mechanical inspector.
   D. None of the above.

1244. When asbestos material is used as a pipe wrapping, the technician should
   A. remove it immediately.
   B. use eye protection, then remove the wrappings.
   C. use eye, hand, and breathing apparatus protection, then remove wrappings.
   D. leave it alone and allow qualified asbestos removal technicians to handle the removal.

SOLDERING AND BRAZING SAFETY

1245. What safety device must always be present when using a torch?
   A. a striker
   B. a wet rag
   C. a fire extinguisher
   D. a respirator

1246. When soldering or brazing, always
   A. adjust the acetylene to 40 PSIG.
   B. adjust the oxygen to 70 PSIG.
   C. use a lighter when igniting the torch.
   D. wear the proper safety glasses.

1247. What is the purpose of a heat shield?
   A. it is used to absorb heat from a filter dryer when brazing
   B. it is used to protect combustible material
   C. it is used to add heat to a joint while brazing
   D. it is used to help cool a joint after it has been soldered or brazed

Section 4 Indoor Air Quality (IAQ) and Safety   199

1248. When using a torch to repair a system that is overhead,
   A. heat the male section of the joint only.
   B. heat the female section of the joint only.
   C. never stand directly under the torch.
   D. heat the center of the joint only.

1249. Brazing materials should not contain
   A. lead.
   B. lead or tin.
   C. flux.
   D. cadmium.

1250. When using torches while repairing copper tubing on a refrigeration system, always
   A. use flux.
   B. use silver solder.
   C. use brazing material.
   D. make sure the work area is well ventilated.

1251. The pressure regulators on the oxygen/acetylene torch should
   A. always be tested for good working condition.
   B. be adjusted out for use.
   C. be adjusted in after each use.
   D. be disassembled and cleaned after each use.

1252. Acetylene pressure should never exceed
   A. 5 PSI.
   B. 10 PSI.
   C. 15 PSI.
   D. 20 PSI.

1253. Soldering or brazing should never be attempted when
   A. the system has been recovered.
   B. the system is under pressure.
   C. the system has been evacuated.
   D. the system has been retrofitted.

1254. Oxygen/acetylene tanks and hoses should
   A. be shut-off and have the hoses bled after each use.
   B. be correctly color-coded.
   C. be correctly labeled and identified.
   D. All of the above.

1255. What is the proper source of ignition for an Oxygen/acetylene torch?
   A. a match
   B. a lighter
   C. a striker
   D. another torch

1256. A system must never be soldered or brazed
   A. when it is at 0 PSIG.
   B. when nitrogen is inside the system.
   C. when it is in a slight vacuum.
   D. when under pressure.

1257. When soldering or brazing a sealed system, first recover the refrigerant, then
   A. never open the system to the atmosphere.
   B. open the system to the atmosphere.
   C. there must be a minimum of 20 PSIG present.
   D. draw a vacuum of 20 inches Hg.

ELECTRICAL SAFETY

1258. After installing or servicing an electrical system,
   A. always inspect the system for loose connections, improper grounding, and defective wiring.
   B. properly color-code all wires.
   C. seal control and line voltage wires in greenfield.
   D. seal control and line voltage wires in conduit.

1259. What must always be observed when servicing electrical equipment?
   A. the brand name of the breaker panel
   B. labels and precautions
   C. the type of fuse being used
   D. the size of the equipment

1260. All electrical installation and repairs made must meet
   A. NFPA guidelines.
   B. building inspector approval.
   C. BOCA approvals.
   D. NEC guidelines.

1261. The technician should always
   A. listen to the customer's diagnosis of a system.
   B. determine the last time the equipment has been serviced.
   C. check the electrical panel for open breakers.
   D. be aware of the voltage rating of the equipment being worked on.

1262. When servicing or replacing electrical components, a technician
   A. must de-energize the equipment.
   B. must wear hand protection.
   C. must use insulated tools only.
   D. must disconnect the primary side of the transformer.

1263. An electrical safety control
   A. must not be permanently bypassed.
   B. must always be installed on the discharge line.
   C. must always be installed on the suction line.
   D. is considered the primary control of the system.

1264. Before repairing electrical equipment,
   A. check with the local electrical inspector.
   B. a voltage test should be done to ensure there is no power to the system.
   C. use an ohmmeter to check all fuses.
   D. an ammeter test is required.

1265. Before starting equipment, the technician should first
   A. check for proper voltage to the equipment.
   B. check superheat.
   C. check refrigerant charge.
   D. check all safety devices.

1266. When servicing large air conditioning equipment, the technician should
   A. always ask for assistance.
   B. always check superheat.
   C. always place a power lock-out / tag-out on the electrical disconnect.
   D. always inform the customer of the equipment's electrical and mechanical condition in great detail.

1267. An electrocution caused by a grounded compressor on an air conditioning system is possible
   A. if the power leads of the system are reversed.
   B. if the fuse protecting the circuit is improperly sized.
   C. if the high side pressure increases causing a current surge.
   D. if grounding wires are improperly connected.

1268. A GFCI is used to
   A. prevent an electrocution caused by a short circuit.
   B. prevent an electrocution caused by an open motor winding.
   C. prevent an electrocution caused by a grounded load.
   D. prevent an electrocution caused by high voltage.

1269. What is commonly used with an extension cord to prevent electrocution when working with power tools?
   A. a GFCI circuit breaker
   B. a GFCI outlet
   C. a GFCI toggle switch
   D. None of the above.

1270. Before testing a capacitor, always
   A. check for the correct MFD rating.
   B. check for the correct voltage rating.
   C. discharge the capacitor.
   D. None of the above.

1271. An electrical shock is what happens when
   A. the body resistance is below 100,000 ohms.
   B. voltage flows through the body.
   C. current flows through the body.
   D. Both B and C.

1272. What type of an electric shock can cause heart damage?
   A. One that is across both arms.
   B. One that is across the right arm and the right leg.
   C. One that is across both legs.
   D. One that is across the hand and the shoulder.

1273. When is a person more susceptible to an electric shock?
   A. when their skin is dry
   B. when a GFCI is used
   C. when their skin is moist
   D. when they are using rubber sole shoes

1274. An electrocution from as low as
   A. 14 amps can kill or cause serious injury.
   B. 10 amps can kill or cause serious injury.
   C. 1 amp can kill or cause serious injury.
   D. .025 amps can kill or cause serious injury.

1275. When replacing a fuse on an HVAC system,
   A. it is safe to use a fuse that is lower than the system's name plate.
   B. it is safe to use a fuse that is higher than the system's name plate.
   C. use a timed delay fuse that is equal to the system's name plate.
   D. use a fast acting fuse that is equal to the system's name plate.

1276. When removing a fuse from a fused disconnect, a technician must use
   A. an insulated fuse puller.
   B. an insulated screwdriver.
   C. insulated needle nose pliers.
   D. an insulated pry bar.

1277. When performing an electrical repair,
   A. always use magnetized hand tools.
   B. always inspect hand tools for adequate insulation.
   C. use an ohmmeter for testing the power supply.
   D. use an ammeter for testing the power supply.

Section 4 Indoor Air Quality (IAQ) and Safety

1278. When working on a blower that has an exposed pulley and belt system, a technician should always
A. test for belt tightness while the system is in operation.
B. inspect the system for proper safety guards.
C. test belt tension when the system is in operation.
D. adjust the belt for a 2 inch slack.

1279. When working on electrical panels, a technician should
A. have rubber sole shoes.
B. wear insulated gloves.
C. wear eye protection.
D. wear a face shield.

1280. To prevent a personal injury,
A. always disconnect the power supply before inspecting fans and motors.
B. the inspection of fans and motors should always be done with power to the system.
C. fans and motors rarely need inspection.
D. None of the above.

1281. Never service machinery
A. when wearing loose clothing.
B. without the use of a safety helmet.
C. unless wearing gloves.
D. None of the above.

1282. If a technician has excessively long hair, he or she should
A. get it cut.
B. do not work on machinery.
C. tie hair back or into a cap before working on machinery.
D. None of the above.

1283. The manufacturers of HVAC/R equipment provide wiring diagrams to
A. aid the owner of the equipment.
B. provide help in case of re-modifying the wiring.
C. indicate the correct wiring and provide location of components.
D. follow federal requirements.

1284. When working on electronic air cleaners,
A. all electrical power should be turned off.
B. the indoor blower motor must be de-energized.
C. the thermostat subbase should be placed on "fan auto."
D. the thermostat subbase should be placed on "fan manual."

1285. All electrical equipment should be
A. tagged "high voltage."
B. turned off before servicing.
C. labeled.
D. color-coded.

EPA CERTIFICATION CORE SECTION

1286. CFCs neither dissolve in water nor break down into compounds that dissolve in water, so they do not rain out of the atmosphere.
A. True
B. False

1287. How many ozone molecules will one chlorine atom destroy?
A. 1
B. 100
C. 100,000

1288. What does ODP stand for?
A. out door pollutants
B. ozone depletion potential
C. ozone distillation process

1289. What is ODP?
A. the measurement of the density of the stratosphere
B. the measurement of the ozone layer
C. the measurement of CFCs and HCFCs ability to destroy ozone

1290. Which refrigerant is the most harmful to the ozone layer?
A. R-12 CFC
B. R-22 HCFC
C. R-134A HFC

1291. The Clean Air Act
A. prohibits venting of CFCs and HCFCs as of November 15, 1995.
B. prohibits venting of CFCs and HCFCs as of July 1, 1992.
C. prohibits venting of CFCs and HCFCs as of November 14, 1994.

1292. The Clean Air Act
A. prohibits venting of HFCs as of November 15, 1995.
B. prohibits venting of HFCs as of July 1, 1992.
C. prohibits venting of HFCs as of November 14, 1994.

1293. The Clean Air Act
    A. requires a service technician to be EPA certified as of November 15, 1995.
    B. requires a service technician to be EPA certified as of July 1, 1992.
    C. requires a service technician to be EPA certified as of November 14, 1994.

1294. When disposing of a refrigerant cylinder,
    A. vent the refrigerant to the atmosphere and remove the top.
    B. recover the refrigerant and render the cylinder useless.
    C. evacuate the cylinder and use as an air tank.

1295. When is it most likely that an appliance will need its refrigerant recovered?
    A. before it can be discarded
    B. after purging the system with dry nitrogen
    C. after the system obtained a leak

1296. What is known as *de minimis* releases?
    A. It is a minimum amount of refrigerant that should not be released to the atmosphere when servicing equipment.
    B. It is a minimum amount of refrigerant that can be released to the atmosphere when servicing equipment.
    C. It is a minimum amount of refrigerant that must be recovered when servicing equipment.

1297. What does the term temperature glide refer to?
    A. the different refrigerant boiling and condensing points of a blended refrigerant
    B. an zeotropic refrigerant as it is charged into a system as a gas
    C. the point at which the refrigerant temperature decreases as it exits an expansion valve

1298. Service technicians who violate the clean air act
    A. may be fined.
    B. may lose certification.
    C. Both A and B.

1299. How much of a fine can be issued for venting refrigerants?
    A. $ 5,000
    B. $ 10,000
    C. $ 25,000

1300. An award of _____ is given to the person that supplies information about the venting of refrigerants.
    A. $ 5,000
    B. $ 10,000
    C. $ 25,000

1301. Fractionation is the separation of blended refrigerants caused by different refrigerant vapor pressures.
    A. True
    B. False

1302. What is the pressure range of the high side gauge that is located on a manifold set?
    A. 30" HG to 300 PSIG.
    B. 0 PSIG to 300 PSIG.
    C. 0 PSIG to 500 PSIG.

1303. Which of the following mixtures are legal to add into a system for the purpose of leak testing?
    A. a trace amount of R-12 and nitrogen
    B. a trace amount of the refrigerant recommended in the system and nitrogen
    C. a trace amount of R-22 and nitrogen

1304. Hygroscopic is oil that
    A. has a high attraction to moisture.
    B. should be stored at ambient temperatures.
    C. can be left open to the atmosphere.

1305. What is the state of the refrigerant leaving the condenser?
    A. low pressure liquid
    B. high pressure liquid
    C. low pressure gas
    D. high pressure gas

1306. What is the color code of refrigerant gauges?
    A. high side is blue, low side is red
    B. low side is blue, high side is red
    C. low side is white, high side is red
    D. high side is white, low side is red

1307. Which of the following is considered recycling?
    A. removing refrigerant from a system into an approved container
    B. removing, filtering, and reusing refrigerant in a system
    C. removing refrigerant from a system and transfering it to a center for a chemical analysis

1308. Refrigerant cylinders should be
    A. free of rust.
    B. non-dented and undamaged.
    C. Both A and B.

1309. Which leak detector will help locate the general area of a systems leak?
    A. electronic / ultrasonic
    B. soap bubbles
    C. halide torch

Section 4 Indoor Air Quality (IAQ) and Safety

1310. Which of the following makes it impossible to reclaim a refrigerant?
  A. having air in the refrigerant
  B. having acid in the refrigerant
  C. having two different refrigerants in one container

1311. Factors that affect the speed of evacuation are
  A. size of equipment.
  B. ambient temperature.
  C. amount of moisture in the system.
  D. All of the above.

1312. To prevent an explosion, never fill a refrigerant cylinder more than
  A. 50% full.
  B. 60% full.
  C. 80% full.
  D. 90% full.

1313. Refrigerant can cause suffocation because
  A. it is lighter than air and it displaces oxygen.
  B. it is heavier than air and it displaces oxygen.
  C. the chlorine causes a reaction to the oxygen molecule.

1314. The Montreal Protocol is
  A. a treaty regulating the production of CFCs.
  B. a method of measuring refrigerant CFCs.
  C. a method of measuring the CFCs in the stratosphere.

1315. What refrigerant is an HCFC?
  A. R-12
  B. R-22
  C. R-134A

1316. What refrigerant is an HFC?
  A. R-12
  B. R-22
  C. R-134A

1317. A ternary blend is
  A. a 3 part blend.
  B. a 2 part blend.
  C. an azeotropic mixture.

1318. What type of oil does a ternary blend use?
  A. polyolester
  B. mineral
  C. alkylbenzene

1319. Which of the following is considered a "System Dependent" (passive) recovery?
  A. using the appliance compressor to remove refrigerant
  B. using an EPA approved recovery machine
  C. None of the above.

1320. Which of the following is considered a "Self-Contained" (active) recovery?
  A. using the appliance compressor to remove refrigerant
  B. using an EPA approved recovery machine
  C. None of the above.

1321. A mechanical float device, an electronic shut-off device, and weight methods are devices used to
  A. charge a system.
  B. evacuate a system.
  C. safely recover refrigerants.

1322. A CFC and an HCFC have the same ozone depletion potential.
  A. True
  B. False

1323. When pressure testing a system, never exceed
  A. high side test pressures.
  B. low side test pressures.
  C. 100 PSIG.

1324. Which agency has to approve refrigerant cylinders before they are transported?
  A. ARI
  B. EPA
  C. DOT

1325. To prevent an explosion, never pressure test a system using
  A. CO_2.
  B. oxygen.
  C. nitrogen.

1326. A system is evacuated to 29.92" Hg. and the pump is turned off. The pressure increases and stops at 25" Hg. What can be said about the system?
  A. It has a leak.
  B. The oil has vaporized.
  C. It has moisture or refrigerant present in the system.

1327. As of _____, the EPA requires technicians performing a sealed system repair to be certified.
  A. November 15, 1993
  B. July 1, 1992
  C. November 14, 1994

TYPE I CERTIFICATION

1328. According to the EPA, a small appliance is
   A. A system containing under 3 lbs. of refrigerant including PTACs.
   B. A system containing 5 lbs. of refrigerant or less including PTACs.
   C. A system containing 5 lbs. of refrigerant or less not including PTACs.

1329. Under conditions of ARI 740-1993, recovery devices manufactured *before* November 15, 1993 must
   A. be able to remove 80% of the refrigerant charge or achieve 4" Hg. vacuum.
   B. be able to remove 90% of the refrigerant charge or achieve 4" Hg. vacuum, if the compressor is in operation.
   C. be able to remove 80% of the refrigerant charge or achieve 4" Hg. vacuum, if the system's compressor is not operating.

1330. Under conditions of ARI 740-1993, recovery devices manufactured *after* November 15, 1993 must
   A. be able to remove 80% of the refrigerant charge or achieve 4" Hg. vacuum.
   B. be able to remove 70% of the refrigerant charge or achieve 4" Hg. vacuum, if the system's compressor is in operation.
   C. be able to remove 80% of the refrigerant charge or achieve 4" Hg. vacuum, if the system's compressor is not operating.

1331. Technicians must certify their recovery equipment with the EPA after
   A. November 15, 1993.
   B. August 12, 1993.
   C. November 14, 1994.

1332. The EPA
   A. does not require a leak in a small appliance to be repaired.
   B. requires a leak in a small appliance to be repaired if it leaks within 6 months.
   C. requires a leak in a small appliance to be repaired if it leaks within 1 year.

1333. While recovering refrigerant, why is it necessary to install a piercing valve on both the high and low side of an appliance that has an inoperative compressor?
   A. it will help drain the oil out of the system while recovering the refrigerant
   B. it will decrease the rate of refrigerant evaporation
   C. it will decrease the time of the refrigerant recovery

1334. When using a system dependent (passive) recovery, the EPA requires
   A. 80% of the refrigerant to be removed if the compressor is operating.
   B. 80% of the refrigerant to be removed if the compressor is not operating.
   C. 90% of the refrigerant to be removed if the compressor is not operating.

1335. The pressure of an R-12 cylinder at 70° F ambient temperature will be
   A. 100 PSIG.
   B. 70 PSIG.
   C. 50 PSIG.

1336. Which of the following refrigerant must not be recovered with a current recovery system?
   A. R-12 CFC
   B. R-502 azeotrope
   C. R-717 ammonia

1337. Which of the following refrigerant is considered a replacement for R-12 refrigerant?
   A. R-22
   B. R-141A
   C. R-134A

1338. Which of the following is a direct replacement for R-12 when no oil change or system adjustments are required?
   A. R-134A
   B. R-141A
   C. None of the above.

1339. When pressurizing a system with nitrogen, it is required to have
   A. at least 50 lbs. of pressure in the system.
   B. a pressure regulator and a relief valve.
   C. the system in operation.

1340. After installing a piercing valve onto a system,
   A. leak test the valve.
   B. remove the valve after servicing the system.
   C. remove the schreader before capping the valve.

1341. If refrigerants are exposed to open flames and hot surfaces,
   A. sulfuric and phosphoric acids are produced.
   B. ammonia and hydrogen are produced.
   C. phosgene gas, and hydrochloric hydrofluoric acids are produced.

1342. What may cause a pungent odor in a CFC/HCFC system?
   A. a burnt compressor
   B. an overcharge
   C. a defective filter dryer

Section 4 Indoor Air Quality (IAQ) and Safety

1343. Which of the following is required if a large amount of R-12 has been released?
   A. safety glasses
   B. a dust mask and safety glasses
   C. a Self-Contained breathing apparatus

1344. At what temperature is R-22 when its pressure is 121 PSIG?
   A. 70° F
   B. 90° F
   C. 105° F

1345. At what pressure is R-502 when its temperature is 80° F?
   A. 40 PSIG
   B. 161 PSIG
   C. 205 PSIG

1346. Relief valves must never be installed
   A. in parallel.
   B. in series.
   C. on liquid receivers.

1347. The color code for refrigerant recovery tanks is
   A. yellow bottoms with gray tops.
   B. yellow tops with gray bottoms.
   C. yellow tops and bottoms with gray in the middle.

TYPE II CERTIFICATION

1348. Type 2 technicians are certified to service
   A. high pressure equipment only.
   B. high pressure and very high pressure equipment, including small appliances and motor vehicles.
   C. high pressure and very high pressure equipment, excluding small appliances and motor vehicles.

1349. After the installation of a refrigeration system, the technician should first
   A. evacuate the system.
   B. pressure test the system using R-22.
   C. pressure test the system using nitrogen.

1350. What maintenance task must always be done to recovery/recycle equipment?
   A. pressure test the system after every system recovery
   B. vacuum test the equipment after every system recovery
   C. change systems oil and filters

1351. What type of leak detection device is used to pinpoint a leak?
   A. an electronic leak detector
   B. a halide leak detector
   C. soap bubbles

1352. An open compressor that has sat for several months is most likely
   A. not to operate because of seized bearings.
   B. to leak from the shaft seal.
   C. to need a hard start kit.

1353. Two indications that a high pressure system has a leak are
   A. oil traces on flared fittings and high superheat.
   B. low superheat and bubbles in the sight glass.
   C. high head pressure and low superheat.

1354. The EPA requires that all comfort cooling systems with more than 50 lbs. of refrigerant be repaired when the annual leak rate exceeds
   A. 15%.
   B. 35%.
   C. 50%.

1355. The EPA requires that all *commercial and industrial process refrigeration* systems with more than 50 lbs. of refrigerant be repaired when the annual leak rate exceeds
   A. 15%.
   B. 35%.
   C. 50%.

1356. Why is it likely for a recovery and recycle machine to overheat when drawing a deep vacuum?
   A. because a deep vacuum causes high head pressure
   B. because the refrigerant is needed to cool the compressor
   C. because the compressor oil is evaporated

1357. What component is directly after the condenser on a system that uses a TXV type metering device?
   A. the TXV
   B. the receiver
   C. the filter dryer

1358. Which refrigerant requires a separate set of gauges, a vacuum pump, and a recovery machine when servicing?
   A. R-12
   B. R-500
   C. R-134A

1359. What methods help to reduce recovery time of a system?
   A. pack ice around the recovery tank, heat the appliance, and recover liquid first
   B. pack ice around appliance, heat the recovery tank, and recover gas first
   C. pack ice around the recovery tank, heat the appliance, and recover gas first

1360. During a recovery, when should the refrigerant be removed from the condenser outlet?
   A. when the condenser is below the receiver
   B. when the condenser is above the receiver
   C. when the condenser is below the compressor

1361. When recovering refrigerant from a system that has a condenser on the roof and the evaporator on the first floor, the recovery should begin from
   A. the suction line entering the evaporator.
   B. the discharge line exiting the compressor.
   C. the liquid line entering the evaporator.

1362. After removing refrigerant from a system, the refrigerant may
   A. be transported to another system on the premises.
   B. be reused in the same system or on another system owned by the same person.
   C. not be reused.

1363. What must be done to an empty recovery cylinder before transferring refrigerant into the cylinder?
   A. It must be evacuated.
   B. It must be charged with nitrogen.
   C. It must be purged.

1364. What is commonly used to aid in minimizing the release of refrigerant to the atmosphere?
   A. low loss fittings, and hand valves on the hoses
   B. pre-purged hoses and schrader valves
   C. vacuum pumps and purge valves

1365. When using recovery and recycling equipment manufactured *after* November 15, 1993, technicians must evacuate an appliance containing more than 200 lbs. of HCFC-22 to what level before making a major repair?
   A. 0 PSIG
   B. 4" Hg.
   C. 10" Hg.

1366. When using recovery and recycling equipment manufactured *before* November 15, 1993, technicians must evacuate an appliance containing less than 200 lbs. of HCFC-22 to what level before making a major repair?
   A. 0 PSIG
   B. 4" Hg.
   C. 10" Hg.

1367. If a leak in an appliance makes the prescribed evacuation level unattainable, the appliance can be evacuated to
   A. 0 PSIG.
   B. 4" Hg.
   C. 10" Hg.

1368. Which of the following repairs are considered a major repair?
   A. the replacement of an evaporator fan
   B. the installation of a low pressure control
   C. the replacement of a condenser coil

1369. When servicing an open system, which of the following should be replaced?
   A. a filter dryer
   B. a sight glass
   C. a pressure control

1370. In a refrigeration system, high pressure liquid exits
   A. the condenser and the metering device.
   B. the receiver and the metering device.
   C. the condenser and receiver.

1371. When evacuating a system, it is required that the vacuum pump pull a minimum of
   A. 500 microns.
   B. 5000 microns.
   C. 1000 microns.

1372. A technician must never
   A. front seat the suction service valve while the compressor is in operation.
   B. front seat the king valve while the compressor is in operation.
   C. operate a hermetic compressor under a deep vacuum.

1373. Non-condensables (air) cause
   A. low superheat.
   B. high head pressure.
   C. low suction pressure.

1374. A technician must never
   A. front seat the discharge service valve while the compressor is in operation.
   B. front seat the king valve while the compressor is in operation.
   C. back seat the discharge service valve while the compressor is in operation.

Section 4 Indoor Air Quality (IAQ) and Safety

1375. Defined by ASHRAE Standard 15, a sensor and alarm are required for A–1 refrigerants (R-12, R-22, R-11)
   A. to sense oxygen depletion.
   B. to sense ozone depletion.
   C. to sense CFCs.

1376. A refrigeration system must always be protected by
   A. a high pressure control.
   B. a safety head.
   C. a pressure relief device.

1377. To prevent freezing when charging R-12 into an evacuated system, it must be in the vapor form to raise the pressure up to
   A. 10" Hg.
   B. 0 PSIG.
   C. 33 PSIG.

1378. The sale of CFC and HCFC refrigerants have been restricted to certified technicians since
   A. November 15, 1993.
   B. July 1, 1992.
   C. November 14, 1994.

1379. Refrigerators built before 1950 often contain
   A. R-12.
   B. R-717.
   C. SO_2.

TYPE III CERTIFICATION

1380. If EPA regulations change after a technician has been certified,
   A. the technician will be grandfathered in to the new changes.
   B. this will only effect technicians that have not been certified.
   C. it is the technician's responsibility to comply with future changes.

1381. Why is it easier to have air enter gaskets and fittings on a low pressure system than a high pressure system?
   A. because air is attracted to R-11 and R-123 more than R-12 and R-22
   B. because a low pressure system operates in a vacuum
   C. because low pressure systems often use centrifugal compressors

1382. At what pressure does a rupture disc open on a low pressure system?
   A. at 10 PSIG
   B. at 15 PSIG
   C. at 20 PSIG

1383. What is the easiest way to leak test a low pressure system?
   A. connect gauges and add 15 PSIG of nitrogen
   B. add warm water over the coil to raise the pressure to 10 PSIG
   C. add R-22 and nitrogen to increase the pressure to 150 PSIG

1384. Why is it important not to exceed 10 PSIG when leak testing a low pressure system?
   A. because the relief valve on the receiver could blow
   B. because the evaporator test pressure is 10 PSIG MAX
   C. because the rupture disc, located on the evaporator, may fail

1385. When using an electronic leak detector, where is a water box leak tested after the water has been removed?
   A. at the drain valve
   B. at the rupture disc
   C. at the inlet valve

1386. What is used when leak testing a water tube on a low pressure chiller?
   A. a hydrostatic tube kit
   B. a hydrometer
   C. a psychrometer

1387. The EPA requires that all comfort cooling systems with more than 50 lbs. of refrigerant must be repaired when the annual leak rate exceeds
   A. 15%.
   B. 35%.
   C. 50%.

1388. The EPA requires that all *commercial and industrial process refrigeration* systems with more than 50 lbs. of refrigerant must be repaired when the annual leak rate exceeds
   A. 15%.
   B. 35%.
   C. 50%.

1389. What pressure is a high-pressure cut-out set for on a recovery system that is used on low-pressure refrigeration systems?
   A. 10 PSIG
   B. 15 PSIG
   C. 20 PSIG

1390. When recovering R-11 or R-123 from a low pressure system, always recover
   A. liquid first.
   B. vapor first.
   C. vapor and liquid first.

1391. What is commonly done to prevent a low pressure system from freezing during evacuation?
   A. stop the water from running through the chiller tubes
   B. run the water through the chiller tubes
   C. remove the water from the chiller tubes

1392. What does the ASHRAE guideline 3-1990 state about a vacuum test?
   A. If a 1 mm Hg. vacuum level rises to 2.5 mm Hg., the system should be leak tested.
   B. If a 1 mm Hg. vacuum level rises to 2.5 mm Hg., the system is not leaking.
   C. If a 1 mm Hg. vacuum level rises to 2.5 mm Hg., the chiller coil should be replaced.

1393. How is refrigerant added to a low pressure chiller system after an evacuation?
   A. by the use of the king valve
   B. by the use of the suction service valve
   C. by the use of the evaporator charging valve

1394. After a chiller system has had a deep system evacuation, what is commonly done to decrease the possibility of system freeze-up prior to a liquid charge?
   A. pressurize the system with liquid up to the 36° saturation temperature
   B. pressurize the system with vapor up to the 36° saturation temperature
   C. pressurize the system with liquid or vapor up to the 20° saturation temperature

1395. When using recovery or recycling equipment manufactured or imported before November 15, 1993, the required level of evacuation for low pressure equipment is
   A. 25 mm Hg. absolute.
   B. 25 inches Hg.
   C. 27 inches Hg.

1396. When using recovery or recycling equipment manufactured or imported on or after November 15, 1993, the required level of evacuation for low pressure equipment is
   A. 25 mm Hg. absolute.
   B. 25 inches Hg.
   C. 27 inches Hg.

1397. What is used on low pressure chillers to remove non-condensables from the system?
   A. a vent system
   B. an aerator
   C. a purge unit

1398. From where does the purge unit remove trapped air in a low pressure system?
   A. the end of the evaporator
   B. the top of the receiver
   C. the top of the condenser

1399. The equipment room that contains a chiller using R-123 group B-1 refrigerant requires a
   A. oxygen deprivation sensor.
   B. ozone depletion sensor.
   C. refrigerant sensor.

1400. It is important to recover the vapor refrigerant in a chiller after the liquid has been removed, since there is approximately _____ lbs. of vapor in a common 350 ton R-11 chiller.
   A. 25
   B. 50
   C. 100

Section 5

Service and Troubleshooting

AIR CONDITIONING TROUBLESHOOTING

1401. What are the symptoms of a TXV air conditioning system if it has lost its charge?
A. the superheat will decrease and the high side pressure will increase
B. the subcooling will increase and the low side pressure will increase
C. the superheat will increase and the low side pressure will decrease
D. the subcooling will increase and the low side pressure will increase

1402. A customer who owns a split air conditioning system complains that his system has been running but his house doesn't seem to be cooling properly. When the service technician arrives on the job, he finds that the indoor blower is operating and the air filter is clean. The condenser fan on the outdoor unit is operating, but he notices that the liquid line is hot to the touch. Before connecting gauges to the system, his next step should be to
A. clean the condenser coil.
B. remove the air filter.
C. connect the TXV sensing bulb to the suction line.
D. make sure the condenser fan is turning counter clockwise.

1403. What are symptoms of an air conditioning system that has a low charge?
A. The evaporator coil will be partially frosted and the discharge pressure will increase.
B. The suction line will sweat and the low-side pressure will decrease.
C. The low-side pressure will decrease and the evaporator coil will be partially frosted.
D. The discharge pressure will decrease and the suction pressure will increase.

1404. A customer calls and says that her air conditioning system is not cooling properly. Upon arrival, a technician notices that the suction line and compressor have frost build up.

Technician "A" says that the problem is probably a low charge. Technician "B" says that it is possibly a dirty air filter.

Which technician is more likely to be correct?
A. technician "A"
B. technician "B"
C. Both technicians are correct.
D. Both technicians are incorrect.

1405. What are the operating pressures of an oil pressure safety switch?
- A. the oil pump pressure and the crankcase pressure
- B. the oil pump pressure and the discharge pressure
- C. the oil pump pressure only
- D. Both B and C.

1406. What is a common discharge temperature of an air cooled air conditioning system that is used on a 90° F day?
- A. 230° F–250° F
- B. 110° F–120° F
- C. 90° F–100° F
- D. 80° F–90° F

1407. What are the symptoms of an air conditioning system that has a bad compressor discharge valve?
- A. high suction pressure, high discharge pressure, and low amp draw
- B. high discharge pressure, low suction pressure, and high amp draw
- C. low amp draw, low suction pressure, and high discharge pressure
- D. low discharge pressure, high suction pressure, and low amp draw

1408. What is the purpose of an EPR valve?
- A. it decreases the pressure to the compressor during an excessive load
- B. it increases pressure in a high temperature evaporator coil when it is used on a dual evaporator system
- C. it decreases pressure in a low temperature evaporator coil when it is used on a dual evaporator system
- D. Both A and C.

1409. Which of the following is a possible cause for a compressor to overheat?
- A. a dirty air filter
- B. a plugged evaporator coil
- C. a low refrigerant charge
- D. Both A and C.

1410. What is a possible cause of a sweat or frost pattern down stream from a sight glass and filter dryer?
- A. an overcharged system
- B. a plugged metering device
- C. a leaking sight glass
- D. a plugged filter dryer

1411. What causes the automatic expansion valve to close?
- A. a decrease in high-side pressure
- B. an increase in low-side pressure
- C. a decrease in low-side pressure
- D. an increase in high-side pressure

1412. A customer calls with an insufficient cooling complaint on a 90° F day. Upon arrival it is found that the system, using R-22, is operating at 50 PSIG on the low side and 150 PSIG on the high side. Further investigation indicates that the evaporator coil is partially frosted on one circuit of the distributor.

Technician "A" says that the system has a low charge and must be leak tested.

Technician "B" says that the capillary tube from the distributor is partially restricted. Which technician is correct?
- A. technician "A"
- B. technician "B"
- C. Both technicians are correct.
- D. Both technicians are incorrect.

1413. What type of problems can be caused by an oversized air conditioning system?
- A. excessive cooling and dehumidification
- B. ice build up at the condenser coil
- C. excessively high suction and discharge pressures
- D. short cycling and insufficient dehumidification

1414. What types of symptoms will be present if the sensing bulb of a TXV is not properly connected?
- A. high suction pressure and low superheat
- B. low suction pressure and high superheat
- C. possible floodback and low discharge pressures
- D. Both B and C.

1415. An air conditioning unit has a suction line that is frosted up to the compressor. What faults could cause this to occur?
- A. a dirty evaporator coil
- B. a plugged air filter
- C. a bad indoor blower relay
- D. All of the above.

1416. If 24 volts are present to the contactor coil on an air conditioning system and the condenser fan or compressor is inoperative, which of the following could be at fault?
- A. a bad cooling thermostat
- B. a blown fuse
- C. a faulty control transformer
- D. All of the above.

Section 5 Service and Troubleshooting

1417. A customer calls with an insufficient cooling complaint. Upon arrival to the service call it is found that the system is R-22 with the high-side pressure at 290 PSIG and the low-side pressure at 85 PSIG. The ambient temperature is 85° F and the indoor WB is 60° F.

Technician "A" says that more than likely the condenser is dirty.

Technician "B" says that the system's pressures are common for this type of day.

Who is correct?
A. technician "A"
B. technician "B"
C. Both technicians are correct.
D. Both technicians are incorrect.

1418. If a TXV sensing bulb had lost its charge, what effect would this have on the air conditioning system?
A. it would cause the high side pressure to increase
B. it would cause the low side pressure to increase
C. it would cause the low side pressure to decrease
D. Both A and C.

1419. A customer calls with a preventative maintenance service call on his R-22 water cooled air conditioning system. The technician notices the high-side pressure is 195 PSIG and the leaving water temperature is 90° F. What is wrong with the system? (Use the pressure temperature chart provided.)
A. The condenser is dirty.
B. The water valve is closed too much.
C. The water valve is opened too much.
D. There is no problem.

1420. What would be the approximate discharge temperature and pressure of an R-22 air cooled condensing unit on an 80° F day? (Use the pressure temperature chart provided.)
A. 80–90° F at 145–170 PSIG
B. 100–110° F at 195–225 PSIG
C. 110–120° F at 225–260 PSIG
D. 120–130° F at 260–295 PSIG

PSIG	Temperature in Degrees F.	
	R-410A	R-22
100	31	58
105	34	62
110	36	64
115	38	66
120	40	68
180	63	94
185	65	96
190	66	98
195	68	100
200	70	102
205	71	104
210	73	107
215	74	108
220	76	109
225	77	110
230	78	112
240	81	114
250	84	118
260	86	120
270	89	122
280	91	126
290	94	128
300	96	132
310	98	134
320	100	136
330	103	138
340	105	142
350	107	144
360	109	146
365	110	147
370	111	148
375	112	149
380	113	150

1421. A customer calls with an insufficient cooling complaint. Upon arrival to the service call it is found that the outdoor condensing unit is operating, but the indoor blower is not in operation.

Technician "A" says that it is possible that it is a bad fuse to the furnace.

Technician "B" says that the control transformer could also be at fault.

Who is correct?
A. technician "A"
B. technicians "B"
C. Both technicians are correct.
D. Both technicians are incorrect.

1422. A party-store owner tried to adjust the water valve on his water-cooled condensing unit, then he called for assistance from an HVAC/R service technician. It was found that when servicing the unit, the outlet water temperature was too low.

Technician "A" says that turning the adjustment stem on water valve clockwise will close the valve, and this will cause an increase in high-side pressure.

Technician "B" says that turning the adjustment stem on the water valve counterclockwise will close the valve, and this will increase the high-side pressure.

Which technician was correct?
  A. technician "A"
  B. technician "B"
  C. Both technicians are correct.
  D. Both technicians are incorrect.

1423. When connecting gauges to an R-410A air conditioning system, on an 85° F day, the pressures read 365 PSIG on the high side and 105 PSIG on the low side. What is a possible problem with the system? Use the pressure temperature chart provided on page 211 to answer this question.
  A. dirty air filter
  B. dirty condenser
  C. insufficient compressor
  D. there is no problem

1424. What are two common symptoms of an overcharged air conditioning system that uses a capillary tube type metering device?
  A. an increase in discharge pressure and a decrease in suction pressure
  B. an increase in suction pressure and an increase in discharge pressure
  C. a decrease in discharge pressure and an increase in suction pressure
  D. a decrease in suction pressure and a decrease in discharge pressure

1425. Name two symptoms that are common to an overcharged air conditioning system and a system that has a dirty condenser.
  A. the subcooling decreases and the amperage increases
  B. the suction pressure increases and the discharge pressure decreases
  C. the suction pressure decreases and the subcooling increases
  D. the discharge pressure increases and the amperage increases

1426. When an air conditioning system has an undercharge, what happens to the superheat and the subcooling of the system?
  A. the superheat increases and the subcooling increases
  B. the superheat increases and the subcooling decreases
  C. the superheat decreases and the subcooling increases
  D. the superheat decreases and the subcooling decrease

1427. If a rooftop air conditioning system does not have any means of low ambient protection, what will be two symptoms on the system when operating under a low ambient condition?
  A. high discharge pressure and low superheat
  B. low discharge pressure and high suction pressure
  C. high amperage and high superheat
  D. high superheat and low discharge pressure

1428. What are two common symptoms of a rooftop air conditioning system that operates with a reverse acting high pressure control stuck in the closed position?
  A. condenser fans operating during a low ambient condition and a low suction pressure
  B. high discharge pressure and high suction pressure
  C. high discharge pressure and condenser fans de-energized
  D. high suction pressure and low superheat

1429. What are two problems that can cause an air conditioning system to have an excessively high superheat?
  A. an oversized metering device and an overcharge
  B. an uninsulated suction line and a restricted metering device
  C. an undersized metering device and a disconnected TXV sensing bulb
  D. a disconnected TXV sensing bulb and a restricted metering device

1430. What are two common problems that can cause a sight glass in an air conditioning system to have bubbles?
  A. an overcharge and a restricted metering device
  B. air in the system and an undercharge
  C. a dirty evaporator coil and a low ambient condition
  D. an undercharge and low ambient conditions

Section 5 Service and Troubleshooting   213

1431. When the CTH is in the closed position and LPS is in the open position, what would the voltage be across terminal 1 at the ATH and T2? Use the Single Phase AC diagram on page 214 to find the answer.
   A. 0 volts
   B. 240 volts
   C. 120 volts
   D. Both A and B.

1432. When the CTH is in the closed position and LPS is in the closed position, what would be the voltage across terminals 1 and 2 of FR1? Use the Single Phase AC diagram on page 214 to find the answer.
   A. 0 volts
   B. 240 volts
   C. 120 volts
   D. None of the above.

1433. On a call for cooling the COMP is in operation but the CF in not operating. A 240 volt reading is obtained across terminal 2 on the ATH and T2. What 2 malfunctions could be present? Use the Single Phase AC diagram on page 214 to find the answer.
   A. a defective ATH and/or a RC
   B. a defective RC and/or a CF
   C. a defective CF and/or an ATH
   D. a defective C2 contact and/or an ATH

1434. On a call for cooling, the CF is operating but the COMP is not. A clamp-on ammeter is placed between T1 and the R terminal on the compressor and a reading of 0 amps is obtained. It is then placed between terminal 1 on the PR and the SC and 0 amps is obtained. What is most likely to be the problem? Use the Single Phase AC diagram on page 214 to find the answer.
   A. a defective SC
   B. an open OL
   C. a defective Run winding
   D. a defective C1 contact

1435. A technician arrives on the job and the customer complains that there is no cooling on his rooftop unit. Troubleshooting the system, the technician finds that the compressor is faulty. He notices that there is another faulty compressor laying next to the unit. A further check shows that the crankcase heater is electrically burnt open. What type of damage is most likely to have been done to the compressor?
   A. broken rings
   B. broken valves
   C. seized bearings
   D. All of the above.

1436. After a compressor motor has been severely burned, what accessory must be added as an aid to rid the air conditioning system of acid?
   A. an acid core liquid line dryer
   B. an acid core receiver
   C. an acid core suction line dryer
   D. All of the above.

1437. When troubleshooting a split system air conditioner, it is found that the compressor motor is humming but not starting.

   Technician "A" says that it could be an open internal overload protector.

   Technician "B" says that it could be a blown fuse.

   Which technician is correct?
   A. technician "A"
   B. technician "B"
   C. Both technicians are correct.
   D. Both technicians are incorrect.

1438. A technician receives an air conditioning service call that states the system is operating but it isn't cooling properly. When the technician arrives, he notices that the compressor has recently been changed and it is an R-22 system. He checks the indoor air filter and the condenser coil. Both are clean and both coils seem to have adequate air flow. When he connects his gauges he notices that the suction pressure is 95 PSIG and the high side pressure is 285 PSIG and the ambient temperature is 80° F. What is most likely to be the problem with the system?
   A. the system was improperly serviced and it has air trapped in the condenser
   B. the unit has been overcharged
   C. the unit has an ice blockage at the metering device caused by improper service procedures
   D. Either A or B.
   E. Either A or C.

1439. An air conditioning system that uses a fan center has an inoperable indoor blower. When the thermostat is set for cooling and dialed below room temperature, a test with the voltmeter shows that there is 24 volts across terminals R and G at the terminal block. What is the problem with the system?
   A. a bad blower
   B. a bad relay
   C. a bad thermostat
   D. Any of the above.

Part 2 Questions

Section 5 Service and Troubleshooting

1440. When testing an indoor blower control circuit and the cooling circuit on a fan center, what terminals should read 24 volts if the air conditioning system is in operation?
   A. (R and Y) and (R and G)
   B. (Y and G) and (R and C)
   C. (R and W) and (R and G)
   D. (Y and C) and (G and C)

1441. Which of the following problems can cause a decrease in a system's pressures and possible icing of the evaporator coil?
   A. a missing air filter
   B. improper rotation of the indoor blower
   C. blower too small
   D. All of the above.

1442. A split system air conditioner that uses a piston type metering device is to be charged with R-22.

   Technician "A" says that the outdoor dry-bulb temperature is needed to determine the proper superheat.

   Technician "B" says that the indoor wet-bulb temperature is needed to determine the proper superheat.

   Which one is correct?
   A. technician "A"
   B. technician "B"
   C. Both technicians are correct.
   D. Both technicians are incorrect.

1443. A 220 volt 30 amp motor circuit keeps tripping the breaker after the motor is started. The starting amperage is 60 amps. When measuring the amperage of the circuit, this is drawn for a brief moment. After start up, the amperage drops to 22 amps and then the breaker trips. What is the problem with the circuit?
   A. a shorted run winding
   B. an open run winding
   C. a grounded motor winding
   D. a defective circuit breaker

1444. A rooftop package system keeps tripping out on the high pressure switch. What are three possible causes for this problem?
   A. condenser air being re-circulated, plugged metering device, and overcharge
   B. excessively bent condenser fins, air in the system, and undercharge
   C. dirty condenser, incorrect rotation of the condenser fan, and bad compressor valves
   D. dirty condenser, bent condenser fins, and air re-circulating over the condenser

1445. What are two common symptoms of refrigerant leaks in an air conditioning system?
   A. oil traces around flared fittings and low superheat
   B. partially frosted evaporator and high superheat
   C. low discharge pressure and high suction pressure
   D. low suction pressure and a high compressor amp draw

1446. A system that is being serviced has been evacuated down to 500 microns. When the gauges are closed and the vacuum pump is turned off, the pressure starts to increase and stops at 400 microns.

   Technician "A" says that the system has a leak and needs to be repaired.

   Technician "B" says that the system has moisture present and it needs to be evacuated longer.

   Which technician is correct?
   A. technician "A"
   B. technician "B"
   C. Both technicians are correct.
   D. Both technicians are incorrect.

1447. What are common methods that are used for leak testing a system?
   A. a vacuum test
   B. oil traces around fittings
   C. soap bubble solution
   D. All of the above.

1448. Using the Three-Phase AC diagram on page 216, determine when the crankcase heater operates?
   A. when the thermostat is in the closed position
   B. when the thermostat is in the open position
   C. as long as there is power to the 240 volt circuit
   D. Both B and C.

1449. If L2 lost its source of power due to a blown fuse, what 240 loads could still operate? Use the Three-Phase AC diagram on page 216 to find the answer.
   A. COMP
   B. CCH
   C. IBM
   D. None of the above.

Section 5   Service and Troubleshooting   **217**

1450. What would be the resistance readings across T1, T2, and T3 on the COMP. if OL were in the open position? Use the Three-Phase AC diagram on page 216 to answer this question.
   A. T1–T2 infinite resistance, T1–T3 zero resistance, and T2–T3 zero resistance
   B. T1–T2 zero resistance, T1–T3 zero resistance, and T2–T3 zero resistance
   C. T1–T2 infinite resistance, T2–T3 infinite resistance, and T2–T3 measurable resistance
   D. T1–T2 infinite resistance, T2–T3 infinite resistance, and T2–T3 infinite resistance

1451. On an 85° F outdoor ambient day the air conditioning systems indoor coil and suction line has frost build up. A voltage check from Y to C and G to C show 24 volts, what are the possible problems with the system? Use the Three-Phase AC diagram on page 216 to answer this question.
   A. a defective C, FR, and/or ATH
   B. a defective RC, CF, and/or ATH
   C. a defective FR, IBM, or RC
   D. a defective FR, C, or LPS

1452. On a 90° F outdoor ambient day the systems condenser fan is inoperable. The discharge pressure is at 350 PSIG and the compressor overheats causing it to cycle on the internal overload protector. A voltage check from Y to C and G to C show 24 volts, what are the possible problems with the system? Use the Three-Phase AC diagram on page 216 to answer this question.
   A. a defective C, FR, and/or ATH
   B. a defective RC, CF, and/or ATH
   C. a defective FR, IBM, or RC
   D. a defective C, CF, or ATH

1453. What is the purpose of the ATH? Use the Three-Phase AC diagram on page 216 to answer this question.
   A. It de-energizes the CF if the outdoor ambient air temperature is too high.
   B. It de-energizes the CF if the outdoor ambient air temperature is too low.
   C. It decreases the speed of the CF if the outdoor ambient air temperature is too high.
   D. It decreases the speed of the CF if the outdoor ambient air temperature is too low.

1454. What is the purpose of a CPR valve?
   A. It is used to increase the pressure in the evaporator coil of a two temp system.
   B. It is used to control refrigerant pressure in the compressor crankcase.
   C. It keeps a constant evaporator pressure on a single pipe system.
   D. It floods the receiver during low ambient temperatures.

1455. What is the common cause of a compressor burnout?
   A. improper system evacuation during installation
   B. short cycling of the system
   C. excessively dirty condenser
   D. All of the above.

1456. Which of the following is a common cause of an excessively low superheat on a TXV system?
   A. a sensing bulb improperly connected to the suction line
   B. a low system charge
   C. the superheat spring adjusted too far in the clockwise position
   D. Both A and C.

1457. What are three methods for controlling a system's high side pressure during low ambient conditions?
   A. cycling condenser fans, EPR valves, and CPR valves
   B. flooding the condenser, cycling condenser fans, and dampering condenser air
   C. pump down system, flooding the condenser, and oversized receivers
   D. reverse acting pressure switch, cycling evaporator fans, and dampering condenser air

1458. What is a common temperature difference between the condensing temperature on an air cooled condensing unit and the outdoor ambient temperature?
   A. 5° to 10° F
   B. 10° to 15° F
   C. 20° to 30° F
   D. 40° to 50° F

1459. What controls are used on an air conditioning or refrigeration system that incorporates an automatic pump down system?
   A. a high pressure switch cycles the liquid line solenoid, and a thermostat controls the contactor coil
   B. a low pressure switch cycles the condenser fans, and a thermostat controls the compressor
   C. a thermostat cycles the condenser fans, and a reverse acting pressure switch controls the compressor
   D. a thermostat cycles the liquid line solenoid, and a low pressure switch controls the compressor

1460. A customer calls and complains that his air conditioning system is not operating properly. It is noticed that the compressor is not operating. A check on the run capacitor shows that there is 33 UF and the rating on the capacitor is 40 UF.

Technician "A" says that the capacitor is good because it is within its normal 20% range.

Technician "B" says that the capacitor is bad because it is not within its 10% range.

Which technician is correct?
A. technician "A"
B. technician "B"
C. Both technicians are correct.
D. Both technicians are incorrect.

1461. A system that is operating with a clear sight glass shows
A. the sight glass is full of liquid
B. the sight glass is full of gas
C. the system is always running efficiently
D. Both "A" and "B" are correct.

1462. Technician "A" says "When charging a system using the superheat method, adding refrigerant to the system will increase the superheat."

Technician "B" says "When charging a system using the superheat method, adding refrigerant will decrease the superheat."

Which technician is correct?
A. technician "A"
B. technician "B"
C. Both technicians are correct.
D. Both technicians are incorrect.

1463. What is the purpose of the TXV?
A. it keeps a constant superheat in the evaporator
B. it keeps a constant pressure in the evaporator
C. it keeps a constant pressure drop through the evaporator
D. Both A and C.

1464. A liquid charged TXV bulb will
A. always have some liquid present in the bulb and diaphragm.
B. have 100% liquid present in the bulb and diaphragm at all times.
C. have liquid present only until the valve is fully open.
D. Both A and B.

1465. When using a vapor charged TXV bulb, it is necessary to have
A. the valve body at a lower temperature than the sensing bulb.
B. the sensing bulb at a lower temperature than the valve body.
C. an external equalizer on the TXV if the evaporator has 1 to 2 PSIG pressure drop.
D. an internal equalizer on the TXV if the evaporator has over 5 PSIG pressure drop.

1466. A customer calls and complains that her high efficiency air conditioning system is operating, but is not cooling properly. Further investigation shows that the system is using a TXV type of metering device. A subcooling check shows that the system appears to have the proper charge. A superheat test shows that the superheat is 30° F when it is supposed to be 15° F. What could be the problem?
A. the TXV valve spring adjustment could be turned counterclockwise too far
B. the sensing bulb could be loose
C. the external equalizer could be plugged
D. the system could have improper evaporator air flow

1467. A system that uses a thermal electric expansion valve
A. must use a liquid line solenoid to pump the system down during the off cycle.
B. can use the thermal electric expansion valve instead of a liquid line solenoid as part of the pump down system.
C. must not use a pump down system during the off cycle.
D. closes the valve as voltage increases and opens the valve as voltage decreases.

1468. What are possible problems that can cause the discharge line temperature on an air conditioning system to be cool to the touch?
A. the system could be undercharged
B. the compressor could have bad valves
C. the system could have a restricted metering device
D. All of the above.

Section 5 Service and Troubleshooting   219

1469. A customer calls to complain that his central air conditioning system is not cooling properly. The technician arrives on the job. He notices that the indoor blower has adequate air flow and the outdoor condenser fan is operating, but the compressor is not operating. After pulling out the electrical disconnect, the technician measures resistance across the compressor motor windings. The readings are as follows Run to Common infinite ohms, Start to Common infinite ohms, and Run to Start 10 ohms. What could have caused this condition?
   A. an open start winding
   B. an open run winding
   C. an open internal overload protector
   D. Both A or B.

1470. A commercial customer calls complaining that her rooftop air conditioning system is not operating. The technician arrives on the job and sets the thermostat to cooling, when he notices that the indoor fan is not operating. He then sets the thermostat to fan on and it still does not operate. What two terminals must be tested for 24 volts to insure that the problem is not the thermostat?
   A. R to C
   B. Y to C
   C. R to G
   D. G to C

1471. A commercial customer calls on a 90° F day, complaining that his rooftop unit is running but not operating properly. The technician arrives on the job and the customer shows him an invoice from the spring season. The invoice states that 2 pounds of R-22 were added to the system. After connecting his gauges, the pressure reads 88 PSIG on the low side and 300 PSIG on the high side. What is the most likely systems problem?
   A. the system has been overcharged
   B. the system has air trapped in the condenser
   C. the system has a small leak
   D. Both A or B.

1472. Which of the following could be the cause for a system's discharge and suction pressure being low?
   A. a restricted metering device
   B. bad compressor valves
   C. an inoperative compressor
   D. Both A and C.

1473. What problem or problems can cause low discharge and suction pressure?
   A. an oversized TXV
   B. an undercharged system
   C. insufficient airflow over the evaporator coil
   D. Both B and C.

1474. What problem can cause an increase of discharge pressure and an increase of suction pressure?
   A. bad compressor valves
   B. an overcharged system
   C. an inoperative compressor
   D. an oversized compressor

1475. What problem or problems can cause the discharge pressure to be low and the suction pressure to be high.
   A. an oversized metering device
   B. bad compressor valve
   C. too low of an outdoor ambient temperature
   D. Both B and C.

1476. Which of the following can cause excessive subcooling?
   A. an undercharged system
   B. an overcharged system
   C. bad compressor valves
   D. Both A and C.

1477. Which of the following can cause low superheat in a TXV system?
   A. an overcharge
   B. an undercharge
   C. a TXV that is turned clockwise too far
   D. an uninsulated TXV sensing bulb

1478. What three things can prevent an air conditioning system from releasing heat at the proper rate?
   A. a dirty evaporator coil, an improperly adjusted metering device, and a dirty condenser
   B. a dirty condenser, re-circulation of condenser air, and a defective condenser fan
   C. a dirty evaporator coil, a defective evaporator fan, and an iced up evaporator coil
   D. All of the above.

1479. What three things can prevent an air conditioning system from absorbing heat at the proper rate?
   A. a dirty evaporator coil, an improperly adjusted metering device, and a dirty condenser
   B. a dirty condenser, re-circulation of condenser air, and a defective condenser fan
   C. a dirty evaporator coil, a defective evaporator fan, and an iced up evaporator coil
   D. All of the above.

1480. What component is most often the cause of a start capacitor failure?
   A. a start relay
   B. a run capacitor
   C. a faulty start winding
   D. a seized compressor

1481. A compressor motor starts and operates for a short period of time and cycles on the external overload protector. What are two possible problems that could cause this?
   A. an open start winding and low voltage
   B. an open start capacitor and a faulty start relay
   C. a shorted run capacitor and an open start capacitor
   D. low voltage and a faulty start relay

1482. Technician "A" says that to manually pump down an air conditioning system with a 2 position service valve, the liquid line must be front seated and the high and low side pressure will decrease to 0 PSIG when the pump down is complete.

   Technician "B" says that to manually pump down an air conditioning system with a 2 position service valve, the liquid line must be front seated and only the low side pressure will decrease to 0 PSIG when the pump down is complete and the high side will drop slightly.

   Which technician is correct?
   A. technician "A" is correct
   B. technician "B" is correct
   C. Both technicians are correct.
   D. Both technicians are incorrect.

1483. Technician "A" says that if you connect 208 volts to the primary side of a 240 / 24 volt transformer, the transformer will burn out.

   Technician "B" says that if you connect 208 volts to the primary side of a 240 / 24 volt transformer, the control circuit voltage will drop.

   Which technician is correct?
   A. technician "A" is correct
   B. technician "B" is correct
   C. Both technicians are correct.
   D. Both technicians are incorrect.

1484. When a high efficiency split air conditioning system has a plugged metering device, what symptoms might occur?
   A. a high amp draw, an increase in high side pressure, and a decrease in low side pressure
   B. a low amp draw, a decrease in high side pressure, and an increase in low side pressure
   C. a high amp draw, a decrease in high side pressure, and a decrease in low side pressure
   D. a low amp draw, a decrease in high side pressure, and a decrease in low side pressure

1485. Why is it unsafe to operate an air conditioning system with a low charge?
   A. it can seize the compressor because of excessive heat
   B. it can seize the compressor because lack of lubrication
   C. it can seize the compressor because of excessive system pressures
   D. Both A and B.

GAS FURNACE TROUBLESHOOTING

1486. On a standing pilot furnace, the thermocouple must be replaced when
   A. it is not positioned properly.
   B. the millivolt reading is 28 or higher.
   C. the millivolt reading is 22 or lower.
   D. the tip of the thermocouple is in the pilot flame $\frac{3}{8}$ of an inch.

1487. Natural gas pressure at the main line is recorded at 7" WC, but at the gas manifold it reads 1.5" WC. What is most likely to be the cause of this?
   A. a defective main line regulator
   B. too large of a gas valve
   C. plugged burner orifices
   D. a defective pressure regulator at the gas valve

1488. A technician servicing a gas fired furnace notices four out of five ribbon burners are firing simultaneously. What could be the problem with this system?
   A. too much gas pressure
   B. too little gas pressure
   C. burners not aligned properly
   D. a defective gas valve

1489. Technician "A" says that if a combination gas valve needs to be replaced if there are 24 volts present to the valve and it will not open.

   Technician "B" says that if a combination gas valve has a new thermocouple and the pilot will not stay lit, the gas valve needs to be replaced. Which technician is correct?
   A. technician "A"
   B. technician "B"
   C. Both technicians are correct.
   D. Both technicians are incorrect.

Section 5 Service and Troubleshooting

1490. What should be the first step that a technician should take if gas pressure at the manifold is too high?
A. adjust the pressure regulator clockwise
B. adjust the pressure regulator counter-clockwise
C. replace the pressure regulator
D. replace the combination gas valve

1491. If the flame is lifting off a burner port, what should a technician do to correct this problem?
A. adjust primary air by closing off the air shutter
B. adjust primary air by opening the air shutter more
C. replace the burner
D. decrease main gas pressure

1492. A step down transformer is reading zero ohms at its secondary winding. What is a possible cause for this problem?
A. an open primary winding
B. an open secondary winding
C. a shorted primary winding
D. a shorted secondary winding

1493. When a fan center relay needs to be tested for 24 volts, what two terminals are used to measure for transformer output voltage?
A. terminals R and W
B. terminals R and G
C. terminals R and Y
D. terminals R and C

1494. On a call for heat, what two terminals should be closed on the thermostat?
A. terminals R and W
B. terminals R and G
C. terminals R and Y
D. terminals R and C

1495. When a thermostat subbase fan selector switch is moved to fan "on," what terminals will close in the thermostat?
A. terminals R and W
B. terminals R and G
C. terminals R and Y
D. terminals R and C

1496. A new thermostat was recently replaced. The technician observes that the burner cycle is not long enough. The most likely cause would be
A. a defective thermostat.
B. low gas pressure.
C. that the anticipator needs adjusting.
D. a defective gas valve.

1497. A customer has replaced a mercury bulb thermostat but complains that the burners will not shut off and the house is getting too hot. The most likely cause would be
A. a defective limit control.
B. a defective gas valve.
C. that the thermostat is not properly level.
D. that the anticipator is set too low.

1498. A bimetal fan control will not turn the indoor blower motor off. As a result, cool air passes through the registers between furnace cycles. What is a possible cause of this problem?
A. The heating circuit is opening on limit.
B. The fan control contacts are stuck open.
C. The fan control contacts are stuck closed.
D. There is a defective gas valve.

1499. Residents are complaining of experiencing symptoms of headaches, watery eyes, and nausea.

Technician "A" says that the furnace heat exchanger possibly is cracked.

Technician "B" says that the furnace may only need cleaning and that the heat exchanger is probably plugged. Which technician is correct?
A. technician "A"
B. technician "B"
C. Both technicians are correct.
D. Both technicians are incorrect.

1500. A belt drive motor has recently been replaced but the furnace is shutting off on limit control. The most likely cause is that
A. the motor is running in the wrong direction.
B. the motor is defective.
C. the motor is too large.
D. the motor mount is not correct.

1501. Upon testing a motor run winding for continuity, the technician records 3 ohms. The winding resistance should read 8 ohms. The technician concludes that
A. the winding is open.
B. the winding is shorted.
C. the winding is grounded.
D. the motor is within satisfactory operating conditions.

1502. A run capacitor is tested with an analog meter. The needle moves towards zero then deflects back to infinity. What is the condition of the capacitor?
A. The capacitor is open.
B. The capacitor is good.
C. The capacitor should not be used.
D. The capacitor needs to be of a higher MFD.

1503. The temperature rise of a particular furnace states that it should be between 35° and 65° F. The furnace is operating at 20° F temperature rise with the proper gas pressure at the manifold. What could be done to solve this problem?
   A. add more outside air to the basement
   B. increase indoor blower speed
   C. decrease indoor blower speed
   D. replace air filter with more efficient type

1504. A belt drive motor has been replaced, but quickly gets hot and stops just short of burning out. Which of the following could cause this problem?
   A. A fan belt that is too tightly adjusted.
   B. A fan belt that has only 1 inch of slack between pulleys.
   C. The pulley on the motor is too small.
   D. None of the above.

1505. A combination fan/limit control has been replaced. The indoor blower motor runs too long and the resident complains of cooler airflow. The technician should
   A. replace the control again.
   B. adjust the limit setting.
   C. lengthen the fan/off setting.
   D. check and readjust the fan on/off differential.

1506. A service call is made to a customer who complains that when his furnace ignites, the flame rolls out in front of the heat exchanger.
   Technician "A" says that it's due to a poor draft, possibly a blocked flue or heat exchanger.
   Technician "B" says that it has to do with the limit control or the fan switch control circuit. Which technician is correct?
   A. technician "A"
   B. technician "B"
   C. Both technicians are correct.
   D. Both technicians are incorrect.

1507. A customer complaint is that of headaches, sleepiness, and nausea.
   Technician "A" finds a crack in the heat-exchanger and says that the furnace should be shut down.
   Technician "B" says that the crack isn't too serious, and that the furnace can operate a while but it must be cleaned. Which technician is correct?
   A. technician "A"
   B. technician "B"
   C. Both technicians are correct.
   D. Both technicians are incorrect.

1508. A technician is servicing a standing pilot gas fired furnace. The complaint is that the burners are not lighting. Where should the technician start troubleshooting the system?
   A. at the transformer secondary and at the combination gas valve
   B. at the transformer primary only
   C. at the primary limit control
   D. at the fan/limit control

1509. On a spark ignition system, the spark for ignition continues with no ignition of gas at the burners.
   Technician "A" concludes that the thermostat is defective.
   Technician "B" concludes that it is possibly the control module. Which technician is correct?
   A. technician "A"
   B. technician "B"
   C. Both technicians are correct.
   D. Both technicians are incorrect.

1510. Which of the following is the correct test for a faulty cable on a direct spark ignition system?
   A. a voltage check
   B. a continuity check
   C. an amperage check
   D. None of the above.

1511. A customer complaint is that burners light on her system, but then immediately shut down. The problem is most likely to be
   A. in the control module.
   B. in the control transformer.
   C. in the limit control.
   D. in the gas valve.

1512. Which of the following would not be a proper procedure to perform on a direct spark ignition system where the spark continues, the burners light, and the flames are immediately extinguished?
   A. make sure the sensor is positioned in the flame correctly
   B. check for proper flame rectification current
   C. check for a proper grounding terminal
   D. make sure there is proper gas flow to the pilot

1513. A hot surface ignition system does not ignite the main burner. After measuring voltage to the igniter, the technician should
   A. perform a resistance test on the igniter.
   B. replace the flame sensor.
   C. replace the main gas valve.
   D. replace the control module.

Section 5 Service and Troubleshooting   223

1514. The burners of a gas furnace that is equipped with a 115 volt hot surface igniter light for approximately 7 seconds and then go out. What is a problem that could cause this symptom?
   A. a faulty igniter
   B. a faulty transformer secondary winding
   C. burners are shutting off on high limit control
   D. a faulty sensor

1515. The burners on a furnace having a hot surface ignition system light up, but quickly shut off.

   Technician "A" states that after testing the gas valve, he has determined that it is shorted.

   Technician "B" visually checked the flame sensor, cleaned it and performed a continuity test. He faults a dirty sensor. Which technician is most likely to be correct?
   A. technician "A"
   B. technician "B"
   C. Both technicians are correct.
   D. Both technicians are incorrect.

1516. The thermostat on a mid-efficiency gas furnace with a hot surface ignition system is placed on a call for heat. The burners do not light, but the igniter does glow. Which of the following would be the next step to take when troubleshooting the system?
   A. test the flame sensor
   B. test the gas valve and gas manifold pressure
   C. test the primary limit control and the fan control
   D. test the thermostat

1517. The draft pressure switch on a mid-efficiency furnace is not allowing the ignition system to energize. Which of the following is a possible problem?
   A. a dirty filter
   B. a defective indoor blower
   C. a defective induced draft motor
   D. a defective flame roll out sensor

1518. A high-efficiency gas furnace will not stay in operation. The unit runs for approximately 5 minutes then is shut down by the main furnace controller with an "open pressure switch" flash code. Which of the following is *not* a proper test to perform in this situation?
   A. check the induced draft motor
   B. check the furnace filter
   C. check for a restricted flue pipe
   D. Both A and C.

1519. A service call is made on a mid-efficiency furnace. On a call for heat, the induced draft motor will not operate. What is the proper troubleshooting procedure that should be taken?
   A. check for control voltage at the "G" and "C" terminals at the circuit board
   B. check for control voltage across terminals "W" and "C" at the circuit board
   C. check for control voltage at the induced draft relay
   D. Both B and C.

1520. A high efficiency furnace that uses a furnace controller will not energize the blower motor after a pre-selected 45 second "time on" setting. Which of the following would be the most correct troubleshooting procedure?
   A. measure the resistance across the furnace controller low speed terminal and common
   B. measure the resistance across the furnace controller high speed terminal and common
   C. measure voltage to the motor and test the run capacitor
   D. measure amperage to the blower motor and test the run capacitor

1521. When servicing a mid-efficiency furnace, the thermostat has been turned to the "Off" position, but the burners remain lit, and the induced draft motor continues to operate. Which of the following would be the most correct troubleshooting procedure?
   A. determine if the thermostat is open and check voltage at the gas valve
   B. check for voltage at the indoor blower motor
   C. replace the control module
   D. check the ignition system

1522. On a high efficiency furnace that uses a furnace controller, the indoor blower motor will not de-energize after its pre-selected "Off Delay" setting. Which of the following is a logical repair for the system?
   A. replace the run capacitor
   B. replace the blower motor
   C. replace the control module
   D. change speed settings on the control module

1523. The burners on a furnace cycle for approximately 3 to 4 minutes. During a call for heat, the system does not provide adequate heat for the home.

Technician "A" has determined that a new thermostat has been installed and the heat anticipator was incorrectly adjusted.

Technician "B" cleans and positions the flame sensor and claims that this was the fault.

Which technician is correct?
A. technician "A"
B. technician "B"
C. Both technicians are correct.
D. Both technicians are incorrect.

1524. A gas furnace has been recently converted to LPG gas. On a call for heat, the furnace is not firing correctly and starting to soot-up. What would be the most likely cause for this problem?
A. low LPG pressure
B. too long of a burner cycle
C. wrong orifice size installed
D. All of the above.

1525. When inspecting a gas furnace, it is noticed that the burner flames are very small in height. What action should be taken by the technician?
A. replace the gas valve
B. replace the orifices
C. check gas pressures and adjust primary air shutters if needed
D. contact the utility company to replace the gas meter

1526. A customer is experiencing loud, delayed ignitions from his furnace.

Technician "A" says that it's possible that the gas valve is not opening or closing properly.

Technician "B" claims that it's possible that after the furnace had been cleaned, the burners were not properly aligned.

Which technician is correct?
A. technician "A"
B. technician "B"
C. Both technicians are correct.
D. Both technicians are incorrect.

1527. A customer called in complaining that his indoor blower would not stop running. What is the most possible cause of this problem?
A. a defective blower motor run capacitor
B. contacts stuck open on the fan control
C. contacts stuck closed on the fan control
D. Both A and C.

1528. After diagnosing a no heat call, it has been determined that the induced draft motor is not operating.

Technician "A" claims that it is the fault of an induced draft pressure switch.

Technician "B" states that the indoor blower relay module is defective. Which technician is more likely to be correct?
A. technician "A"
B. technician "B"
C. Both technicians are correct.
D. Both technicians are incorrect.

1529. The burners on a gas fired furnace are continuously lit, and setting the thermostat to the "OFF" position does not make any difference. What is a possible problem for this fault?
A. a defective gas valve
B. a defective fan control
C. a limit control stuck in the closed position
D. a shorted secondary winding on the transformer

1530. A customer complains that he has no heat and that the furnace does nothing at all. He says that he has a mid-efficient furnace with an induced draft motor and a spark ignition system. Which of the following would be a likely place to start troubleshooting?
A. check the fan door switch
B. check for a blown fuse
C. check for a defective furnace switch
D. All the above.

1531. On a hot surface ignition system, the igniter does not glow. What would be the most likely cause of this fault?
A. a bad flame sensor
B. a bad gas valve
C. a bad closed pressure switch
D. a bad igniter

1532. On a spark ignition system, the spark is occurring but its burners will not fire. Which of the following would you check?
A. test terminals PV and MV/PV for 24 volts
B. test for continuity of PV coil
C. test for gas to the pilot line
D. All the above.

1533. The thermocouple of a furnace has just been replaced, but the pilot light continues to go out. What test should a technician perform next?
A. ensure that the pilot flame is engulfing the tip of thermocouple by $\frac{3}{8}$"
B. test continuity of the main valve
C. test for voltage to the main gas valve
D. check for an open limit control

Section 5 Service and Troubleshooting

1534. An existing thermocouple has been tested with a millivolt meter and has registered at 13 millivolts. Which of the following would be the correct action to take?
A. make sure that the thermostat is calling for heat
B. replace the thermocouple
C. replace the pilot gas valve
D. there is no problem with the thermocouple, and it should be left alone

1535. What is an acceptable voltage of a thermocouple?
A. 30 MV
B. 20 MV
C. 10 MV
D. 5 MV

1536. An indoor blower motor will run, but it cycles on an internal overload protector. Using a digital meter, technician "A" states that the meter read zero when placing it across the motor's run capacitor.

Technician "B" states that the meter read "OL" when placing it across the run capacitor. Which technician would have the correct diagnosis?
A. technician "A"
B. technician "B"
C. Both technicians are correct.
D. Both technicians are incorrect.

1537. A furnace has been operating well during mild winters. During the very cold days, the customer complains that the three-year-old furnace does not provide continuous heat to satisfy the demand. The technician does a Btu input measurement by clocking the gas meter and finds the system within specifications. The temperature rise is also within systems limits. Which of the following would be the most correct diagnosis?
A. The furnace was undersized in heating capacity.
B. The gas pressure needs to be increased.
C. The ductwork needs redesigning.
D. None of the above.

1538. During furnace operations, the thermostat is quickly satisfied. The new furnace heats the entire house very fast, then shuts off on thermostat demand. Which would be the most correct statement of this situation?
A. The thermostat needs replacing.
B. The blower speed should be increased.
C. The furnace heating capacity is oversized.
D. The duct system is too large.

1539. Two gas furnaces were installed. Both furnaces were 150 MBtu/h (150 thousand Btus per hour). Both furnaces may operate at the same time. When one furnace is running, all operations are satisfactory. When both furnaces are operating at the same time, the burner flames of both furnaces decrease in size. Which procedure would be most likely correct?
A. replace heating thermostat
B. increase the gas pressure regulator of the first furnace
C. replace both gas valves
D. determine if the gas piping layout is properly sized

1540. On an LPG furnace installation, technician "A" states that only the burner orifices need changing.

Technician "B" states that the orifice's gas valve regulator spring and pilot orifice should be changed. Which technician is correct?
A. technician "A"
B. technician "B"
C. Both technicians are correct.
D. Both technicians are incorrect.

1541. A draft test has been performed and it is indicating that there is an improper draft for the gas furnace that is being used. Which of the following would be the correct procedure to perform?
A. increase the gas pressure
B. increase the induced draft motor speed
C. check for chimney obstructions
D. replace the heat exchanger

1542. On an LPG furnace installation, the burner flames are low and the new furnace is sooting up. Which of the following would be the most correct procedure to solve this problem?
A. replace the gas valve
B. measure gas pressure and install a low gas pressure switch
C. call the LPG fuel provider for service
D. clean heat exchanger and operate the furnace

1543. A crack has been detected on a heat exchanger. The furnace is only five years old. Which of the following would be the most correct procedure to perform?
A. replace the entire furnace
B. replace the heat exchanger under manufacturers warranty
C. replace the complete ductwork and furnace
D. None of the above.

1544. On a belt drive indoor blower motor, a fan belt needs replacing.

Technician "A" says that if you adjust the belt to ride higher on the pulley, you will increase the blower speed.

Technician "B" says that if you adjust the belt to ride lower into the pulley, you will increase the blower speed. Which technician is correct?
A. technician "A"
B. technician "B"
C. Both technicians are correct.
D. Both technicians are incorrect.

1545. A technician replaces a fan belt and a fan motor. The new motor is operating with too high of an amp draw. What is most likely the cause of this fault?
A. The belt was adjusted too tight.
B. The belt was adjusted too loose.
C. The motor horsepower is too low.
D. Both A and C.

1546. On a no heat call, a fan motor is not operating during the heating cycle.

Technician "A" measures voltage at the fan relay's normally open contacts.

Technician "B" measures voltage at the bimetal fan control. Which technician performed the correct procedure?
A. technician "A"
B. technician "B"
C. Both technicians are correct.
D. Both technicians are incorrect.

1547. The amperage of an indoor blower motor must be determined during the heating operation of a furnace. Where should this reading be taken?
A. after the N.O. contacts at the fan relay
B. before the N.O. contacts at the fan relay
C. at the high speed on the motor
D. at the red or white wire on the motor

1548. Where should a resistance reading be taken when measuring the low speed of an indoor blower motor?
A. at the red wire and green ground wire on the motor
B. at the red wire and the white wire on the motor
C. at the black wire and the white wire of the motor
D. at the yellow wire and the green ground wire on the motor

1549. A frequently used fireplace satisfies a thermostat located in the same room as the fireplace. This shuts down the call for heat while the greater portion of the home remains cold. What should the technician do to solve this problem?
A. compensate by turning the thermostat higher
B. damper down the fireplace control
C. relocate the thermostat
D. increase gas pressure at the furnace

1550. What will an extremely dirty air filter do to furnace operation?
A. cause the furnace to shut off on the limit control
B. cause the indoor blower motor to overheat
C. cause an increase of furnace temperature rise
D. All the above.

1551. Where can a limit control be wired in a gas fired furnace?
A. in the high voltage circuit
B. in the low voltage circuit
C. in the high or low voltage circuit
D. None of the above.

1552. How are limit controls and flame roll out sensors wired into their circuits?
A. in series with each other
B. in parallel with each other
C. in series parallel with each other
D. None of the above.

1553. A limit control should never be
A. wired in the high voltage circuit.
B. wired in the low voltage circuit.
C. bypassed permanently.
D. None of the above.

1554. Most limit controls that are wired on the high voltage side of a circuit are in series with the transformer. Where would a voltage check be made to see if there is proper source voltage to the transformer?
A. after the limit control and at the common of transformer
B. before the limit control and at the common of transformer
C. at both leads of the transformer
D. Both A and C.

1555. Adjusting a heat anticipator to a higher setting will
A. provide a longer burner cycle.
B. provide a shorter burner cycle.
C. cause the indoor fan motor short cycle.
D. prematurely shut off the gas valve.

Section 5 Service and Troubleshooting

1556. A service call is made to a customer who complains that the furnace circuit breaker will not stay in the on position.

Technician "A" says that the problem can be in the blower motor or the primary side of the transformer.

Technician "B" says that the problem is most likely in the heating controls circuit. Which technician is most likely correct in his diagnosis?
- A. technician "A"
- B. technician "B"
- C. Both technicians are correct.
- D. Both technicians are incorrect.

1557. An induced draft motor is not operating when the thermostat is calling for heat. What would be the most correct test to make?
- A. an amperage check at the motor
- B. a resistance check of the motor
- C. a voltage check on circuit board at induced draft motor terminals
- D. a voltage check at the primary of transformer

1558. Which of the following residential furnace components would be checked for 24 volt operation?
- A. inducer motor
- B. blower motor
- C. pressure switch
- D. primary of transformer

1559. Technician "A" says, "To ensure that a heating thermostat is in the closed position, a voltage check at "R" and "W" should read 24 volts."

Technician "B" says, "To ensure that a heating thermostat is in the closed position, a voltage check at "C" and "W" should read 24 volts." Which technician is correct?
- A. technician "A"
- B. technician "B"
- C. Both technicians are correct.
- D. Both technicians are incorrect.

1560. Which of the following is not a thermostat problem?
- A. improper anticipator setting
- B. subbase is not level
- C. furnace switch turned off
- D. loose connection at subbase

1561. Which of the following can cause a limit control to open?
- A. a dirty filter
- B. too high of supply air static pressure
- C. not enough return air
- D. All the above.

1562. A direct short to ground of a primary winding on the transformer would most likely cause
- A. a circuit breaker to trip.
- B. a transformer to become defective.
- C. a gas valve to become defective.
- D. a thermostat to become defective.

1563. To correctly set a heat anticipator, the control circuit camperage reading should be
- A. performed at the gas valve.
- B. performed at the secondary of the transformer.
- C. performed at the thermostat subbase.
- D. All of the above.

1564. When a natural gas furnace has flames that are more yellow than blue, what corrective action should a technician take?
- A. replace the gas valve
- B. increase gas pressure
- C. open the primary air ports
- D. close the primary air ports

1565. What are the normally open set of contacts used for on a fan center relay?
- A. high-speed operation of the blower motor
- B. low-speed operation of blower motor
- C. humidifier circuit switching
- D. isolation contacts for the heating circuit

1566. A customer complaint is that the furnace is not operating. After inspection of the system, the technician notices a small hairline crack in the ceramic of the sensing probe. What action must the technician take to repair this problem?
- A. repair the probe by sealing the crack with silicone
- B. replace the sensing probe
- C. no action; sensing probes can handle small cracks
- D. None of the above.

1567. How should a limit safety check be performed on a gas furnace?
- A. temporarily jump out the limit control for at least 10 minutes then remove the jumper
- B. stop the blower operation for at least 10 minutes or until the limit control opens
- C. block off cold air return for 5 minutes, or until the limit control closes
- D. jump out thermostat for 15 minutes until the furnace cycles on the limit control

1568. How is an induced draft pressure switch tested for proper operation?
   A. jump out the pressure switch after setting the thermostat to the heat position and test the control with a manometer
   B. decrease indoor motor speed and test the control with a manometer
   C. de-energize induced draft motor and test the switch with a manometer
   D. jump out the pressure switch while testing it with a manometer

1569. Which of the following are input signals to a high efficiency furnace control board?
   A. the thermostat, the flame roll out sensor, and the ignitor
   B. the pressure switch, the pilot valve, and the limit control
   C. the flame sensor circuit, the main gas valve, and the thermostat
   D. the thermostat, the flame sensor circuit, and the limit control

1570. Which of the following are output signals to a high efficiency furnace control board?
   A. the thermostat, the main gas valve, and the ignitor
   B. the main gas valve, the pilot valve, and the limit control
   C. the flame sensor circuit, the main gas valve, and the ignitor
   D. the ignitor, the pilot gas valve, and the main gas valve

1571. Where do high efficiency condensing furnaces get their combustion air?
   A. from outside of the home only
   B. from inside of the home only
   C. from inside or outside of the home
   D. None of the above.

1572. Which of the following is not a cause of insufficient low gas input?
   A. a plugged orifice
   B. a faulty regulator
   C. a faulty heat exchanger
   D. a bad gas valve

1573. A customer complains that there is not enough supply air coming through the registers.

   Technician "A" checked the furnace temperature rise, inspected the air filter, and adjusted a loose fan belt.

   Technician "B" inspected the air filter, observed the blower motor operation, checked for duct obstructions, and made sure that the balancing dampers were in the correct position.

   Which technician performed a proper procedure?
   A. technician "A"
   B. technician "B"
   C. Both technicians are correct.
   D. Both technicians are incorrect.

1574. A technician was called because not enough heat was being generated at the furnace. The technician checked the heat exchanger, flue pipe, and burners. He also tested the inlet and outlet gas pressure to the furnace. His measurement was 4" WC to the main gas valve and 2 inches out of the gas valve. Which of the following could be the problem with the system?
   A. a faulty gas valve
   B. a faulty gas meter regulator
   C. a faulty heat exchanger
   D. a faulty gas control circuit

1575. Which of the following would *not* be a cause of a faulty thermostat on a standing pilot furnace?
   A. The gas valve will not de-energize.
   B. The gas valve will not energize.
   C. The furnace is short cycling.
   D. The pilot will not stay lit.

1576. A new high efficiency furnace has been installed and the furnace seems to come on too often. Which of the following would most likely be the problem?
   A. a defective fan control
   B. an improperly adjusted heat anticipator
   C. a defective blower motor
   D. too much air flow

1577. Which of the following would not cause the burners of a gas furnace to shut down?
   A. excessive air flow
   B. a defective gas valve
   C. a faulty thermostat
   D. a defective induced draft motor

Section 5 Service and Troubleshooting

1578. Which of the following is *not* a cause for the burners of a gas furnace to short cycle?
   A. a dirty air filter
   B. poor thermostat location
   C. heat anticipator set too low
   D. fan control contacts stuck closed

1579. A customer complains that her house is too warm. The technician arrives on the job and notices that the gas valve is stuck open. The technician taps the side of the valve and the gas flow stops. What action should the technician take?
   A. replace the gas valve
   B. lubricate the sticking valve
   C. turn the valve on and off several times to free the problem
   D. Any of the above.

1580. A pilot light has been established, but it is too small and soon goes out. Which of the following procedures would be correct to remedy this condition?
   A. replace the pilot valve
   B. repair the pilot valve coil
   C. clean the pilot orifice
   D. replace entire gas valve

1581. What is the sequence of operation of a mid-efficiency furnace using a hot surface indirect ignition system?
   A. The thermostat calls for heat, the ignition system lights the pilot, the inducer motor is energized, the pressure switch closes, the main gas valve is energized, and the blower motor is energized.
   B. The thermostat closes, the inducer motor is energized, the main burner is lit from the ignitor, and the blower motor is energized.
   C. The thermostat closes, the inducer motor is energized, the pilot is lit from the ignitor, the pilot is proved, the main gas valve is energized, and the blower motor is energized.
   D. The thermostat opens, the inducer motor is energized, the main gas valve is energized, and the blower motor is energized.

1582. On an intermittent spark ignition system, the pilot valve is energized, but there is no spark from the ignitor. Which of the following could be the problem?
   A. a faulty ignition cable
   B. a faulty control module
   C. a bad ground
   D. All of the above.

1583. On a direct-spark ignition system, the ignitor is sparking, but the main burners will not light. Which of the following is an *incorrect* test to perform?
   A. check for 24 volts across the gas terminals
   B. check for correct position of spark gap
   C. check for a cracked insulation on the ignitor ceramic
   D. check from terminals W to C to make sure that the thermostat is closed

1584. On a direct hot surface ignition system, the ignitor is not glowing, but the main gas valve is energized during a trial for ignition. Which of the following is an *incorrect* test to perform?
   A. measure voltage to the ignitor
   B. measure resistance on the ignitor
   C. measure resistance of the limit control
   D. measure input voltage to the control module

1585. A furnace that uses a direct hot surface ignition system will not ignite the burners.

   Technician "A" measures 50 ohms on the hot surface ignitor, and says that the ignitor is defective.

   Technician "B" also measures 50 ohms on the ignitor, and says that the ignitor is not defective. The problem is probably a bad control module.

   Which technician is correct?
   A. technician "A"
   B. technician "B"
   C. Both technicians are correct.
   D. Both technicians are incorrect.

1586. What terminals should measure 24 volts if the thermostat is calling for heat? Use the Bimetal Flame Sensor Pilot Ignition System diagram on page 230 to answer this question.
   A. LIM 1 and LIM 2
   B. R and W
   C. GAS 1 and GAS 2
   D. GC and C

1587. What terminals should measure 24 volts if the pilot has been proven? Use the Bimetal Flame Sensor Pilot Ignition System diagram on page 230 to answer this question.
   A. WH at the sensor and GAS 2
   B. YEL at WH at the sensor
   C. GAS 3 and GAS 1
   D. R and W

Bimetal Flame Sensor Pilot Ignition System

1588. Using the Bimetal Flame Sensor Pilot Ignition System diagram, what two terminals energize the fan motor during the heating cycle?
   A. terminals EAC 1 and EAC 2
   B. terminals 4 and C
   C. terminals 1 and C
   D. terminals R and GC

1589. Using the Bimetal Flame Sensor Pilot Ignition System diagram, what terminals should always measure 24 volts if power is connected to the system?
   A. terminals Gas 1, Gas 2, Sec 1, and Sec 2
   B. terminals Sec 1, Sec 2, R, and W
   C. terminals Sec 1, Sec 2, R, and C
   D. All of the above.

1590. Where should a measurement of 24 volts be when the heating thermostat is in the closed position? Use the Indirect Spark Ignition System diagram on page 231 to answer this question.
   A. at the "TH" terminal and terminal before "PS"
   B. at the "W" terminal and any of the "HL" terminals
   C. at the "W" terminal and "SL" terminal
   D. at the "TH" terminal and "TR" terminal

1591. The indoor blower motor will not start on a call for heat. The technician suspects a faulty set of "FR" contacts. Where should a voltage reading occur if the contacts are in good condition? Use the Indirect Spark Ignition System diagram on page 231 to answer this question.
   A. across the FR coil and from the load side of the normally open FR contacts to the common side of the motor
   B. across the FR coil and across the normally open contacts
   C. across the FR coil and across the FS contacts
   D. across the FR coil and from the load side of the normally open contacts to the line side of the motor

1592. What are the correct terminals to measure voltage to the main gas valve? Use the Indirect Spark Ignition System diagram on page 231 to answer this question.
   A. "TH" and "TR"
   B. "PV" and "C"
   C. "TH" and "C"
   D. "MV" and "C"

Section 5   Service and Troubleshooting

Indirect Spark Ignition System

Legend

C	Contractor
DS	Door Switch
FR	Fan Relay
FS	Fan Switch
HL	High Limit
IBM	Indoor Blower Motor
IDR	Induced Draft Relay
MV	Main Valve
PS	Pressure Switch
PV	Pilot Valve
SL	Secondary Limit
TH	Thermostat

OIL FURNACE TROUBLESHOOTING

1593. What is the Btu input rating of an oiled fired furnace that uses a nozzle rated at .85 GPM?
 A. 16,470 Btu/h
 B. 93,333 Btu/h
 C. 119,000 Btu/h
 D. 210,000 Btu/h

1594. What are two problems that can cause and oil fired gun burner to operate below 100 PSIG?
 A. insufficient oil and improperly adjusted electrodes
 B. a plugged nozzle and a bad oil pump
 C. a plugged inline filter and a bad ignition transformer
 D. a bad oil pump and a plugged inline filter

1595. Technician "A" says that the input voltage of an ignition transformer used on a gun type oil burner is 208V, and the output voltage is 15,000V.

Technician "B" says that the input voltage of an ignition transformer used on a gun type oil burner is 120V, and the output is 10,000V.

Which technician is correct?
 A. technician "A"
 B. technician "B"
 C. Both technicians are correct.
 D. Both technicians are incorrect.

1596. Which of the following can cause a weak spark in a high pressure gun burner?
 A. a cracked electrode
 B. a bad ignition transformer
 C. a bad thermocouple
 D. Both A and B.

1597. When adjusting the electrode assembly, what is the proper measurement from the center of the nozzle to the tip of the electrode?
 A. $\frac{1}{8}"$ to $\frac{1}{4}"$
 B. $\frac{5}{16}"$ to $\frac{1}{4}"$
 C. $\frac{1}{2}"$ to $\frac{9}{16}"$
 D. None of the above.

1598. The capacity of a high pressure oil burner nozzle is rated in
   A. GPH (gallons per hour).
   B. GPM (gallons per minute).
   C. Btu/h (British thermal units per hour).
   D. PPG (pounds per gallon).

1599. What type of spray pattern does a residential oil fired furnace commonly use?
   A. solid
   B. semi-hollow
   C. uniform
   D. hollow

1600. What would the spray angle be on a residential oil fired furnace?
   A. 30°
   B. 45°
   C. 60°
   D. 80°

1601. Technician "A" says that the motor that is commonly used on a gun type oil burner is a capacitor start motor.
   Technician "B" says that the motor that is commonly used on a gun type oil burner is a split phase motor.
   Which technician is correct?
   A. technician "A"
   B. technician "B"
   C. Both technicians are correct.
   D. Both technicians are incorrect.

1602. When replacing a burner motor that is used on a high pressure gun burner, is critical to have the proper speed. Burner motors have two common speeds that are used in the field. What are they?
   A. 650 and 1300 RPM
   B. 750 and 1400 RPM
   C. 1725 or 3450 RPM
   D. 2025 and 2525 RPM

1603. Technician "A" says that the size of a fill pipe on an oil storage tank is 2 inches in diameter.
   Technician "B" says that the size of a fill pipe on an oil storage tank is $2\frac{1}{2}$ inches in diameter.
   Which technician is correct?
   A. technician "A"
   B. technician "B"
   C. Both technicians are correct.
   D. Both technicians are incorrect.

1604. Technician "A" says that the size of an air vent on an oil storage tank is 2 inches in diameter.
   Technician "B" says that the size of an air vent on an oil storage tank is $1\frac{1}{2}$ inches in diameter.
   Which technician is correct?
   A. technician "A"
   B. technician "B"
   C. Both technicians are correct.
   D. Both technicians are incorrect.

1605. A technician performs a combustion efficiency test on an oil fired furnace. He notices that the CO_2 reading is high. What could cause this problem?
   A. not enough fuel in the fuel air mixture
   B. too much air in the fuel air mixture
   C. not enough air for the fuel air mixture
   D. Both A and B.

1606. A technician is servicing an oil fired furnace and he notices that the single stage oil pump has to be replaced. What type of system is he servicing?
   A. an above ground one pipe system
   B. an underground one pipe system
   C. an above ground two pipe system
   D. an underground two pipe system

1607. A technician is servicing an oil fired furnace that uses a cad cell primary control. He notices that the control locks out after approximately 45 seconds. Which of the following can cause this to happen?
   A. the cad cell is measuring above 100,000 ohms of resistance in the dark
   B. the cad cell is measuring below 1,600 ohms of resistance in the light
   C. the cad cell is measuring above 1,600 ohms of resistance in the light
   D. the cad cell is measuring below 100,000 ohms in the dark

1608. When troubleshooting an oil fired furnace that uses a cad cell primary control, what two wires should be tested for line voltage that feeds the burner motor and ignition transformer?
   A. the orange and white wires
   B. the blue and orange wires
   C. the black and blue wires
   D. the white and black wires

Section 5 Service and Troubleshooting 233

1609. A technician is servicing an oil fired furnace that uses a stack mount relay. He notices that the burner motor is not operating. What two terminals must he place the voltmeter across to test for output voltage from the control to the burner motor?
A. terminals 1 and 2
B. terminals 1 and 3
C. terminals 2 and 3
D. terminals 3 and 4

1610. A service technician gives a combustion efficiency test to an oil fired furnace. He notices that there is too much of a draft. What can the technician do to decrease the amount of draft in the system?
A. he can close the barometric damper
B. he can increase the size of the flue pipe
C. he can add a chimney cap
D. he can open the barometric damper

1611. A technician is servicing an oil fired furnace and he notices that there is a small amount of oil that drips from the nozzle when the system is de-energized. What can be added to the system to decrease the amount of dripping during shutdown?
A. a new nozzle
B. a delayed oil valve
C. a new inline filter
D. a new nozzle assembly

1612. How is a delayed oil valve wired in a heating circuit?
A. It is connected across the black and white wire on the cad-cell relay.
B. It is connected across terminals 2 and 3 on a stack mount relay.
C. It is connected across terminals 1 and 2 on a stack mount relay.
D. It is connected across the limit control.

1613. Technician "A" says that a common CO_2 measurement for an oil fired furnace is from 8.5% to 9.5%.

Technician "B" says that a common CO_2 for an oil fired furnace is from 10% to 12%.

Which technician is correct?
A. technician "A"
B. technician "B"
C. Both technicians are correct.
D. Both technicians are incorrect.

1614. A technician arrives on a service call for an oil fired furnace. He notices that the overfire draft reading is +.01" WC. He then measures the breech draft and it is a -.04" WC. What could be the problem with the furnace?
A. a plugged chimney
B. a plugged heat exchanger
C. insufficient primary air
D. All of the above.

1615. A common breech draft for an oil furnace is
A. -1" WC to -2" WC
B. -.05" WC to -.1" WC
C. -.03" WC to -.06" WC
D. -.01" WC to -.02" WC

1616. When servicing an oil fired furnace, a technician finds that the furnace is giving a smoke reading of #4. What is a possible problem that can cause this?
A. excessive combustion air
B. insufficient combustion air
C. there is no problem; the system should produce a smoke reading of #4
D. the barometric damper open too much

1617. What does a smoke reading of #0 indicate on an oil fired furnace?
A. not enough of an over the fire draft
B. not enough primary air
C. too much primary air
D. no problem, good working condition

1618. What is a possible problem if an oil fired furnace has a stack temperature reading over 800° F?
A. the nozzle is too large
B. excessive combustion air
C. too high of an oil pump pressure
D. All of the above.

1619. Using the Cad-Cell Relay diagram on page 234, answer the following question. The system will not in operate, the disconnect switch, the limit switch, and the thermostat are closed. There is a voltage reading of 120 volts across terminals BK and WH. A slight leak is noticed in the combustion chamber. Why will the system not operate?
A. The light leak in the combustion chamber causes the resistance of the cad-cell to decrease and not fire.
B. The light leak in the combustion chamber causes the resistance of the cad-cell to increase and not fire.
C. The primary control must be condemned and replaced.
D. Both B and C.

Cad-Cell Relay Diagram

Legend
BM — Blower Motor
Disc — Disconnect
FS — Fan Switch
IBM — Indoor Blower Motor
IT — Ignition Transformer
LIM — Limit
TH — Thermostat

1620. Using the Cad-Cell Relay diagram, a technician notices that the system energizes for about 2 minutes, and then shuts down. A reading across terminals 1 and 2 on the LIM shows 120V. Another reading across terminals 1 and 2 on FS also shows 120V. What is the possible problem with the system?
   A. The LIM is defective.
   B. The cad-cell relay is defective.
   C. The IBM is defective.
   D. The FS is defective.

1621. Using the Cad-Cell Relay diagram, a technician sets the thermostat to call for heat and closes the disconnect switch. The system is sparking but does not fire. A further check shows that the burner motor is not in operation. What could be the problem with the system?
   A. The cad-cell relay could be defective.
   B. The cad-cell could be defective.
   C. The burner motor could be defective.
   D. Both A and B.

1622. The disconnect switch, the limit switch and the thermostat are in the closed position. There is spark at the electrodes, but no ignition. A further check shows that the burner motor is not operating. Where should a voltage reading be taken to further troubleshoot the system? Use the Stack Mount Relay diagram on page 235 to answer this question.
   A. across terminals 1 and 2
   B. across terminals 1 and 3
   C. across terminals 1 and 4
   D. across terminals 3 and 2

1623. A technician notices that the system energizes for about 45 seconds and shuts down. Before it energizes, there is a satisfactory flame present in the combustion chamber. A reading across terminals 1 and 2 on the LIM shows 0V. Another reading across terminals 1 and 2 on the stack mount relay also shows 120V. What is the problem? Use the Stack Mount Relay diagram on page 235 to answer this question.
   A. The LIM is defective.
   B. The stack mount relay is defective.
   C. The IBM is defective.
   D. The FS is defective.

Section 5 Service and Troubleshooting

Stack Mount Relay Diagram

1624. A technician sets the thermostat to call for heat and closes the disconnect switch. The system is sparking but does not fire. A further check shows that the burner motor is not in operation. What could be the problem with the system? Use the Stack Mount Relay diagram to answer this question.
   A. The stack mount relay could be defective.
   B. The limit control could be defective.
   C. The burner motor could be defective.
   D. Both A and C.

1625. What two problems on a high pressure gun burner can stop the flow of oil to the nozzle?
   A. a defective ignition transformer and an inoperative oil pump
   B. a broken coupling and a plugged inline filter
   C. an oil pump with too high of a pressure adjustment and a defective burner motor
   D. a bad run capacitor on the oil pump and a plugged nozzle

1626. Technician "A" says that the distance the electrodes should be placed in front of a nozzle with an 80° spray angle is $\frac{1}{4}"$.

   Technician "B" says that the distance the electrodes should be placed in front of a nozzle with an 80° spray angle is $\frac{1}{8}"$.

   Which technician is correct?
   A. technician "A"
   B. technician "B"
   C. Both technicians are correct.
   D. Both technicians are incorrect.

1627. On a stack mount relay, what position should the cold contacts be in on the start up of an oil fired furnace?
   A. the normally closed position
   B. the normally open position
   C. they can be in the open or closed position
   D. there are no cold contacts that are located on a stack mount relay

1628. When troubleshooting an oil fired furnace which has an ignition problem, what are three tests that can be performed?
  A. test the ignition transformer, the burner motor, and the nozzle assemble
  B. test the ignition transformer, the electrode assemble, and the blast tube assembly
  C. test the ignition transformer, the nozzle assembly, and the blast tube assembly
  D. test the ignition transformer, the electrode assembly, and the electrode adjustment

1629. When troubleshooting an oil fired furnace which has a oil delivery problem, what are three tests that can be performed?
  A. test oil pump pressure, burner motor operation, and the nozzle assembly
  B. test oil pump pressure, the electrode adjustment, and the blast tube assembly
  C. test oil pump pressure, the electrode assembly, and the nozzle assembly
  D. test oil pump pressure, the burner motor operation, and the blast tube assembly

1630. If a high pressure gun type oil burner has a sufficient spark and the oil pump pressure is adequate, what can cause the system not to fire properly?
  A. a broken coupling
  B. a plugged nozzle
  C. a plugged inline oil filter
  D. a defective ignition transformer

1631. What could cause a high pressure gun type oil burner to have a weak or insufficient spark?
  A. cracked ceramic electrodes
  B. improper electrode adjustment
  C. faulty ignition transformer
  D. All of the above.

1632. Which of the following could cause an oil burner to have insufficient oil?
  A. plugged nozzle
  B. bad oil pump
  C. empty oil tank
  D. All of the above.

HEAT PUMP TROUBLESHOOTING

1633. On an air to air heat pump system, the suction pressure and discharge pressure are low during the cooling months. What are two possible problems?
  A. a dirty air filter and a dirty condenser
  B. a low charge and a dirty condenser
  C. a restricted indoor metering device and a dirty air filter
  D. a restricted outdoor metering device and a restricted indoor check valve

1634. On an air to air heat pump system, the discharge pressure and suction pressure are high during the heating months. What are two possible problems?
  A. a restricted indoor metering device and a restricted outdoor check valve
  B. a dirty outdoor coil and a restricted outdoor metering device
  C. a defective defrost system and a dirty outdoor coil
  D. a dirty indoor filter and non-condensables in the system

1635. A service call is made on an air to air heat pump system during the winter. When servicing the system, it is found that the outdoor coil has frost build up and the system is operating on emergency heat.

Technician "A" says that the system might have a defrost problem.

Technician "B" says that the reversing valve might be leaking.

Which technician is correct?
  A. technician "A"
  B. technician "B"
  C. Both technicians are correct.
  D. Both technicians are incorrect.

1636. A service call is made on an air to air heat pump system during the winter months. The technician arrives on the job and notices that the discharge pressure is low and the suction pressure is high. What two problems could cause this symptom?
  A. a restricted outdoor metering device and a defective defrost system
  B. an inefficient compressor and a dirty indoor air filter
  C. an excessive heat load and a dirty condenser
  D. a leaking reversing valve and bad compressor valves

Section 5   Service and Troubleshooting

1637. How can a system be tested to determine if the reversing valve solenoid is leaking internally?
   A. A temperature measurement can be taken at the inlet and the outlet of the valve in the heating and cooling mode.
   B. The valve can only be changed since it is impossible to test the valve for internal leaks.
   C. A temperature drop reading across the indoor coil during the cooling season will determine if the reversing valve is leaking.
   D. The reversing valve often has a high pitched whistle if it is leaking internally.

1638. What component must always be de-energized during the defrost cycle on a heat pump?
   A. the reversing valve solenoid
   B. the outdoor fan
   C. the indoor fan
   D. the defrost relay coil

1639. When an air to air heat pump is in the cooling mode, what can cause a decrease in high side and low side pressure?
   A. a restricted indoor metering device
   B. a defective indoor coil check valve
   C. bad compressor valves
   D. a defective outdoor metering device

1640. Using the Heat Pump diagram (cooling mode), what is connected to the top tube of the reversing valve?
   A. the compressor discharge line
   B. the compressor suction line
   C. the indoor coil
   D. the outdoor coil

1641. Using the Heat Pump diagram (cooling mode), what is connected to the bottom left tube of the reversing valve?
   A. the compressor discharge line
   B. the compressor suction line
   C. the indoor coil
   D. the outdoor coil

Heat Pump Diagram
Heating Mode

1642. Using the Heat Pump diagram (heating mode), what is connected to the bottom left tube of the reversing valve?
   A. the compressor discharge line
   B. the compressor suction line
   C. the indoor coil
   D. the outdoor coil

1643. What is connected to the bottom center tube of the reversing valve? Use the Heat Pump diagram (heating mode) to answer this question.
   A. the compressor discharge line
   B. the compressor suction line
   C. the indoor coil
   D. the outdoor coil

1644. Technician "A" says that the check valves in a heat pump system are located in series with both metering devices.

   Technician "B" says that the check valves in a heat pump system are located in parallel with both metering devices.

   Which technician is correct?
   A. technician "A"
   B. technician "B"
   C. Both technicians are correct.
   D. Both technicians are incorrect.

1645. A service technician notices that a check valve needs to be replaced in a heat pump system. Since it's a ball type check valve, it is critical for him to install the valve
   A. horizontally, with the seat on the outlet of the valve.
   B. vertically, with the seat on the bottom of the valve.
   C. horizontally, with the seat on the inlet of the valve.
   D. vertically, with the seat on the top of the valve.

1646. After troubleshooting a heat pump system, the technician concludes that the check valve is defective. He connects a recovery system and removes the refrigerant from the unit. He then proceeds to unsweat the old check valve and remove it from the system. He cleans the area where the new valve will be installed, places it into position, and brazes it into the system while not protecting the valve. After a thorough evacuation, he starts the system and notices that the new valve is defective. What could have been the problem with the new valve?
   A. A heat sink was not used and the valve warped.
   B. The valve may have been defective from the factory.
   C. The problem was misdiagnosed.
   D. All of the above.

Section 5 Service and Troubleshooting

1647. Technician "A" says that the auxiliary heaters found on an air to air heat pump are used to increase the system's pressure during a defrost cycle.

Technician "B" says that the auxiliary heaters found on an air to air heat pump are used to increase supply air temperature for comfort during a defrost cycle.
A. technician "A"
B. technician "B"
C. Both technicians are correct.
D. Both technicians are incorrect.

1648. When a heat pump switches into the defrost cycle, it must
A. remain in defrost until the temperature rise across the indoor coil reaches 50° F.
B. remain in defrost until the indoor coil has reached a pre-determined temperature.
C. remain in defrost until the outdoor coil has reached a pre-determined temperature.
D. remain in defrost until the indoor thermostat is satisfied.

1649. A defrost termination thermostat is commonly located
A. on the vapor line, and opens at 55° F.
B. on the liquid line, and closes at 55° F.
C. on the indoor coil, and closes at 55° F.
D. on the outdoor coil, and opens at 55° F.

1650. Temperature split refers to
A. the temperature difference of the air across the outdoor coil.
B. the temperature difference between the inside and the outside of the home.
C. the temperature difference of the air across the indoor coil.
D. the temperature difference between the air entering a coil and the temperature of the coil.

1651. Technician "A" says that the outdoor ambient thermostat commonly de-energize the compressor during the second stage of heating in an air to air heat pump.

Technician "B" says that the outdoor ambient thermostat used with an air to air heat pump commonly energizes the auxiliary strip heaters during the first stage of heating.

Which technician is correct?
A. technician "A"
B. technician "B"
C. Both technicians are correct.
D. Both technicians are incorrect.

1652. If there is malfunction in a hold back thermostat that is used with an air to air heat pump, how would this affect the system?
A. The system would not hold back the compressor operation during the defrost cycle. This could cause excessive power consumption.
B. The system would not stay energized during defrost if the room thermostat satisfied. This could cause freezing of the outdoor coil.
C. The system would not de-energize the auxiliary strip heaters during the defrost cycle. This could cause excessive heat build up.
D. The system would not energize the auxiliary strip heaters during the heating cycle. This could cause insufficient heating.

1653. A technician arrives on a service call for an air to air heat pump during the heating season. He notices that the system operates normally until the defrost circuit is energized. The system stays in defrost for approximately 15 minutes when there is no ice or frost build up on the coil. What could be the problem with the system?
A. the defrost timer is defective
B. the defrost termination switch is stuck closed
C. the defrost termination switch is stuck open
D. the reversing valve solenoid is stuck open

1654. A technician receives a service call for an air to air heat pump. The customer is complaining about insufficient heat during low ambient conditions. Upon arrival to the job, he notices that the mechanical heat pump system is operating properly but the auxiliary heat is not operating. What circuit should be checked next?
A. the first stage heating circuit
B. the second stage heating circuit
C. the air pressure control circuit
D. the auxiliary contactor circuit

1655. A technician arrives at a home on a service call for an air to air heat pump. After measuring the air temperature over the indoor coil and finding a temperature rise of 35° F, he determines that
A. the indoor air filter might be dirty
B. the indoor blower is running too fast
C. the outdoor coil might be dirty
D. the temperature rise is sufficient

Part 2 Questions

Heat Pump Electrical Diagram

Legend for Heat Pump Elecrtical Diagram

ADR–Auxiliary Defrost Relay	DR–Defrost Relay	HT2–Heating Thermostat 2nd Stage
APS–Air Pressure Switch	FR–Fan Relay	IFM–Indoor Fan Motor
C–Contactor	FU–Fusible Link	LIM–Limit Switch
CCH–Crankcase	H1–Auxiliary Strip Heater #1	ODT–Outdoor Thermostat
CF–Condenser Fan	H2–Auxiliary Strip Heater #2	RV–Reversing Valve
Comp.–Compressor	H3–Auxiliary Strip Heater #3	TD1–Timed Delay Relay #1
CR–Control Relay	Heater HP–High Pressure Switch	TD2–Timed Delay Relay #2
CT1–Cooling Thermostat 1st Stage	HR–Holding Relay	TM–Timer Motor
CT2–Cooling Thermostat 2nd Stage	HT1–Heating Thermostat 1st Stage	

Section 5 Service and Troubleshooting

1656. With ODT closed between terminals 2 and 3 and the thermostat in the heating mode with W2 in the closed position, what voltage would be expected across terminal 2 on LIM and terminal 1 on FU in the H3 circuit? Use the Heat Pump Electrical diagram on page 240 to answer this question.
A. 0 volts
B. 24 volts
C. 120 volts
D. 240 volts

1657. What should be the resistance across terminals 1 and 2 on TM1 during a normal heating operation? Use the Heat Pump Electrical diagram on page 240 to answer this question.
A. 0 ohms
B. 50 ohms
C. 1000 ohms
D. infinite ohms

1658. What voltage should be across terminal 1 on LIM and terminal 1 on FU in the H1 circuit during the first stage of heating? Use the Heat Pump Electrical diagram on page 240 to answer this question.
A. 0 volts
B. 24 volts
C. 120 volts
D. 240 volts

1659. If FU was in the open position in the H2 circuit when the first stage heating was energized, what voltage would be across terminals 1 and 2? Use the Heat Pump Electrical diagram on page 240 to answer this question.
A. 0 volts
B. 24 volts
C. 120 volts
D. 240 volts

1660. What will happen to the system if the HP switch opens? Use the Heat Pump Electrical diagram on page 240 to answer this question.
A. The system will remain de-energized until HP closes and then the system will energize again.
B. The system will short cycle on the HP control.
C. The system will be locked out until the power is disconnected and then reconnected.
D. Both A and B.

1661. What are symptoms of an air to air heat pump system that has a low charge when in the cooling mode?
A. The indoor coil will be partially frosted and the discharge pressure will increase.
B. The suction line will sweat and the low-side pressure will decrease.
C. The low-side pressure will decrease and the indoor coil will be partially frosted.
D. The discharge pressure will decrease and the suction pressure will increase.

1662. A customer calls and says that her air to air heat pump system is not cooling properly. Upon arrival, a technician notices that the suction line and compressor has frost build up.

Technician "A" says that the problem is probably a low charge.

Technician "B" says that it is possibly a dirty air filter.

Which technician is more likely to be correct?
A. technician "A"
B. technician "B"
C. Both technicians are correct.
D. Both technicians are incorrect.

1663. Name two symptoms that are common to an overcharged air to air heat pump system with a system that has a dirty condenser.
A. The subcooling decreases and the amperage increases.
B. The superheat increases and the discharge pressure increases.
C. The suction pressure increases and the subcooling increases.
D. The discharge pressure increases and the amperage increases.

1664. When an air to air heat pump system has an undercharge, what happens to the superheat and the subcooling of the system?
A. The superheat increases and the subcooling increases.
B. The superheat increases and the subcooling decreases.
C. The superheat decreases and the subcooling increases.
D. The superheat decreases and the subcooling decreases.

AIR DISTRIBUTION TROUBLESHOOTING

1665. An air conditioning service call is made and the complaint is that the system does not seem to be cooling properly. It is noticed that the suction line and the evaporator coil are frosted up. The filter is checked and it is clean. It is suspected that the indoor blower is not operating at the proper speed.

Technician "A" says that the speed can be verified with a strobe tachometer to be sure that the RPMs are as written on the nameplate.

Technician "B" says that the airflow can be measured by taking a static pressure reading.

Which technician is correct?
A. technician "A"
B. technician "B"
C. Both technicians are correct.
D. Both technicians are incorrect.

1666. An air flow measurement taken in a duct system by drilling small holes in a vertical and horizontal section of a rectangular duct to find an average air velocity reading is known as a
A. pitot tube traverse.
B. anenometer test.
C. air quantity test.
D. a static pressure test.

1667. To calculate velocity pressure when using a manometer, which of the following would be the proper formula?
A. static pressure + total pressure = velocity pressure
B. static pressure − total pressure = velocity pressure
C. total pressure − static pressure = velocity pressure
D. total pressure + static pressure = velocity pressure

1668. What is the primary purpose of an isolation collar?
A. It is used to isolate the duct system from the furnace when it needs to be replaced.
B. It is used to eliminate furnace vibration through the duct system.
C. It is used to increase the air velocity in a duct system.
D. It is used to decrease the air velocity in a duct system.

1669. During the cooling season, a complaint is made that a two story home is too cool on the first floor and too warm on the second floor. What can be done to help the system cool properly?
A. The first floor inline dampers can be adjusted parallel to the duct.
B. The first floor inline dampers can be adjusted perpendicular to the duct.
C. The second floor inline dampers can be adjusted perpendicular to the duct.
D. None of the above.

1670. When installing sections of ducts such as plenums, trunk sections, and duct transitions, it is important to
A. attach the sections using drive cleats only.
B. attach the sections using S-slips only.
C. connect the duct sections using the proper S-slips and drive cleats loosely to prevent duct vibration.
D. connect the duct sections using the proper S-slips and drive cleats tightly to prevent duct vibration.

1671. A service call is made stating that a family's furnace and duct system has recently been installed by a friend of the family. The problem is that there doesn't seem to be enough airflow out of the registers. After arriving to the job, the technician notices that the furnace is a 75,000 Btu/h and the air conditioning system is 2 tons of cooling. After further investigation, the technician notices that the main supply duct system is 60" x 8" by 25' long. What is the problem with the system?
A. The system's static pressure is too low.
B. The system's main duct is too small.
C. The system's static pressure is too high.
D. The system's main duct is too large.
E. Both A and D.

1672. Technician "A" measures the airflow in a main supply duct by using a Pitot Tube Traverse. His measurement is 1200 FPM. The area of the duct at the point of measurement is 8" x 30". He says that he has 2000 CFM.

Technician "B" measures the airflow in a main supply duct by using a Pitot Tube Traverse. His measurement is 1200 FPM. The area of the duct at the point of measurement is 10" x 30". He says that he has 2500 CFM.

Which technician is correct?
A. technician "A"
B. technician "B"
C. Both technicians are correct.
D. Both technicians are incorrect.

Section 5 Service and Troubleshooting

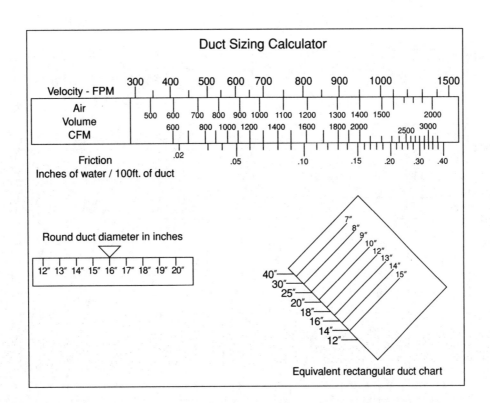

1673. When should the dampers and registers of a new system be 100% open?
A. after the system has been fully balanced
B. on initial start up of the system
C. during the heating season only
D. during the cooling season only

1674. When using the Duct Sizing Calculator, Technician "A" uses 1800 CFM as the air quantity with 900 FPM air velocity. He says that the static pressure is .10 inches of water column.

When using the Duct Sizing Calculator, Technician "B" uses 1600 CFM as the air quantity with 1000 FPM air velocity. He says that the static pressure is .10 inches of water column.

Which technician is correct?
A. technician "A"
B. technician "B"
C. Both technicians are correct.
D. Both technicians are incorrect.

1675. When using the Duct Sizing Calculator, the required volume of air is 1600 CFM and a velocity is 1000 FPM. What size of round and rectangular duct is needed?
A. 18" and 8" x 30"
B. 16" and 10" x 20"
C. 20" and 12" x 18"
D. 16" and 14" x 18"

1676. Using the Duct Sizing Calculator, what is the air velocity if the air quantity is 1400 CFM?
A. 800 FPM
B. 950 FPM
C. 1800 FPM
D. 2000 FPM

1677. Technician "A" says that an 8" x 12" register that has the velocity of 300 FPM with the free area of 75% has an air quantity of 150 CFM.

Technician "B" says that an 8" x 12" register that has the velocity of 300 FPM with the free area of 75% has an air quantity of 250 CFM.

Which technician is correct?
A. technician "A"
B. technician "B"
C. Both technicians are correct.
D. Both technicians are incorrect.

Part 2 Questions

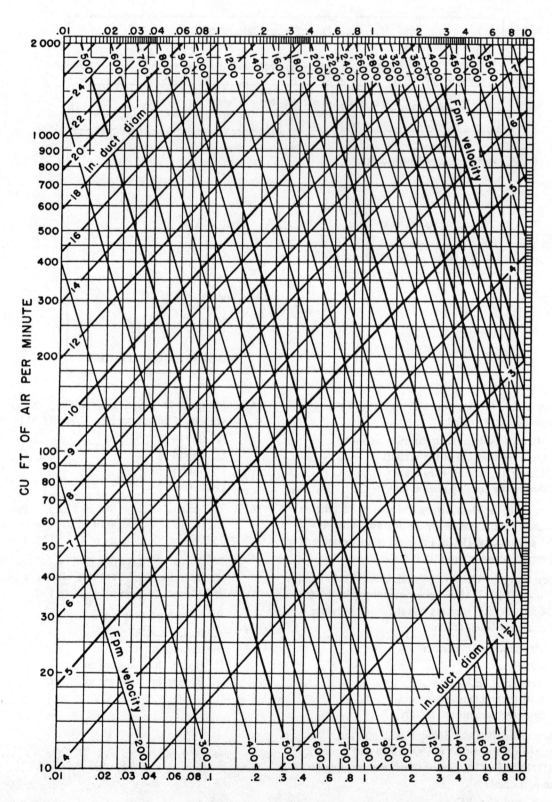

FRICTION LOSS IN INCHES OF WATER PER 100 FT

(Based on Standard Air of 0.075 lb per cu ft density flowing through average, clean, round, galvanized metal ducts having approximately 40 joints per 100 ft.) Caution: Do not extrapolate below chart.

Section 5 Service and Troubleshooting

1678. A service call is made on the complaint of not enough warm air out of the supply registers. The technician does an air flow calculation by measuring the temperature difference across the heat exchanger, which is 100° F. He then determines the furnace output, which is 90,000 Btu/h. With this information, he can determine what the quantity of airflow out of the system. Which of the following formulas must be used?
   A. CFM = Output (Btu/h) / 1.08 × temperature rise
   B. CFM = Output (Btu/h) × 1.08 / temperature rise
   C. CFM = Output (Btu/h) × temperature rise / 1.08
   D. CFM = Output (Btu/h) / 1.08 + temperature rise

1679. On the previous question, what is the quantity of air that is flowing in the system?
   A. 833 CFM
   B. 972 CFM
   C. 8333 CFM
   D. 890 CFM

1680. What would be the velocity of the air at a 0.2 friction loss per 100 feet and 50 CFM of air? Use the Friction Loss Chart on page 244 to help answer this question.
   A. 85 FPM
   B. 200 FPM
   C. 500 FPM
   D. 600 FPM

1681. What would the diameter of the duct have to be to move 300 CFM at the velocity of 600 FPM? Use the Friction Loss Chart on page 244 to help answer this question.
   A. 6 inches
   B. 8 inches
   C. 10 inches
   D. 12 inches

1682. What would be the friction loss per 100 feet if the speed of the air through an 8 inch duct were 600 FPM? Use the Friction Loss Chart on page 244 to help answer this question.
   A. .05
   B. .08
   C. 1.0
   D. .15

1683. Using the Friction Loss Chart on page 244, technician "A" says that to keep the static pressure of a duct at .1" WC, with 250 CFM in the duct, the size of the duct must be 8 inches round.

Using the Friction Loss Chart on page 244, technician "B" says that to keep the velocity of a 6 inch duct to 500 FPM, the air quantity must be 100 CFM. Which technician is correct?
   A. technician "A"
   B. technician "B"
   C. Both technicians are correct.
   D. Both technicians are incorrect.

1684. Which of the following formulas are used to calculate the area of a round duct?
   A. Diameter × circumference
   B. Radius × 3.14
   C. $3.14 \times radius^2$
   D. $3.14 \times diameter^2$

1685. Technician "A" says that a blower over 5 horsepower is commonly a PSC type of motor.

Technician "B" says that a blower over 5 horsepower is commonly a three phase type of motor.

Which technician is correct?
   A. technician "A"
   B. technician "B"
   C. Both technicians are correct.
   D. Both technicians are incorrect.

1686. What is commonly used to cycle condenser fan motors on HVAC rooftop equipment during low ambient conditions?
   A. a high ambient pressure switch
   B. a low pressure switch
   C. a high pressure switch
   D. a reverse acting high pressure switch

1687. Why is it necessary to cycle the condenser fans during low ambient conditions?
   A. to prevent an increase in high side pressure and the flooding of the evaporator coil
   B. to prevent a decrease in high side pressure and the flooding of the evaporator coil
   C. to prevent an increase in high side pressure and the starving of the evaporator coil
   D. to prevent a decrease in high side pressure and the starving of the evaporator coil

1688. A belt drive fan motor that runs at 1050 RPM uses a 6 inch pulley and operates a blower that has a 9 inch pulley. What is the RPM of the blower?
   A. 1500
   B. 1575
   C. 775
   D. 700

1689. What quantity of air must the indoor blower provide per ton of refrigeration effect?
    A. 300 CFM
    B. 400 CFM
    C. 600 CFM
    D. 800 CFM

1690. When the static pressure is increased in a duct system, how does this effect the blower motor?
    A. The amperage increases.
    B. The amperage decreases.
    C. The air quantity increases.
    D. The air quantity decreases.

1691. Which of the following can cause an increase (more positive) of supply air duct static pressure and an increase (less negative) of return air duct static pressure?
    A. a dirty air filter
    B. too small of a return air duct
    C. too small of a supply air duct
    D. Both A and B.

1692. Which of the following can cause a decrease (less positive) of supply air duct static pressure and a decrease (more negative) of return air duct static pressure?
    A. too large of a supply air duct
    B. too small of a return air duct
    C. too small of a supply air duct
    D. too large of a return air duct

1693. Which of the following can cause an increase (more positive) of supply air duct static pressure and a decrease (more negative) of return air duct static pressure?
    A. too large of a supply air duct and too small of a return duct
    B. too small of a return air duct and too small of a supply duct
    C. too small of a supply air duct and too large of a return duct
    D. too large of a return air duct and too large of a supply duct

1694. Which of the following can cause a decrease (less positive) of supply air duct static pressure and an increase (less negative) of return air duct static pressure?
    A. too large of a supply air duct and too small of a return duct
    B. too small of a return air duct and too small of a supply duct
    C. too small of a supply air duct and too large of a return duct
    D. too large of a return air duct and too large of a supply duct

1695. The minimum total area of a supply air duct used on a forced warm air furnace
    A. must be at least 1 square inch per 1000 Btu/h approved output rating.
    B. must be at least 2 square inches per 1000 Btu/h approved output rating.
    C. must be at least 4 square inches per 1000 Btu/h approved output rating.
    D. must be at least 6 square inches per 1000 Btu/h approved output rating.

1696. The minimum total area of a supply air duct used on a heat pump
    A. must be at least 1 square inch per 1000 Btu/h approved output rating.
    B. must be at least 2 square inches per 1000 Btu/h approved output rating.
    C. must be at least 4 square inches per 1000 Btu/h approved output rating.
    D. must be at least 6 square inches per 1000 Btu/h approved output rating.

1697. Technician "A" says that a smaller area in the supply air trunk will increase the air velocity.

    Technician "B" says that a smaller area in the supply air trunk increases the air quantity.

    Which technician is correct?
    A. technician "A"
    B. technician "B"
    C. Both technicians are correct.
    D. Both technicians are incorrect.

1698. Which of the following will cause a decrease of supply air static pressure?
    A. a section of the return air trunk being disconnected
    B. a higher fan speed
    C. a section of the supply air trunk being disconnected
    D. Both B and C.

1699. Technician "A" says to increase the speed on a belt drive blower, the adjusting pulley should be closed more.

    Technician "B" says to increase the speed on a belt drive blower, the adjusting pulley should be opened more.

    Which technician is correct?
    A. technician "A"
    B. technician "B"
    C. Both technicians are correct.
    D. Both technicians are incorrect.

1700. What is the definition of "friction loss" in a duct system?
    A. the resistance to airflow
    B. the resistance to air pressure
    C. the resistance to static pressure
    D. None of the above.

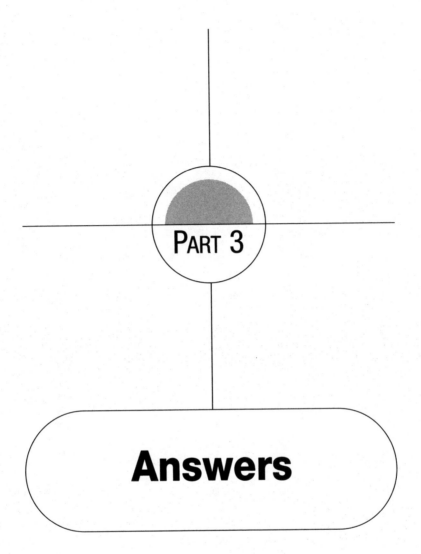

Section 1 System Components and Tools

Section 1
System Components and Tools

HAND TOOLS

1. D
2. B
3. C
4. A
5. D
6. A
7. D
8. B
9. D
10. A
11. B
12. C
13. D
14. B
15. B
16. A
17. C
18. D
19. A
20. B
21. D
22. A
23. B
24. C
25. D
26. A
27. A
28. A
29. C
30. A
31. B
32. A
33. C
34. B

BASIC CONSTRUCTION

35. C
36. B
37. A
38. D
39. B
40. B
41. C
42. C
43. D
44. C
45. A
46. D
47. B
48. B
49. A

REFRIGERATION THEORY

50. A
51. C
52. B
53. D
54. B
55. D
56. A
57. C
58. A
59. B
60. A
61. A
62. C
63. A

64. D
65. B
66. C
67. C
68. C
69. D
70. A
71. B
72. D
73. A
74. B
75. C
76. A
77. B
78. D
79. B
80. C
81. D
82. B
83. B
84. D
85. A
86. C
87. B
88. A

AIR CONDITIONING COMPONENTS

89. A
90. B
91. D
92. C
93. C
94. C
95. D
96.
    1. B
    2. H
    3. C

4. I
5. E
6. J
7. F
8. G
9. D
10. A
97. A
98.
1. A
2. D
3. B
4. C
5. H
6. F
7. I
8. E
9. J
10. G
99. B
100. C
101. A
102. C
103. C
104. A
105. C
106. D
107. A
108. A
109. D
110. C
111. B
112. D
113. A
114. C
115. B
116. C
117. A
118. A
119. B
120. C
121. C
122. B
123. C
124. D
125. B

REFRIGERANTS

126. B
127. B
128. D
129. C
130. B
131. A
132. D
133. C
134. A
135. A
136. D
137. B
138. D
139. B
140. B
141. C
142. A
143. B
144. D
145. B
146. A
147. B
148. C
149. A
150. B
151. C
152. B
153. D
154. B

LUBRICANTS

155. A
156. D
157. B
158. A
159. D
160. C
161. A
162. C
163. A
164. B
165. D
166. A
167. C
168. A
169. C
170. A
171. B
172. C
173. B

COMPRESSORS

174. C
175. B
176. D
177. D
178. B
179. B
180. C
181. B
182. C
183. C

Section 1 System Components and Tools

184. A	213. C	243. C
185. C	214. D	244. C
	215. C	
	216. B	
	217. C	
AIR COOLED CONDENSERS	218. A	**SERVICE VALVES**
186. B		245. B
187. D		246. A
188. D	**INDOOR EVAPORATOR COILS**	247. B
189. C		248. C
190. C	219. B	249. B
191. B	220. B	250. A
192. D	221. A	251. B
	222. D	252. D
	223. B	253. C
	224. D	254. C
	225. C	255. D
	226. B	256. A
WATER COOLED CONDENSERS	227. C	257. B
193. C	228. A	
194. B	229. B	
195. A	230. C	
196. D	231. D	**LEAK DETECTION**
197. B	232. B	258. A
198. C		259. B
199. D		260. A
200. C		261. C
201. C	**METERING DEVICES**	262. B
202. D	233. C	263. D
203. C	234. B	264. C
204. D	235. A	265. D
205. A	236. D	266. D
206. D	237. C	267. A
207. C	238. C	268. C
208. B	239. D	269. B
209. A	240. C	270. C
210. C	241. B	271. C
211. A	242. A	272. D
212. B		

RECOVERY EQUIPMENT

273. A
274. A
275. C
276. D
277. A
278. B
279. D
280. B
281. D
282. A
283. D
284. C
285. B

SOLDERING AND BRAZING

286. C
287. B
288. C
289. A
290. B
291. D
292. C
293. A
294. A
295. D
296. C
297. C
298. B
299. A
300. D
301. A
302. D
303. C
304. D
305. B

306. C
307. A

MANIFOLD GAUGES

308. C
309. D
310. B
311. B
312. C
313. C
314. A
315. D
316. A
317. B
318. C

SYSTEM EVACUATION

319. A
320. C
321. B
322. C
323. C
324. D
325. A
326. C
327. C
328. B
329. B
330. B

REFRIGERANT CHARGING

331. A
332. B

333. B
334. C
335. D
336. A
337. C
338. A
339. C
340. D
341. A

DEFROST SYSTEMS

342. C
343. B
344. A
345. A
346. D
347. D
348. D
349. B
350. C
351. A
352. D
353. C
354. A
355. B
356. A
357. A
358. D
359. C
360. D
361. D
362. D
363. A
364. B
365. C
366. D
367. C

Section 1   System Components and Tools

FUNDAMENTALS OF GAS COMBUSTION

368. B
369. B
370. D
371. D
372. A
373. B
374. A
375. D
376. B
377. A
378. C
379. C

GAS FURNACE CONTROLS AND COMPONENTS

380. A
381. D
382. C
383. B
384. C
385. A
386. B
387. B
388. D
389. B
390. C
391. D
392. D
393. A
394. B
395. B
396. D
397. D
398. D
399. C
400. B
401. A
402. B
403. C
404. A
405. A
406. C
407. A
408. B
409. B
410. C
411. D
412. C
413. B
414. B
415. D
416. A
417. C
418. C
419. C
420. B
421. D
422. D
423. C
424. C

COMBUSTION AIR

425. C
426. D
427. C
428. C
429. D
430. B
431. C
432. A
433. B
434. B
435. C
436. B
437. B
438. B
439. D
440. A
441. B
442. C
443. C
444. A
445. C
446. A
447. C
448. A
449. D

VENTS AND CHIMNEYS

450. A
451. C
452. D
453. A
454. B
455. B
456. B
457. D
458. A
459. D
460. A
461. D
462. C
463. B
464. A
465. D
466. A
467. B
468. B
469. D
470. C

471. D
472. C
473. B
474. C
475. C
476. B
477. D
478. C
479. D
480. A
481. C
482. B
483. C
484. A
485. A
486. B
487. A
488. B
489. A
490. D
491. B
492. D
493. A
494. B
495. A
496. B
497. B
498. A
499. C

GAS PIPING

500. A
501. B
502. C
503. A
504. D
505. A
506. B
507. B
508. A
509. C
510. B
511. C
512. D
513. D
514. D
515. D
516. B
517. B
518. A
519. D
520. D
521. A
522. D
523. D
524. B
525. C
526. D
527. A
528. A
529. B
530. C

OIL FURNACES

531. C
532. D
533. B
534. C
535. A
536. A
537. B
538. D
539. C
540. C
541. B
542. B
543. C
544. A
545. C
546. B
547. C
548. A
549. D
550. B
551. A
552. D
553. D
554. B
555. B
556. C
557. B
558. B
559. C
560. A
561. B
562. A
563. D
564. C
565. B
566. B
567. D
568. C
569. C
570. D
571. A

HEAT PUMPS

572.
    1. C
    2. H
    3. E
    4. B
    5. D
    6. F
    7. G
    8. A

Section 1 System Components and Tools

573. C		643. B
574. A	**Section 2**	644. C
575. B	**Electrical Theory**	
576. C	**and Application**	
577. C		**DIRECT CURRENT**
578. D		645. A
579. B	**ELECTRICAL THEORY**	646. C
580. A	614. B	647. C
581. A	615. A	648. C
582. D	616. B	649. B
583. B	617. D	650. A
584. C	618. C	651. B
585. A	619. C	652. A
586. C	620. C	653. D
587. D	621. A	654. C
588. C	622. A	655. C
589. B	623. B	656. B
590. D	624. D	657. C
591. D	625. D	658. A
592. B	626. B	659. B
593. B	627. C	660. C
594. B	628. C	661. C
595. B	629. D	662. B
596. C	630. B	663. D
597. B	631. A	664. D
598. D	632. B	665. A
599. D	633. C	666. A
600. D	634. A	667. A
601. B	635. A	668. A
602. D		669. D
603. C		
604. A		
605. D	**MAGNETISM**	
606. B	636. B	**ALTERNATING CURRENT**
607. C	637. A	
608. D	638. D	670. B
609. C	639. A	671. C
610. D	640. C	672. A
611. B	641. A	673. D
612. A	642. C	674. C
613. D		

675. B
676. D
677. A
678. D
679. C
680. B
681. A
682. B
683. C
684. C
685. D
686. A
687. B
688. C
689. C
690. B
691. D
692. B
693. A
694. D
695. D

POWER DISTRIBUTION

696. B
697. A
698. A
699. B
700. D
701. C
702. B
703. D
704. B
705. C
706. B
707. A
708. C
709. C
710. D

AC MOTORS

711. B
712. C
713. B
714. D
715. A
716. B
717. B
718. C
719. A
720. C
721. C
722. A
723. C
724. B
725. A
726. A
727. D
728. C
729. D
730. D
731. D
732. B
733. C
734. A
735. C
736. D
737. A
738. B
739. A
740. C
741. C
742. A
743. D
744. B
745. D
746. B
747. C
748. C

749. D
750. C
751. B
752. D
753. A
754. B
755. D
756. D
757. C
758. A
759. A
760. C
761. A
762. B
763. C
764. A
765. B
766. A
767. D
768. B
769. C
770. B
771. A
772. D
773. C

ELECTRICAL CONTROLS

774. C
775. D
776. B
777. C
778. D
779. A
780. B
781. C
782. B
783. A
784. C

Section 2 Electrical Theory and Application

785. D	826. D	**WIRING LAYOUTS AND ELECTRICAL DIAGRAMS**
786. A	827. C	
787. A	828. B	
788. B	829. D	861. D
789. B	830. A	862. A
790. C	831. D	863. D
791. D	832. C	864. B
792. D	833. A	865. A
793. B	834. D	866. B
794. C	835. B	867. C
795. C	836. C	868. D
796. A	837. A	869. A
797. A	838. B	870. B
798. B	839. C	871. B
799. C	840. C	872. C
800. D	841. A	873. B
801. B	842. B	874. C
802. B	843. D	875. B
803. C	844. B	876. D
804. D	845. D	877. A
805. A	846. B	878. C
806. C	847. D	879. A
807. B	848. B	880. C
808. C	849. B	881. B
809. D	850. C	882. D
810. D	851. C	883. C
811. A	852. A	884. D
812. C		885. B
813. D		886. C
814. A		887. A
815. B		888. C
816. B	**SOLID-STATE ELECTRONICS**	889. B
817. C		890. C
818. B	853. C	891. C
819. C	854. D	892. A
820. D	855. A	893. C
821. A	856. D	894. A
822. C	857. A	895. B
823. B	858. B	896. A
824. A	859. C	897. D
825. B	860. D	898. C

899. C
900. D
901. C
902. B
903. A
904. D
905. C

ELECTRICAL METERS

906. D
907. B
908. B
909. A
910. C
911. D
912. B
913. C
914. A
915. C
916. B
917. C
918. A
919. B
920. A
921. B
922. B
923. B
924. A
925. D
926. A
927. C
928. C
929. C
930. B
931. D

ELECTRICAL TROUBLESHOOTING

932. C
933. B
934. D
935. B
936. C
937. B
938. D
939. B
940. C
941. B
942. A
943. D
944. C
945. A
946. C
947. A

Section 3
Air Flow Components and Duct Fabrication

AIR FLOW TOOLS AND MEASUREMENTS

948. B
949. B
950. C
951. A
952. D
953. B
954. A
955. D
956. B

957. B
958. C
959. B
960. C
961. A
962. D
963. B
964. A
965. C
966. A
967. B
968. C
969. C
970. A
971. D
972. B
973. C
974. B
975. B
976. C
977. C
978. A
979. D
980. B
981. D

BLOWERS AND FANS

982. C
983. B
984. B
985. C
986. A
987. A
988. B
989. B
990. D
991. D

Section 3 Air Flow Components and Duct Fabrication

992. C	1027. C	1059. A
993. B	1028. B	1060. A
994. C	1029. C	1061. C
995. B	1030. A	1062. B
996. D	1031. D	1063. C
997. B	1032. A	1064. A
998. A	1033. D	1065. C
999. D	1034. B	1066. D
1000. B	1035. C	1067. C
1001. A	1036. A	1068. D
1002. A	1037. B	1069. D
	1038. D	1070. B
	1039. C	1071. B
	1040. D	1072. A
	1041. C	1073. B
	1042. A	1074. C
	1043. B	1075. B
		1076. C
		1077. B
		1078. C
		1079. D

AIR DUCT SYSTEMS AND FABRICATION

1003. C
1004. B
1005. D
1006. A
1007. B
1008. A
1009. D
1010. A
1011. D
1012. B
1013. D
1014. D
1015. A
1016. C
1017. B
1018. D
1019. C
1020. D
1021. A
1022. C
1023. D
1024. B
1025. D
1026. A

FLEXIBLE (FLEX) DUCT

1044. B
1045. B
1046. D
1047. D
1048. C
1049. D
1050. B
1051. A

FIBERGLASS DUCTBOARD

1052. C
1053. D
1054. D
1055. C
1056. B
1057. A
1058. B

AIR SIDE COMPONENTS

1080. B
1081. D
1082. A
1083. B
1084. D
1085. B
1086. D
1087. D
1088. D
1089. D
1090. C
1091. A
1092. D
1093. B

AIR FILTERS

1094. A
1095. A
1096. C
1097. B
1098. C
1099. A
1100. C
1101. C
1102. C
1103. D
1104. B
1105. B
1106. A
1107. C

HUMIDIFIERS

1108. A
1109. D
1110. D
1111. A
1112. D
1113. A
1114. A
1115. B
1116. B
1117. C
1118. A
1119. D
1120. C
1121. C

TEMPERATURE AND HUMIDITY

1122. C
1123. A
1124. A
1125. B
1126. B
1127. C
1128. A
1129. C
1130. B
1131. D
1132. D
1133. B
1134. A

Section 4
Indoor Air Quality (IAQ) and Safety

INDOOR AIR QUALITY (IAQ)

1135. A
1136. D
1137. D
1138. C
1139. A
1140. B
1141. D
1142. C
1143. A
1144. A
1145. C
1146. B
1147. B
1148. D
1149. D
1150. C
1151. A
1152. C
1153. B
1154. C
1155. D
1156. C
1157. D
1158. A
1159. A
1160. D
1161. B
1162. D
1163. B
1164. A
1165. C
1166. B
1167. D
1168. D
1169. D
1170. C
1171. A
1172. D
1173. A
1174. A
1175. A
1176. D
1177. B
1178. D
1179. A
1180. D
1181. A
1182. A
1183. D

PERSONAL SAFETY

1184. D
1185. A
1186. C
1187. C
1188. A
1189. B
1190. D

Section 4 Indoor Air Quality (IAQ) and Safety

1191. C	1225. D	**ELECTRICAL SAFETY**
1192. D	1226. C	1258. A
1193. A	1227. A	1259. B
1194. C	1228. B	1260. D
1195. D	1229. A	1261. D
1196. B	1230. C	1262. A
1197. D	1231. C	1263. A
1198. A	1232. C	1264. B
1199. C	1233. A	1265. A
1200. A	1234. D	1266. C
1201. B	1235. C	1267. D
	1236. C	1268. C
	1237. A	1269. B
	1238. B	1270. C
	1239. B	1271. C
	1240. C	1272. A
REFRIGERANT SAFETY	1241. A	1273. C
	1242. C	1274. D
1202. C	1243. B	1275. C
1203. D	1244. D	1276. A
1204. B		1277. B
1205. D		1278. B
1206. D		1279. A
1207. B		1280. A
1208. C		1281. A
1209. A	**SOLDERING AND BRAZING SAFETY**	1282. C
1210. C		1283. C
1211. D	1245. C	1284. A
1212. C	1246. D	1285. B
1213. D	1247. B	
1214. B	1248. C	
1215. D	1249. D	
1216. C	1250. D	**EPA CERTIFICATION CORE SECTION**
1217. A	1251. A	
1218. C	1252. C	1286. A
1219. A	1253. B	1287. C
1220. B	1254. D	1288. B
1221. A	1255. C	1289. C
1222. C	1256. D	1290. A
1223. D	1257. B	1291. B
1224. C		

1292. A
1293. C
1294. B
1295. A
1296. B
1297. A
1298. C
1299. C
1300. B
1301. A
1302. C
1303. C
1304. A
1305. B
1306. B
1307. B
1308. C
1309. A
1310. C
1311. D
1312. C
1313. B
1314. A
1315. B
1316. C
1317. A
1318. C
1319. A
1320. B
1321. C
1322. B
1323. B
1324. C
1325. B
1326. C
1327. C

TYPE I CERTIFICATION

1328. B
1329. A
1330. C
1331. B
1332. A
1333. C
1334. B
1335. B
1336. C
1337. C
1338. C
1339. B
1340. A
1341. C
1342. A
1343. C
1344. A
1345. B
1346. B
1347. B

TYPE II CERTIFICATION

1348. C
1349. C
1350. C
1351. C
1352. B
1353. A
1354. A
1355. B
1356. B

1357. B
1358. C
1359. A
1360. A
1361. C
1362. B
1363. A
1364. A
1365. C
1366. A
1367. A
1368. C
1369. A
1370. C
1371. A
1372. C
1373. B
1374. A
1375. A
1376. C
1377. C
1378. C
1379. C

TYPE III CERTIFICATION

1380. C
1381. B
1382. B
1383. B
1384. C
1385. A
1386. A
1387. A

Section 4 Indoor Air Quality (IAQ) and Safety

1388. B	1418. C	1459. D
1389. A	1419. D	1460. B
1390. A	1420. B	1461. D
1391. B	1421. D	1462. B
1392. A	1422. B	1463. A
1393. C	1423. D	1464. A
1394. B	1424. B	1465. B
1395. B	1425. D	1466. C
1396. A	1426. B	1467. B
1397. C	1427. D	1468. D
1398. C	1428. A	1469. C
1399. C	1429. B	1470. D
1400. C	1430. D	1471. D
	1431. A	1472. A
	1432. A	1473. C
	1433. B	1474. B
	1434. A	1475. B
	1435. D	1476. B
	1436. C	1477. D
	1437. D	1478. B
	1438. D	1479. C
	1439. C	1480. A
	1440. D	1481. D
	1441. D	1482. A
	1442. C	1483. B
	1443. D	1484. D
	1444. D	1485. D
	1445. B	
	1446. B	
	1447. D	
	1448. B	

Section 5
Service and Troubleshooting

AIR CONDITIONING TROUBLESHOOTING

1401. C	1442. C	
1402. A	1443. D	
1403. C	1444. D	
1404. B	1445. B	
1405. A	1446. B	
1406. B	1447. D	
1407. D	1448. B	
1408. B	1449. C	
1409. C	1450. D	
1410. D	1451. C	
1411. B	1452. B	
1412. B	1453. D	
1413. D	1454. B	
1414. A	1455. D	
1415. D	1456. A	
1416. B	1457. B	
1417. A	1458. C	

GAS FURNACE TROUBLESHOOTING

1486. C
1487. D
1488. C
1489. C
1490. B
1491. A
1492. D
1493. D

1494. A	1534. B	1574. B
1495. B	1535. A	1575. D
1496. C	1536. A	1576. B
1497. C	1537. A	1577. A
1498. C	1538. C	1578. D
1499. C	1539. D	1579. A
1500. A	1540. B	1580. C
1501. B	1541. C	1581. C
1502. B	1542. B	1582. D
1503. C	1543. B	1583. D
1504. A	1544. A	1584. C
1505. D	1545. D	1585. B
1506. A	1546. B	1586. C
1507. A	1547. D	1587. A
1508. A	1548. B	1588. B
1509. B	1549. C	1589. C
1510. B	1550. D	1590. D
1511. A	1551. C	1591. A
1512. D	1552. A	1592. D
1513. A	1553. C	
1514. D	1554. D	
1515. B	1555. A	
1516. B	1556. A	**OIL FURNACE TROUBLESHOOTING**
1517. C	1557. C	
1518. B	1558. C	1593. C
1519. D	1559. C	1594. D
1520. C	1560. C	1595. B
1521. A	1561. D	1596. D
1522. C	1562. A	1597. C
1523. A	1563. C	1598. A
1524. D	1564. C	1599. D
1525. C	1565. A	1600. D
1526. B	1566. B	1601. B
1527. C	1567. C	1602. C
1528. D	1568. D	1603. A
1529. A	1569. D	1604. B
1530. D	1570. D	1605. C
1531. D	1571. A	1606. A
1532. D	1572. C	1607. C
1533. A	1573. C	1608. A

Section 5 Service and Troubleshooting

1609.	C	1639.	A	1669.	B
1610.	D	1640.	A	1670.	D
1611.	B	1641.	C	1671.	E
1612.	B	1642.	C	1672.	C
1613.	A	1643.	B	1673.	B
1614.	B	1644.	C	1674.	B
1615.	C	1645.	B	1675.	B
1616.	B	1646.	D	1676.	B
1617.	C	1647.	B	1677.	A
1618.	D	1648.	C	1678.	A
1619.	A	1649.	D	1679.	A
1620.	D	1650.	D	1680.	D
1621.	C	1651.	D	1681.	C
1622.	D	1652.	B	1682.	B
1623.	B	1653.	B	1683.	C
1624.	D	1654.	B	1684.	C
1625.	B	1655.	D	1685.	B
1626.	B	1656.	D	1686.	D
1627.	A	1657.	D	1687.	D
1628.	D	1658.	D	1688.	D
1629.	A	1659.	D	1689.	B
1630.	B	1660.	C	1690.	A
1631.	D	1661.	C	1691.	C
1632.	D	1662.	B	1692.	B
		1663.	D	1693.	B
		1664.	B	1694.	D
				1695.	B
				1696.	D

HEAT PUMP TROUBLESHOOTING

1633.	C
1634.	D
1635.	A
1636.	D
1637.	A
1638.	B

AIR DISTRIBUTION TROUBLESHOOTING

1665.	A	1697.	A
1666.	A	1698.	D
1667.	C	1699.	A
1668.	B	1700.	A